# Next-generation Sequencing and Bioinformatics for Plant Science

https://doi.org/10.21775/9781910190654

Edited by

Vijai Bhadauria

Crop Development Centre and Department of Plant Sciences
University of Saskatchewan
Saskatoon, SK
Canada

Caister Academic Press

Caister Academic Press
Norfolk, UK

www.caister.com

British Library Cataloguing-in-Publication Data
A catalogue record for this book is available from the British Library

ISBN: 978-1-910190-65-4 (paperback)
ISBN: 978-1-910190-66-1 (ebook)



Cover design adapted from an image provided by Dr Xifeng Chen and Dr Bojun Ma.

# Contents

# Current Books of Interest

| | |
|---|---|
| Bacterial Evasion of the Host Immune System | 2017 |
| Illustrated Dictionary of Parasitology in the Post-Genomic Era | 2017 |
| The CRISPR/Cas System: Emerging Technology and Application | 2017 |
| Brewing Microbiology: Current Research, Omics and Microbial Ecology | 2017 |
| Metagenomics: Current Advances and Emerging Concepts | 2017 |
| *Bacillus*: Cellular and Molecular Biology (Third Edition) | 2017 |
| Cyanobacteria: Omics and Manipulation | 2017 |
| Foot-and-Mouth Disease Virus: Current Research and Emerging Trends | 2017 |
| Brain-eating Amoebae: Biology and Pathogenesis of *Naegleria fowleri* | 2016 |
| *Staphylococcus*: Genetics and Physiology | 2016 |
| Chloroplasts: Current Research and Future Trends | 2016 |
| Microbial Biodegradation: From Omics to Function and Application | 2016 |
| Influenza: Current Research | 2016 |
| MALDI-TOF Mass Spectrometry in Microbiology | 2016 |
| *Aspergillus* and *Penicillium* in the Post-genomic Era | 2016 |
| The Bacteriocins: Current Knowledge and Future Prospects | 2016 |
| Omics in Plant Disease Resistance | 2016 |
| Acidophiles: Life in Extremely Acidic Environments | 2016 |
| Climate Change and Microbial Ecology: Current Research and Future Trends | 2016 |
| Biofilms in Bioremediation: Current Research and Emerging Technologies | 2016 |
| Microalgae: Current Research and Applications | 2016 |
| Gas Plasma Sterilization in Microbiology: Theory, Applications, Pitfalls and New Perspectives | 2016 |
| Virus Evolution: Current Research and Future Directions | 2016 |
| Arboviruses: Molecular Biology, Evolution and Control | 2016 |
| *Shigella*: Molecular and Cellular Biology | 2016 |
| Aquatic Biofilms: Ecology, Water Quality and Wastewater Treatment | 2016 |
| Alphaviruses: Current Biology | 2016 |
| Thermophilic Microorganisms | 2015 |
| Flow Cytometry in Microbiology: Technology and Applications | 2015 |
| Probiotics and Prebiotics: Current Research and Future Trends | 2015 |
| Epigenetics: Current Research and Emerging Trends | 2015 |
| *Corynebacterium glutamicum*: From Systems Biology to Biotechnological Applications | 2015 |
| Advanced Vaccine Research Methods for the Decade of Vaccines | 2015 |
| Antifungals: From Genomics to Resistance and the Development of Novel Agents | 2015 |
| Bacteria-Plant Interactions: Advanced Research and Future Trends | 2015 |

Full details at www.caister.com

# Editorial

With the advent of high-throughput sequencing platforms, such as Illumina's Genome Analyzer, HiSeq, MiSeq and NextSeq, Roche/454's Genome Sequencer FLX, Thermo-Fisher Scientific's SOLiD, Ion Torrent and Ion Proton, PacBio's Real-Time Sequencer and more recently Oxford Nanopore's MinION, it has become feasible to sequence entire genomes and transcriptomes at an exponential pace. This has huge implications in plant breeding and genetics. The reviews presented in this volume summarize recent developments in next-generation sequencing (NGS) and bioinformatics tools and their application in understanding and improving agronomic traits.

Next-generation sequencing (NGS) coupled with high-performance computing have revolutionized the field of plant breeding and genetics (Bhadauria and Banniza, 2014; Bhadauria *et al.*, 2016). This volume compiles recent advances in the NGS as well as the application of NGS in understanding and improving agronomics traits such as yield, drought tolerance and disease resistance.

In this volume, the review by Sharma *et al.* (Chapter 1) outlines the evolution of DNA sequencing techniques and platforms, including the first generation (Sanger's chain-termination method and Maxam–Gilbert's chemical cleavage method), second-generation (Illumina's GA, HiSeq and MiSeq as well as Roche/454's GS FLX), third-generation single molecule real-time sequencing (PacBio's SMRT RS) and the more recent fourth-generation sequencing platform Oxford Nanopore's MinIon. The application of NGS in the genome sequencing and evolution of rice, maize and sorghum (Chapter 2) and orchids (Chapter 3) shed light on the challenges in the post-NGS era, such as genome assembly and annotation. Gomez-Casati and colleagues, in Chapter 5, describe the application of omics (gene expression and regulation as well as quantitative proteomics approaches, such as iTRAQ) in fruit development and ripening, and plant disease resistance. Bioinformatics resources for plant genomics are detailed and discussed in two reviews (Chapters 4 and 5). The NGS-based genotyping-by-sequencing approaches are useful to map polymorphism in experimental populations and germplasm, which then can be used to track genomic regions controlling quantitative traits, such as fusarium head blight and stripe rust resistance in wheat (Chapater 6) and rice (Chapter 7). Sequencing of the transcriptome from infected plant tissues (dual RNA-seq) can provide molecular insight into host defence and pathogen virulence during incompatible and compatible interactions, thereby facilitating in designing crops with improved resistance (Chapter 8). In addition to genome and transcriptome sequencing, the NGS can also be used in sequencing of small RNA (20–24 nucleotides; Chapter 9) and transcription factor binding sites (ChIP-seq; Chapter 10) in genomes. Structural variations, such as abnormal chromosome number, chromosomal rearrangement, copy number variation, presence or absence variation, mobile element insertion and deletion and homologous exchange play key role in phenotypic diversity of agronomic traits, such as biotic and abiotic stress tolerance. Zhang and colleagues looked into the application of NGS in mapping such structural variations.

## References

Bhadauria, V., and Banniza, S. (2014). What lies ahead in post-genomics era: a perspective on genetic improvement of crops for fungal disease resistance? Plant Signal Behav. *9*, e28503.

Bhadauria, V., Wong, M.M., Bett, K.E., and Banniza, S. (2016). Wild help for enhancing genetic resistance in lentil against fungal diseases. Curr. Issues Mol. Biol. *19*, 3–6.

Vijai Bhadauria
Crop Development Centre and Department of Plant Sciences,
University of Saskatchewan, Saskatoon, SK,
Canada

Lucia Popescu
Department of Soil Science,
University of Saskatchewan, Saskatoon, SK,
Canada

https://doi.org/10.21775/9781910190654.01

# Status and Prospects of Next-generation Sequencing Technologies in Crop Plants

1

Tilak R. Sharma*, Basavantraya N. Devanna, Kanti Kiran, Pankaj K. Singh, Kirti Arora, Priyanka Jain, Ila M. Tiwari, Himanshu Dubey, Banita K. Saklani, Mandeep Kumari, Jyoti Singh, Rajdeep Jaswal, Ritu Kapoor, Deepak V. Pawar, Shruti Sinha, Deepak S. Bisht, Amolkumar U. Solanke and Tapan K. Mondal

ICAR-National Research Centre on Plant Biotechnology, Pusa Campus, New Delhi, India.

*Correspondence: trsharma1965@gmail.com and trsharma@nrcpb.org

https://doi.org/10.21775/9781910190654.02

### Abstract
The history of DNA sequencing dates back to 1970s. During this period, the two first-generation nucleotide sequencing techniques were developed. Subsequently, Sanger's dideoxy method of sequencing gained popularity over Maxam and Gilbert's chemical method of sequencing. However, in the last decade, we have observed revolutionary changes in DNA sequencing technologies leading to the emergence of next-generation sequencing (NGS) techniques. NGS technologies have enhanced the throughput and speed of sequencing combined with bringing down the overall cost of the process over a time. The major applications of NGS technologies being genome sequencing and resequencing, transcriptomics, metagenomics in relation to plant–microbe interactions, exon and genome capturing, development of molecular markers and evolutionary studies. In this review, we present a broader picture of evolution of NGS tools, its various applications in crop plants, and future prospects of the technology for crop improvement.

### Introduction
The overall growth, development and behavioural characteristics of every living creature are largely determined by its genetic constitution. Subsequent to the famous double-helix model of DNA, proposed by Watson and Crick (1953), scientists began to find the ways and means to determine the nucleotide sequence of DNA. The first significant breakthrough in this area was achieved in late 1970s when two groups working independently reported two different approaches for DNA sequencing (Maxam and Gilbert, 1977; Sanger *et al.*, 1977). Though Maxam and Gilbert's approach for DNA sequencing was preferred initially, it was Sanger's sequencing technology which subsequently got popularized among the scientific community. The classical genome sequencing projects such as the Human Genome Project (HGP), the *Arabidopsis* Genome Initiative and the International Rice Genome Sequencing Project were successfully completed using Sanger's sequencing approach. Subsequently, many plant genomes were sequenced using this sequencing technology. Though Sanger's dideoxy sequencing method is considered as gold standard with respect to genome sequencing, there

are many shortcomings in this approach. The important shortcomings of Sanger's sequencing method are that it is time-consuming, low throughput and high cost, and needs more labour, *in vivo* cloning of DNA fragments, etc. Therefore, scientists and bioengineers tried to develop new sequencing techniques also known as second-generation (2GS) or next-generation sequencing (NGS) technologies. The success in this direction was reported in 2005, when the first NGS system was developed by 454 and commercialized by Roche as the GS20 (Margulies *et al.*, 2005). Subsequently, many NGS/2GS systems have been reported, which include Solexa GA2 (now Illumina), Applied Biosystem's SOLiD and Ion Torrent www.appliedbiosystems.com; www.illumina.com; www.iontorrent.com).

The developments in the field of NGS technologies have led to a revolution in the field of genetics and genomics. In spite of their inherent limitation of producing shorter reads with higher error rates than Sanger sequencing, NGS technologies are still gaining popularity, largely due to their ability to produce massive quantities of data at a relatively low cost and in a short time period. Further developments in NGS technologies have consistently led to increased read lengths and this is an active area of research in the future of NGS technologies. The present review compiles the different NGS technologies and their applications in the field of plant biology with special emphasis on agricultural plants (Fig. 1.1). In this review we focused primarily on the crop genomes giving weightage to genome sequencing techniques implied and their challenges, applications of NGS in plant biology with emphasis on agricultural crops and conclude with the future strategies and perspective in the area of plant genomics.

## Evolutions of DNA sequencing technology

### First generation sequencing technologies

The double helix structure of DNA proposed by Watson and Crick (1953) was the milestone

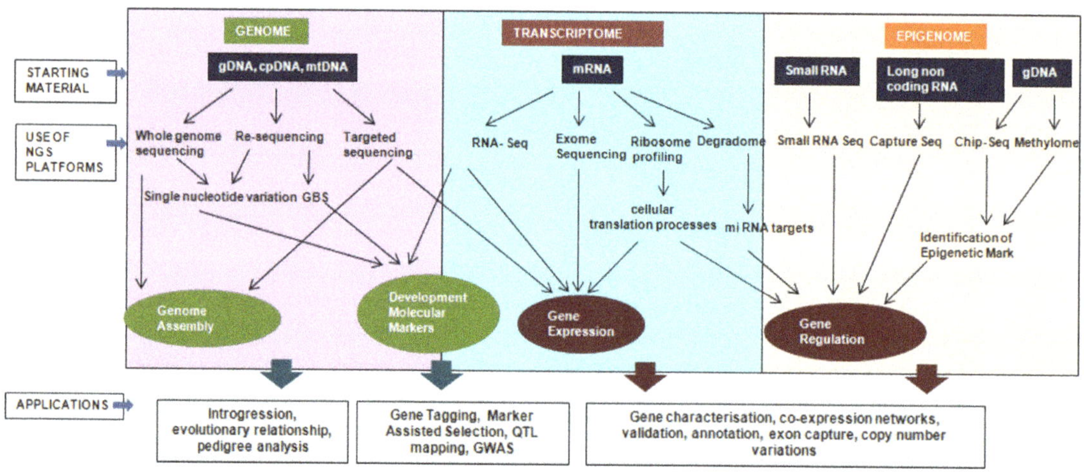


**Figure 1.1** NGS platforms as a tool for the plant genomic research. The phenotype of an organism is a manifestation of controlled expression of the underlying gene. The gene(s) expression is largely regulated at three stages: (a) genomic, (b) transcriptomic and (c) epigenetic. For unravelling the complex cellular machinery and its regulatory network the NGS platforms, with suitable modifications, can be utilized at any of these three stages (each stage is represented by different colour boxes). Genomic DNA, chloroplast DNA and mitochondrial DNA are the primary sources of genetic information which can be divulged by NGS assisted whole genome sequencing, high throughput re-sequencing and targeted sequencing. For example, high sequencing coverage obtained by NGS platforms is of great utility in determining the single nucleotide variations (SNVs) in genome which are of immense importance in establishing genotype phenotype relationships. In addition, NGS tools like MethylC-Seq are a 'gold standard' technique for studying genome wide methylation pattern. Similarly, sequencing of mRNA, small RNAs, long non coding RNAs, degradome and exome helps in understating the expression pattern of genes, their spatial and temporal regulations, co-expression networks and association with trait.

in molecular biology research. Subsequently, interest was shifted to decipher the nucleotide sequence of DNA molecules. The first major success in this field was reported simultaneously in 1977 by two independent groups of researchers when two chemistries for DNA sequencing were published (Maxam and Gilbert, 1977; Sanger *et al.*, 1977). Broadly referred to as first-generation sequencing technologies, these two methods utilized different chemistries for DNA sequencing. Due to its complex procedure and low resolution Maxam and Gilbert's method did not gain wide acceptance. Sanger's original dideoxynucleotide (ddNTPs) chain-termination sequencing method was comparatively less cumbersome and relatively accurate but required radiolabelled ddNTPs, and the chemistry involved four separate base specific reactions and autoradiography in order to detect the sequence of DNA molecule. Sanger's method was later on modified and improved. An account of improvement and evolution of Sanger's method can be found in previous reviews (França *et al.*, 2002; Ansorge, 2009; Mardis, 2013).

## Second-generation sequencing technologies

Second-generation sequencing technologies were capitalized by Illumina Genome Analyzer, Roche/454 FLX, Life technologies SOLiD and Ion Torrent PGM. Second-generation sequencers omitted the need of *in vivo* cloning, and here DNA can directly be fragmented to produce sequencing libraries of appropriate sizes *in vitro* by adapter ligated amplification using a PCR-based system. PCR-based amplification is required to produce millions of copies of the original DNA fragment, required to produce signal intensity sufficient to detect the incorporated bases during sequencing steps. Second-generation sequencing technologies are divided in two categories on the basis of reaction chemistry they use: (a) sequencing by synthesis used by Illumina, Roche/454 and Ion Torrent and (b) sequencing by ligation used by ABI's SOLiD sequencers.

## Third-generation sequencing technology: single molecule sequencing

Third-generation sequencing (3GS) technology differs from second-generation sequencing (2GS) technologies as it does not require the amplification step, leading to elimination of inclusion errors incurred by polymerase during library preparation. Second-generation sequencers use a cyclic wash-and-scan method, and as the number of washing and imaging cycles increases, addition of nucleotides becomes more asynchronous which increases sequencing errors and signal to noise ratio. This particularly limits the read length produced by second-generation sequencers. The advantage of some single molecule sequencing platforms is the ability to detect epigenetic modifications in the genome, which are largely diluted due to amplification step involved in 2GS techniques (Munera *et al.*, 2012). Also, the short reads generated by the 2GS technologies imposes problems in accurate *de novo* genome assembly, repetitive elements and large structural variation analyses (Fan *et al.*, 2010; Delcher *et al.*, 2010). The longer reads generated by third-generation sequencer helps in *de novo* genome sequencing and assembly, improving old assemblies, allowing more accurate analysis of structural variation within the genome and more contiguous reconstruction of genomes which were limited in second-generation sequencers (Carneiro *et al.*, 2013). At present, the major challenge with the single molecule sequencers is inherent higher error rate due to limited ability of detectors to identify and interpret very low levels of signal generated from individual molecules. The 3GS techniques are Pacific Biosciences Single Molecule Real Time sequencers (SMRT) and Helicos Genetic analysis system True single molecule sequencing (tSMS). Pacific Biosciences SMRT was the first 3GS technology to hit the market in 2010. With an improved chemistry (C4), the average read length of PacBio RS II system is over 10 kb and can generate 1 Gb of data per run. In sequel platform PacBio has increased the ZMWs density of loading of samples due to which the data output has increased drastically up to 5–10 Gb per run. Since this technology uses single DNA polymerase per ZMW, total length of a read is dependent on lifespan of the polymerase. Single molecule sequencers suffers from higher raw data error rate, which accounts for up to 15% of incorporated bases, however, this error rate can be minimized by multiple sequencing of same template. The tSMS from Helicos Genetic analysis system was the first commercially available true single molecule sequencing system based on sequencing by synthesis chemistry. This platform can produce average read length of 30 bases. This platform suffers from higher deletion rate ranging from 1 to 5%, which can be reduced by sequencing from both the ends of a fragment (Buzby *et al.*, 2008).

### Fourth-generation sequencing technologies: nanopore sequencing

Fourth-generation sequencing (4GS) technologies omit the use of labelled nucleotides and does not rely on an optical system to detect incorporated nucleotides and also there is no need for the synchronous reagent addition and wash steps. This technology uses the electronic or chemical properties of each nucleotide to determine the sequence of DNA as it is threaded through a nanopore. Current nanopore sequencing technologies use the ionic current blockage method. Currently two nanopore-based systems, biological and solid-state, are being developed and refined. Oxford Nanopore technology has released MinION for test users which will be launched for commercial application. It is small, USB powered, easy to carry equipment. It can produce average read length of 5.4 kb and go up to 10 kb (www.nature.com/news/data-from-pocket-sized-genome-sequencer-unveiled-1.14724).

## NGS and plant genomics

Plants being the ultimate source of food and metabolic energy for nearly all animals, cannot manufacture their own food, but also provide sustenance, shelter, clothing, medicines, fuels, and the raw materials from which innumerable other products are made. The plant kingdom is filled with amazingly incomparable diversity and significance (Schatz *et al.*, 2010) pertaining to which sequencing of plant genomes is of great importance. Out of the hundreds of thousands of plant species around the world, only few of them have been sequenced (Michael and Van Buren, 2015; Table 1.1).

### Genomics of plants

Advanced high-throughput genome sequencing techniques have proved to be a boon in providing practical solutions to the challenges in the field of genomics, especially crops. The plant kingdom is majorly divided into spore bearing vascular plants and seed bearing vascular plants. The former is further classified into algae, mosses/liverworts and ferns, and the latter into non flowering (gymnosperms) and flowering (all angiosperms) plants. Since the first published genome sequence of *Arabidopsis thaliana* (Arabidopsis Genome Initiative, 2000), a large number of plant genomes belonging to different phylum have been sequenced and published using both first-generation and NGS technologies (Turktas *et al.*, 2015).

### Sequenced plant genomes: links from vascular plants to angiosperms

The genome sequences of *Chlamydomonas reinhardtii* (120 Mb) and *Volvox carteri* (138 Mb) are amongst the first multi- and unicellular algae species, respectively (Merchant *et al.*, 2007; Prochnik *et al.*, 2010). Amongst the mosses, the genome of *Physcomitrella* with an assembled genome size of 500 Mb (Rensing *et al.*, 2008) and *Selaginella moellendorffii*, the first non-seed plant genome reported with an assembled genome size of 212.6 Mb (Banks *et al.*, 2011), are breakthrough model organisms in order to study the evolution of vascular land plants and their divergence. Further for a better understanding of the evolution of plants, the genome sequence information from other taxa, especially charophytes, ferns and gymnosperms, could serve as key references. In this context the three gymnosperm (conifer) genome sequences published recently, viz. the genome of Norway spruce (*Picea abies*) with an assembled size of 19.6 Gb (Nystedt *et al.*, 2013) is the first available gymnosperm sequence followed by the loblolly pine (*Pinus taeda*) genome (Zimin *et al.*, 2014; Neale *et al.*, 2014) and white spruce (*Picea glauca*) genome (Birol *et al.*, 2013) with an assembled genome size of 22 Gb and 20.8 Gb, respectively. These genomes helped to understand the divergence of angiosperms and gymnosperms (350 Myr ago) (Jiao *et al.*, 2011).

The development of water-conducting xylem cells and the reproductive development are the two major differences between angiosperms and gymnosperms. Nystedt *et al.* (2013) compared seven genomes, two from the basal plants and five from angiosperms and identified 1021 *P. abies*-specific gene families. *P. abies* homologs present in the angiosperms revealed that gymnosperms lack orthologues of flowering-responsible phosphatidylethanolamine-binding protein (PEBP) protein. Sequencing of angiosperm

**Table 1.1** Details of plant genomes sequenced and published during 2015 to 2016

| Number | Scientific name | Common name | Genome size (Mb) | Type | Reference |
|---|---|---|---|---|---|
| 1 | *Vigna angularis* | Adzuki bean | 538 | Dicot | Kang *et al.*, 2015 |
| 2 | *Thlaspi arvense* | Field pennycress | 539 | Dicot | Dorn *et al.*, 2015 |
| 3 | *Primula veris* | Cowslip | 480 | Dicot | Nowak *et al.*, 2015 |
| 4 | *Hordeum vulgare* | Tibetan hulless barley | 4480 | Dicot | Zeng *et al.*, 2015 |
| 5 | *Vaccinium corymbosum* | American blue berry | 500 | Dicot | Gupta *et al.*, 2015 |
| 6 | *Arabis alpina* | Alpine rockcress | 375 | Dicot | Willing *et al.*, 2015 |
| 7 | *Ipomea trifida* | Wild sweet potato | 520 | Dicot | Hirakawa *et al.*, 2015 |
| 8 | *Gossypium hirsutum* | Upland cotton | 2340 | Dicot | Li *et al.*, 2015 |
| 9 | *Boea hygrometrica* | | 1690 | Dicot | Xiao *et al.*, 2015 |
| 10 | *Solanum commersonii* | Wild potato | 830 | Dicot | Aversano *et al.*, 2015 |
| 11 | *Catharanthus roseus* | Madagascar periwinkle | 738 | Dicot | Kellner *et al.*, 2015 |
| 12 | *Ocimum sanctum* | Holy basil | 386 | Dicot | Rastogi *et al.*, 2015 |
| 13 | *Moringa oleifera* | Drumstick tree | 315 | Dicot | Tian *et al.*, 2015 |
| 14 | *Zizania latifolia* | Jiaobei | 590 | Dicot | Guo *et al.*, 2015 |
| 15 | *Cymbomonas tramitiformis* | | 850 | Algae | Burns *et al.*, 2015 |
| 16 | *Gossypium barbadense* | Sea island cotton | 2470 | Dicot | Liu *et al.*, 2015 |
| 17 | *Lolium perenne* | Perennial ryegrass | 2000 | Monocot | Byrne *et al.*, 2015 |
| 18 | *Chlorella pyrenoidosa* | | 57 | Monocot | Fan *et al.*, 2015 |
| 19 | *Anana comosus* | Pineapple | 526 | Monocot | Ming *et al.*, 2015 |
| 20 | *Oropetium thomaeum* | | 245 | Monocot | Van Buren *et al.*, 2015 |
| 21 | *Lemna minor* | Common duckweed | 481 | Monocot | Van Hoeck *et al.*, 2015 |
| 22 | *Trifolium pratense* | Red clover | 420 | Dicot | De Vega *et al.*, 2015 |
| 23 | *Salvia miltiorrhiza* | Chinese red sage | 641 | Dicot | Zhang *et al.*, 2015 |
| 24 | *Parachlorella kessleri* | | 63 | Algae | Ota *et al.*, 2016 |
| 25 | *Zostera marina* | Common eelgrass | 238 | Monocot | Olsen *et al.*, 2016 |
| 26 | *Dendrobium catenatum* | Chained dendrobium | 1110 | Monocot | Zhang *et al.*, 2016 |
| 27 | *Arachis duranensis* | Wild peanut A | 1250 | Dicot | Bertioli *et al.*, 2016 |
| 28 | *Arachis ipaensis* | Wild peanut A | 1560 | Dicot | Bertioli *et al.*, 2016 |
| 29 | *Cynara cardunculus* | Globe artichoke | 1084 | Dicot | Scaglione *et al.*, 2016 |
| 30 | *Rosa roxburghii* | Chestnut rose | 481 | Dicot | Lu *et al.*, 2016 |
| 31 | *Zoysia japonica* | | 390 | Monocot | Tanaka *et al.*, 2016 |
| 32 | *Zoysia matrella* | Manila grass | 380 | Monocot | Tanaka *et al.*, 2016 |
| 33 | *Zoysia pacifica* | | 370 | Monocot | Tanaka *et al.*, 2016 |
| 34 | *Fagopyrum esculentum* | Common buckwheat | 1300 | Dicot | Yasvi *et al.*, 2016 |
| 35 | *Gonium pectorale* | | 149 | Algae | Hanschen *et al.*, 2016 |
| 36 | *Rubus accidentalis* | Black raspberry | 293 | Dicot | Van Buren *et al.*, 2016 |
| 37 | *Petunia oxillaris* | White moon petunia | 1400 | Dicot | Bombarley *et al.*, 2016 |
| 38 | *Pogostemon cablin* | Patchouli | 1570 | Dicot | He *et al.*, 2016 |
| 39 | *Lepidium meyenii* | Maca | 751 | Dicot | Zhang *et al.*, 2016 |
| 40 | *Daucus carota* | Carrot | 473 | Dicot | Iorizzo *et al.*, 2016 |
| 41 | *Juglans regia* | Common walnut | 606 | Dicot | Martinez-Garcia *et al.*, 2016 |
| 42 | *Drosera capensis* | Cape sundew | 293 | Dicot | Butts *et al.*, 2016 |

**Table 1.1** Continued

| Number | Scientific name | Common name | Genome size (Mb) | Type | Reference |
|---|---|---|---|---|---|
| 43 | *Zostera muelleri* | Dwarf grass wrack | 390 | Dicot | Lee *et al.*, 2016 |
| 44 | *Chenopodium quinoa* | Quinoa | 1500 | Dicot | Yasvi *et al.*, 2016 |
| 45 | *Artocarpus camansi* | Breadnut | 669 | Dicot | Gardner *et al.*, 2016 |
| 46 | *Citrus paradisi* × *Poncirus trifoliata* | Swingle citrumela | 380 | Dicot | Zhang *et al.*, 2016 |
| 47 | *Musa itinerans* | Yunnan banana | 615 | Monocot | Wu *et al.*, 2016 |
| 48 | *Cicer reticulatum* | Wild chickpea | 817 | Dicot | Gupta *et al.*, 2016 |
| 49 | *Trifolium subterraneum* | Subterranean clover | 540 | Dicot | Hirakawa *et al.*, 2016 |
| 50 | *Quercus lobata* | Valley oak | 725 | Dicot | Sork *et al.*, 2016 |
| 51 | *Brassica nigra* | Black mustard | 591 | Dicot | Yang *et al.*, 2016 |
| 52 | *Brassica juncea* | Chinese mustard | 922 | Dicot | Yang *et al.*, 2016 |
| 53 | *Rhazya stricta* | Harmal | 274 | Dicot | Sabir *et al.*, 2016 |

genomes thus additionally served and immensely helped in understanding their divergence from rest of the plant genomes. With over 100 plant genome sequence data already published (Michael and Van Buren, 2015; https://genomevolution.org/wiki/index.php/Sequenced_plant_genomes), it is possible to study their complex life cycle, evolutionary history and genome structural organization. The crops and model plant genomes such as *Arabidopsis*, *Brachypodium distachyon*, *Physcomitrella patens* and *Setaria italica*, *Oryza sativa*, *Populus trichocarpa*, *Zea mays*, *Glycine max*, *Solanum lycopersicum* and *Pinus taeda* are extremely vital not just as crops but as functional models to enable genome-wide studies of various key genes/gene families, pathways and important traits (Arabidopsis Genome Initiative, 2000; Vogel *et al.*, 2010; Rensing *et al.*, 2008; Bennetzen *et al.*, 2012; Zhang *et al.*, 2012a; Goff *et al.*, 2002; Yu *et al.*, 2002; Tuskan *et al.*, 2006; Schnable *et al.*, 2009; Schmutz *et al.*, 2010; Tomato Genome Consortium, 2012; Zimin *et al.*, 2014). Nevertheless, non-functional models and non-crop plants serve as sources to explore the evolution of flowering plants and an in-depth understanding of plant genome architecture. The genome sequences of non-functional model plants like *Utricularia gibba* (bladderwort; 77 Mb), *Genlisea aurea* (corkscrew; 43.4 Mb), an aquatic non-grass monocot *Spirodela polyrhiza* (greater duckweed; 158 Mb), *Selaginella moellendorffii* (212.6 Mb), and *Amborella trichopoda* (870 Mb) are available (Banks *et al.*, 2011; Albert *et al.*, 2013; Ibarra-Laclette *et al.*, 2013; Leushkin *et al.*, 2013; Wang *et al.*, 2014). These can help to identify the whole evolutionary link between basal vascular plants and the most complicated and diverse angiosperms. Most importantly, it can help in identifying plant-specific gene families responsible for the specialized characteristic of each family and genus of the plant kingdom.

## Role of third-generation sequencing platforms in decoding plant genomes

We have seen a generation of large volumes of sequencing information through NGS and assembly platforms. Although NGS systems producing up to 100 Gb of data per run have advanced genome information (Imelfort and Edwards, 2009; Mardis, 2008; Schatz *et al.*, 2010; The 1000 Genomes Project Consortium, 2010), the diversity and variations of different genomes and short read lengths from NGS make it difficult to achieve a complete published genome information. Therefore, taking DNA sequencing to a further level and dramatically reducing costs, several companies have hit the market with the third-generation sequencing (3GS) technologies. The remarkable quality of genome sequences produced by 3GS and mapping technologies are hundreds to thousands of times more contiguous and enable improved analysis of nearly every aspect of the genome. The genomes are more complete and show accurate representation of genes, regulatory regions and other important genomic elements, as well as better resolution of the overall chromosome organization (Burton *et al.*, 2013; Roberts *et al.*, 2013; Ross *et al.*, 2013; Cao *et al.*, 2014; Pendleton *et al.*, 2015; Lee *et al.*, 2016).

## Next-generation sequencing and plant transcriptome analysis

The rapid developments in NGS technologies have revolutionized the field of plant transcriptomics, specifically in those plants where no genome sequence information is available. The NGS-based RNA sequencing technologies are generally called mRNA-seq tools. These technologies provide a novel method for identifying, mapping, and quantifying transcriptomes under different developmental stages and stress conditions. Deep RNA sequencing is a powerful tool for comparing gene expression analysis and discovering the full length 5′ and 3′ untranslated regions, novel splice junctions and transcripts, alternative transcription start sites, and rare transcripts (Cloonan *et al.*, 2008; Mortazavi *et al.*, 2008; Wilhelm *et al.*, 2008; Zhang *et al.*, 2010; Fig. 1.2). RNA-seq data with both technical and biological replicates show good level of reproducibility (Cloonan *et al.*, 2008). The transcribed RNA from genomic DNA, besides coding for proteins, also has potential non-coding RNAs. Owing to their significance in regulation of gene expression, these non-coding RNA have attracted great attention in recent past.

## Applications of transcriptome profiling and long non-coding RNA (lncRNA) in plant

### Cereals

A large number of studies are carried out to generate transcriptomes from cereal crops. RNA-seq was used to understand the transcriptional regulatory network of *Opaque2* (*O2*). *O2* transcription factor in maize plays an important role as a central regulatory molecule during maize endosperm development by regulating multiple regulatory pathways (Li *et al.*, 2015). The RNA-seq analysis found 1605 differentially expressed genes and 383 differentially expressed long, non-coding RNAs between wild-type and *O2* endosperms, 15 days after pollination. Transcriptome sequencing in rice was used to study differentially expressed genes

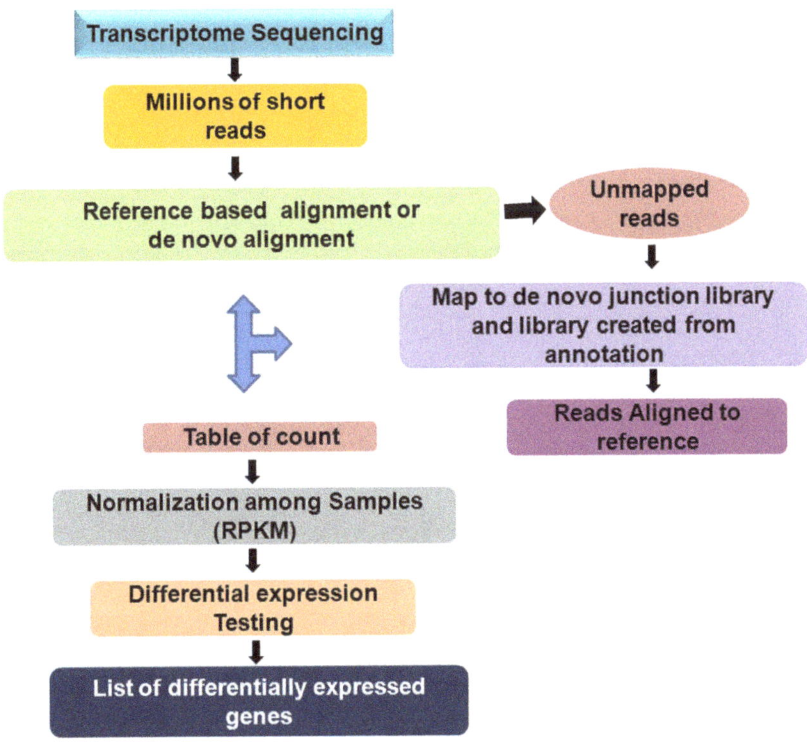


**Figure 1.2** Pipeline for transcriptome sequencing using NGS techniques.

(DEGs) upon infection with *Magnaporthe oryzae* and to probe the molecular response of rice to *Ustilaginoidea virens* infection (Kawahara *et al.*, 2012; Bagnaresi *et al.*, 2012; Chao *et al.*, 2014). Similarly, genome-wide transcription profiling of primed (Selenium and salicylic acid priming) and non-primed seedlings of rice was reported (Hussain *et al.*, 2016). Transcriptome of bread wheat provided a methodology for homeolog-specific sequence assembly (Schreiber *et al.*, 2012) and deep transcriptome sequencing of a wheat genome was used to construct a fine transcriptome map of the chromosome 3B (Pingault *et al.*, 2015). A NGS tool was also used in barley, finger millet and sorghum to study transcriptomes under different conditions (Bedada *et al.*, 2014; Rahman *et al.*, 2014; Tombuloglu *et al.*, 2015; Abdel-Ghany *et al.*, 2016).

Though many studies were conducted to profile the mRNA transcriptome of cereals using NGS technology, very few studies are available wherein NGS has been used for profiling of lncRNA. These studies include genome-wide analysis of non-coding parts of transcriptomes uncovering lncRNA in maize as well as rice and discovered long intergenic non-coding RNAs (lincRNA) which contain SNP associated with agriculturally important traits in maize and two lincRNAs in rice (Kim *et al.*, 2012; Wang *et al.*, 2015a).

### Pulses and oilseeds

Many studies of NGS applications in pulse crops are being reported in last few years. RNA-seq is successively used in chickpea, lentil, *Medicago truncatula* and *Vigna unguiculata* to identify developmental stage-specific transcriptomes, drought and salinity responsive transcriptomes, development of large scale genomics resources in lentil such as unigenes and 2393 EST-SSRs, to identify osmotic stress related lncRNAs and to generate a gene expression atlas of different plant tissues (Singh and Jain, 2014; Wang *et al.*, 2015b; Garg *et al.*, 2016; Yao *et al.*, 2016)

NGS technology has been used in oil-producing plants to study the transcriptome dynamics under different conditions. In soybean, RNA-seq atlas was built using NGS (Severin *et al.*, 2010). Transcriptome profiling of peanut, *Brassica napus* and sunflower was performed to identify various biochemical pathways. Further novel lncRNAs of *Brassica napus* in response to *Sclerotenia sclerotiorum* infection were identified. NGS technology was also used to understand the molecular basis of a higher rate of recombination in cultivated genotypes during meiosis as compared to their wild ancestors in sunflower (Joshi *et al.*, 2016; Florez-Zapata *et al.*, 2016).

### Horticulture and ornamental crops

Li *et al.* (2012) identified 92 DEGs that were associated with the metabolism and anthocyanin synthesis during fruit ripening. Rowland *et al.* (2012) analysed transcriptomes of different blueberry tissues to identify genes that were associated with cold acclimation and fruit development. Various other transcriptome studies involving fruits, flowers and vegetables crops are given Table 1.2.

### Commercial crops

Cotton is an important fibre crops in India that has high commercial value. In the study aimed at analysing the lncRNAs in cotton, Wang *et al.* (2015c) characterized lncRNAs in *Gossypium* species. Recently, Lu *et al.* (2016) identified 10,820 lncRNAs that were associated with different abiotic stresses. Similarly, Zou *et al.* (2016) also identified a total of 5996 lncRNAs that were reported rapid and dynamic changes during early fibre and rapid elongation stages. NGS techniques were used for discovery of genes, regulatory sequences and non-coding RNA for the improvement of sugarcane as a biofuel crop. In a recent study, six genotypes of sugarcane were studied and a total of 72,269 unigenes were identified. Out of which more than 28,788 of these unigenes showed significant similarity to sorghum proteins indicating that both species share a higher degree of genetic lineage. In another study, transcriptome analysis of sugarcane during smut pathogen infection (*Sporisorium scitamineum*) identified 65,852 unigenes and most of the DEGs were related to metabolic pathways in resistant and susceptible genotypes (Que *et al.*, 2014).

Recently, lncRNAs have been found to play an important role in gene regulation by acting as endogenous target mimics. The size of lncRNAs is usually more than 200 nucleotides in length (Chen, 2009; Rinn and Chang, 2012). In plants, identification of long non-coding RNA is more recent and not as comprehensive compared to other eukaryotes (Ulitsky and Bartel, 2013; Zhang *et al.*, 2014). Non-coding RNA regulates the expression

**Table 1.2** List of mRNA and lncRNA studies and NGS platforms used for sequencing in different crops

| Number | Crop | Platform | Validation method | Condition | Reference |
|---|---|---|---|---|---|
| 1 | *Glycine max* | Illumina | – | Fourteen diverse tissues | Severin *et al*., 2010 |
| 2 | *Cucumis sativus* | Roche-454 | – | Flower buds | Guo *et al*., 2010 |
| 3 | *Lens culinaris* | Roche 454 | – | Developmental stages | Kaur *et al*., 2011 |
| 4 | *Daucus carota* | Illumina | – | Root and leaf tissues | Iorizzo *et al*., 2011 |
| 5 | *Oryza sativa* | Illumina | qRT-PCR | Biotic stress | Kawahara *et al*., 2012 |
| 6 | *Oryza sativa* | Illumina | qRT-PCR | Biotic stress | Bagnaresi *et al*., 2012 |
| 7 | *Triticum aestivum* | Illumina | – | Root and shoot tissue from seedlings | Schreiber *et al*., 2012 |
| 8 | *Arachis hypogaea* | Illumina | qRT-PCR | Immature seeds | Zhang *et al*., 2012 |
| 9 | *Cyanococcus* | Illumina | qRT-PCR | Fruit skin and pulp | Li *et al*., 2012 |
| 10 | *Cyanococcus* | Roche 454 | qRT-PCR | Cold acclimation | Rowland *et al*., 2012 |
| 11 | *Carnation* | GS FLX 454 pyrosequencing | – | Flower bud, flowers, leaves and stem (ethylene treated and control) | Tanase *et al*., 2012 |
| 12 | *Allium sativum* | Illumina | – | Vegetative buds | Sun *et al*., 2012 |
| 13 | *Saccharum* | Illumina | – | Leaves | Vicentini *et al*., 2012 |
| 14 | *Chrysanthemum* | Illumina | – | Stems and leaves | Wang *et al*., 2013 |
| 15 | *Chrysanthemum* | Illumina | – | Dehydration stress | Xu *et al*., 2013 |
| 16 | *Allium cepa* | Roche-454 | – | Vernalized bulbs, leaves, unopened umbels, bulbs, and roots | Duangjit *et al*., 2013 |
| 17 | *Oryza sativa* | Illumina | qRT-PCR | Biotic stress | Chao *et al*., 2014 |
| 18 | *Oryza sativa* | Illumina | qRT-PCR | Anthers before flowering, pistils before flowering, spikelets and shoots | Zhang *et al*., 2014 |
| 19 | *Hordeum vulgare* | Roche 454 | – | Drought stress | Bedada *et al*., 2014 |
| 20 | *Eleusine coracana* | Ion torrent | qRT-PCR | Salinity stress | Rahman *et al*., 2014 |
| 21 | *Cicer arietinum* | Illumina | – | Vegetative and reproductive tissues | Singh and Jain, 2014 |
| 22 | *Mangifera indica* | Illumina | qRT-PCR | Hot water treatment | Luria *et al*., 2014 |
| 23 | *Oryza sativa* | Illumina | qRT-PCR | Developmental stages | Wang *et al*., 2015b |
| 24 | *Zea mays* | Illumina | qRT-PCR and polyclonal antibody | Endosperm development | Li *et al*., 2015 |
| 25 | *Oryza sativa* and *Zea mays* | Illumina | qRT-PCR | Flower buds, flowers, flag leaves, roots, (before and after flowering stage), shoot and root tissues | Wang *et al*., 2015a |
| 26 | *Oryza sativa* and *Zea mays* | Illumina | qRT-PCR | Flower bud, milk grain and mature seed | Wang *et al*., 2015a |
| 27 | *Triticum aestivum* | Illumina | – | Root, leaf, stem, spike, and grain at three developmental stages | Pingault *et al*., 2015 |
| 28 | *Hordeum vulgare* | Illumina | qRT-PCR | Boron treatment | Tombuloglu *et al*., 2015 |
| 29 | *Medicago truncatula* | Illumina | qRT-PCR | Osmotic and salt stress | Wang *et al*., 2015 |

**Table 1.2** Continued

| Number | Crop | Platform | Validation method | Condition | Reference |
|---|---|---|---|---|---|
| 30 | *Mangifera indica* | Illumina | qRT-PCR | Fruit ripening | Dautt-Castro *et al.*, 2015 |
| 31 | *Gossypium* | Illumina | qRT-PCR | Cotton fibre | Wang *et al.*, 2015c |
| 32 | *Oryza sativa* | Illumina | qRT-PCR | Selinium and salicylic acid priming | Hussain *et al.*, 2016 |
| 33 | *Sorghum bicolor* | SMRT(single-molecule real-time) cells (Pacific Biosciences) | qRT-PCR, cloning and sequencing | Seedlings | Abdel-Ghany *et al.*, 2016 |
| 34 | *Sorghum bicolor* | Pacific Biosciences SMRT long-read isoform sequencing | RT-PCR | 8 day old seedlings | Abdel-Ghany *et al.*, 2016 |
| 35 | *Helianthus annulus* | Illumina | qRT-PCR | Disc florets at R2 development stage | Florez-Zapata *et al.*, 2016 |
| 36 | *Arabidopsis thaliana* | Illumina | qRT-PCR, RACE, western blot | Biotic stress | Gao *et al.*, 2016 |
| 37 | *Actinidia deliciosa* | Illumina | qRT-PCR | Fruit samples from five 5-year-old plants | Tang *et al.*, 2016 |
| 38 | *Raphanus sativus* | Illumina | qRT-PCR | Taproot sample at developmental stages | Yu *et al.*, 2016 |
| 39 | *Cicer arietinum* | Illumina | qRT-PCR | Salinity and drought stress | Garg *et al.*, 2016 |
| 40 | *Mangifera indica* | Roche 454 and Illumina | qRT-PCR | Fruit ripening | Srivastava *et al.*, 2016 |
| 41 | *Actinidia chinensis* | Illumina | RT-PCR and qRT-PCR | Fruit samples from 5-year-old plant and 20, 120 and 127 days after pollination (DAP) | Tang *et al.*, 2016 |
| 42 | *Rosa* | Illumina Hiseq | qRT-PCR | Cold stress | Zhang *et al.*, 2016 |
| 43 | *Allium cepa* | Illumina | qRT-PCR | Floral inflorescences | Liu *et al.*, 2016a |
| 44 | *Raphanus sativus* | Illumina | qRT-PCR | Bolting and flowering stage root, stem, leaf, flower, floral buds | Nie *et al.*, 2016 |
| 45 | *Raphanus sativus* | Illumina | qRT-PCR | Salt stress | Sun *et al.*, 2016 |
| 46 | *Brassica napus* | Illumina | qRT-PCR | Biotic stress | Joshi *et al.*, 2016 |
| 47 | *Helianthus* | Illumina | – | Non-meiosis versus meiosis-specific cell | Flórez-Zapata *et al.*, 2016 |
| 48 | *Gossypium* | Illumina | qRT-PCR | Drought stress | Lu *et al.*, 2016 |
| 49 | *Gossypium* | Illumina | qRT-PCR | Cotton fibres and leaves | Zou *et al.*, 2016 |

level of target genes via various molecular mechanisms (Quan *et al.*, 2015; Zhu and Wang, 2012). lncRNAs act as a regulator in important biological processes by enhancing target site accessibility to RNA polymerases, formation of RNA-dsDNA triplex, inhibition of RNA polymerase activities as well as regulation of transcription factors (Lipshitz *et al.*, 1987; Nguyen *et al.*, 2001; Willingham *et al.*, 2005; Martianov *et al.*, 2007; Hirota *et al.*, 2008; Mariner *et al.*, 2008). They also have a role in post-transcriptional modulations of mRNA. lncRNAs are found to have role in regulating complex gene regulatory networks involved in plant development and stress management (Crespi *et al.*, 1994;

Ding *et al.*, 2012; Heo and Sung, 2011; Zhang *et al.*, 2014).

## Identification of small non-coding RNA by NGS technology

Small RNAs (sRNAs) are another important class of RNA molecules of 21–24 bases long and include microRNAs (miRNAs), short interfering RNAs (siRNAs), transacting siRNAs (ta-siRNAs) and cis/trans-natural antisense small-interfering RNAs (nat-siRNAs). These small RNAs are non-coding, essential entities that regulate gene expression in epigenetic processes almost in every domain of life (Ruiz-Ferrer and Voinnet, 2009; Zhang *et al.*, 2011). The sequence alterations in miRNA or other small non coding RNAs can be mapped by genomic sequencing or RNA-seq, while the expression levels can be determined by RNA-seq or deep sequencing or expression microarrays (Singh *et al.*, 2012). The alliance of gene expression data with a small RNA data set will help to understand how different biological processes coordinate together in a cellular context (Jain, 2012). In recent times, a large number of studies involving NGS technology have been reported primarily to identify novel as well as conserve miRNA, along with their targets under different stresses or developmentally regulated conditions (Table 1.3) sRNAome analysis of *Arabidopsis thaliana* revealed while expression of miR156, miR399, miR778, miR827, and miR2111 was induced, yet, expression of miR169, miR395, and miR398 was repressed under phosphate (Pi) deficiency indicating that these miRNAs may be involved in Pi uptake in *Arabidopsis* (Hsieh *et al.*, 2009). In *Brachypodium distachyon*, a model crop, three conserved and 25 novel miRNAs showed significant change of expression in response to cold stress and differential expression profiling of 94 conserved miRNAs from 28 families during vegetative and reproductive tissues (Zhang *et al.*, 2009; Wei *et al.*, 2009). A review describing the study of various aspects of miRNA using deep sequencing tools was documented for further information (Yang and Li, 2012).

### Agricultural crops

Rice being a major cereal crop, various transcriptome studies to identify non-coding RNAs have been reported. NGS was used to categorize different classes of small RNA regulatory elements from mature rice grain and seedlings (Heisel *et al.*, 2008; Guo *et al.*, 2012; Campo *et al.*, 2013; Li *et al.*, 2014). These studies together signified the role of miRNAs in rice blast disease and stripe virus resistance. Besides biotic factors, a study involving abiotic stresses has also been reported and elucidated 294 known and 539 novel heat-responsive miRNAs during heat stress (Li *et al.*, 2015b).

In maize, the epigenome was critically scrutinized by NGS technologies for its relationships to mRNA and small RNA transcriptomes (Wang *et al.*, 2009; Liu *et al.*, 2014; Zhou *et al.*, 2016) to understand the molecular aspects underlying maize ear development and to elucidate the rice *black-streaked dwarf* virus-responsive pathway in maize.

Wheat being hexaploid, its genome is considered to be genetically complex. In an earlier NGS-based study on wheat, 58 miRNAs comprising 43 miRNA families were identified, of which 20 families were conserved and 23 were found to be novel (Yao *et al.*, 2007). Further, in another study, 170 conserved miRNAs representing 25 families and 23 novel miRNAs were also identified (Wei *et al.*, 2009). Further, in wheat, a total of 2076 small RNAs were identified (Yao *et al.*, 2010). Similarly a set of wheat miRNAs which play roles in regulating wheat response to powdery mildew pathogen *Erysiphe graminis* f. sp. *tritici* and heat stress were also identified using high-throughput sequencing (Xin *et al.*, 2010). Tang *et al.* (2012) analysed the male sterility system of thermosensitive genic male sterile (TGMS) lines of wheat and concluded that miR167 and tasiRNA-ARF play a role in the developmental response to cold stress and the regulatory pathways of sRNA that were linked with male sterility. A brief summary of work related to wheat is given in Table 1.3.

Though initial efforts were concentrated to study the small RNAs in cereals, of late efforts are also being made to understand the role of these RNA molecules in pulse, oilseeds and other crops of economic importance. In *Medicago truncatula*, an important model legume, a small RNA library was generated and conserved and novel small RNAs (miR1507, miR2118, miR2119 and miR2199) were identified. They also identified three novel transcripts encoding TIR-NBS-LRR disease resistance (Jagadeeswaran *et al.*, 2009). The miRNA (miR408) could induce drought tolerance in chickpea (Jain *et al.*, 2014; Srivastava *et al.*, 2015;

**Table 1.3** Small-RNA studies and NGS platforms used for sequencing in different crops

| Number | Crop | Platform used | Validation method | Experimental conditions | Reference |
|---|---|---|---|---|---|
| 1 | *Triticum aestivum* L. line 3338 w | 454 Life Sciences™ technology | RNA gel blot analysis and RT-PCR | Wheat and monocot-specific miRNAs | Yao *et al.*, 2007 |
| 2 | *Oryza sativa* spp. *japonica* cv. Nipponbare | 454 Life Sciences platform | Small RNA blotting and 5′RACE | Mature rice grain and seedlings | Heisel *et al.*, 2008 |
| 3 | *Solanum lycopersicum* | 454 Life Sciences™ technology | Northern blotting and 5′RACE | Fruit ripening | Moxon *et al.*, 2008 |
| 4 | *Arabidopsis thaliana* ecotype Columbia | Solexa sequencing technology (Illumina) | RLM-5′RACE | Phosphate deficiency | Hsieh *et al.*, 2009 |
| 5 | *Medicago truncatula* Gaertner cv. Jemalong | 454 Life Sciences™ technology | Small RNA blotting and 5′RACE | Legume specific | Jagadeeswaran *et al.*, 2009 |
| 6 | *Zea mays* inbred line B73 | Solexa Sequencing | Computational validation | Tissue-specific miRNA distribution | Wang *et al.*, 2009 |
| 7 | *Brachypodium distachyon* BD21-3 cv. Chinese Spring | Solexa Sequencing | Northern blotting, RT-PCR and 5′RACE | vegetative and reproductive growth stage | Wei *et al.*, 2009 |
| 8 | *Brachypodium distachyon* | Illumina-Solexa 1 G Genetic Analyzer | RNA gel blot analysis | Cold stress | Zhang *et al.*, 2009 |
| 9 | *Triticum aestivum* L line JD8-Pm30 | Solexa sequencing | RNA gel blot analysis and qRT-PCR | Powdery mildew infection and heat stress | Xin *et al.*, 2010 |
| 10 | *Triticum aestivum* L | 454 Life Sciences™ technology | Northern blotting and RT-PCR | heat, cold, salt and dehydration stress | Yao *et al.*, 2010 |
| 11 | *Arachis hypogaea* L. cultivar Huayu19 | Solexa 1G Genome Analyzer | Stem–loop RT-PCR and qRT-PCR | Tissue- and/or growth stage specific expression | Chi *et al.*, 2011 |
| 12 | *Glycine max* inbred line of 'HJ-1' | Solexa sequencing | Northern blot analysis and qRT-PCR | Drought, salinity, and alkalinity stress | Li *et al.*, 2011 |
| 13 | *Carthamus tinctorius L.* | Illumina Solexa Genome Analyzer | Computational validation | Tissue specific (Seed, leaf and petal) | Li *et al.*, 2011 |
| 14 | *Glycine max* cv. Heinong44 | Solexa sequencing | RT-PCR and RLM-5′RACE | miRNA biogenesis | Song *et al.*, 2011 |
| 15 | *Brassica rapa* ssp. *chinensis* | Illumina Genome Analyzer | Northern blot analysis, qRT-PCR, and cRT-PCR | Heat responsive | Wang *et al.*, 2011 |
| 16 | *Brassica rapa* ssp. *chinensis* | Illumina Genome Analyzer | Northern Blotting, RT-PCR and 5′RACE | Heat stress | Yu *et al.*, 2012 |
| 17 | *Oryza sativa* L. *japonica*. cv. Nipponbare | Illumina Solexa sequencing | sRNAs gel blot analysis and qRT-PCR | Virus infected | Guo *et al.*, 2012 |
| 18 | *Triticum aestivum* TGMS line BS366 | Illumina Genome Analyzer | RLM-5′RACE and In situ hybridisation | Cold stress | Tang *et al.*, 2012 |
| 19 | *Brassica napus* line, Westar | Solexa sequencing (Illumina) | qRT-PCR and RLM-5′RACE | miRNAome | Xu *et al.*, 2012 |
| 20 | *Oryza sativa* L. cv. Nipponbare, *Oryza glaberrima* and wild rice species | 454 Life Sciences Technology and Microarray | Northern Blotting, qRT-PCR | *Magnaporthe oryzae* responsive | Campo *et al.*, 2013 |

**Table 1.3** Continued

| Number | Crop | Platform used | Validation method | Experimental conditions | Reference |
|---|---|---|---|---|---|
| 21 | *Carthamus tinctorius L.* | Illumina HiSeq™ 2000 | Northern blot hybridization, stem–loop RT-PCR | Developing seeds with high linoleic and oleic acid content | Cao *et al.*, 2013 |
| 22 | *Saccharum* spp. cultivars RB867515 | Solexa platform | Stem–loop RT-PCR and qRT-PCR | Drought responsive | Gentile *et al.*, 2013 |
| 23 | *Brassica napus* L. cv.DH12075 | SOLiD v3 sequencing | qRT-PCR | Seed maturation | Huang *et al.*, 2013 |
| 24 | *Triticum aestivum* L. | Pyrosequencing | Small RNA blot analysis | siRNA-mediated silencing of TAS3 transcripts | Li *et al.*, 2013 |
| 25 | *Gossypium hirsutum* | Illumina Genome Analyzer | qRT-PCR and RLM-5′RACE | Somatic embryogenesis | Yang *et al.*, 2013 |
| 26 | *Solanum melongena* L. | Illumina/Solexa | qRT-PCR | *Verticillium dahliae* infection | Yang *et al.*, 2013 |
| 27 | *Solanum tuberosum* group Andigena (line ADG573) | Illumina GAIIX sequencer | Computational validation | miRNAome | Zhang *et al.*, 2013 |
| 28 | Miniature Tomato cv. Micro-Tom | Illumina Solexa system | qRT-PCR | Cucumber mosaic virus responsive | Feng *et al.*, 2014 |
| 29 | *Cicer arietinum* L. genotype ICC4958 | Illumina Genome Analyser | qRT-PCR | miRNAome profiling | Jain *et al.*, 2014 |
| 30 | *Solanum tuberosum* cv. Kufri Chandramukhi | Illumina GAIIx | qRT-PCR and RLM-RACE | Different developmental stages | Lakhotia *et al.*, 2014 |
| 31 | *Oryza sativa* | Illumina sequencing | sRNAs gel blot analysis, overexpressing transgenic and qRT-PCR | *Magnaporthe oryzae* infected (0, 12, 24, 48, 72 hpi) | Li *et al.*, 2014 |
| 32 | *Zea mays* inbred lineB73 | Solexa Sequencing | qRT-PCR | Maize ear development | Liu *et al.*, 2014 |
| 33 | *Coffea canephora* | Illumina HiSeq 2000 | Computational validation | miRNAome | Loss-Morrais *et al.*, 2014 |
| 34 | *Brassica napus* L. | Solexa sequencing (Illumina) | qRT-PCR and RLM-5′RACE | *Verticillium longisporum* infection | Shen *et al.*, 2014 |
| 35 | Broccoli (*Brassica oleracea* var. *italica*) | Illumina HiSeq | qRT-PCR | Salinity stress | Tian *et al.*, 2014 |
| 36 | *Poncirus trifoliata* (L.) Raf. | Illumina Genome Analyser | qRT-PCR | Cold responsive | Zhang *et al.*, 2014 |
| 37 | *Solanum tuberosum* tetraploid cultivar 'Zihuabei'. | Solexa sequencing technology | qRT-PCR | Drought stress | Zhang *et al.*, 2014 |
| 38 | *Solanum linnaeanum*, brinjal | Illumina GAIIx | qRT-PCR | Salt stress | Zhuang *et al.*, 2014 |
| 39 | *Vigna unguiculata* | Illumina Genome Analyzer | Northern blot hybridization | Drought stress | Barrera-Figueroa *et al.*, 2011 |
| 40 | *Salicornia europaea* | Illumina sequencing | qRT-PCR and RLM-5′RACE | Salt stress | Feng *et al.*, 2015 |

**Table 1.3** Continued

| Number | Crop | Platform used | Validation method | Experimental conditions | Reference |
|---|---|---|---|---|---|
| 41 | *Brassica napus* L. cv.DH12075 | SOLiD v3 sequencing system | qRT-PCR | Seed maturation | Hayzadeh *et al.*, 2015 |
| 42 | *Rosa multiflora* Thunb. | Solexa-Illumina platform | RT-PCR | Pathogen stress (Virus and Viroids) | He *et al.*, 2015 |
| 43 | *Cajanus cajan* L. | Whole genome shotgun sequencing | Computational validation | Plant growth and development | Kompelli *et al.*, 2015 |
| 44 | *Oryza sativa* cultivar Nagina 22 | Ion Proton Sequencer | qRT-PCR | Heat stress | Li *et al.*, 2015 |
| 45 | *Cymbidium ensifolium* 'Tiegusu' | Solexa technology | RT-qPCR | Floral development | Li *et al.*, 2015 |
| 46 | Turnip cultivar 'Chang Huang Man Jing' | Illumina HiSeqTM 2000 | qRT-PCR | Tuberous root development | Li *et al.*, 2015 |
| 47 | *Raphanus sativus* L. advanced inbred line 'NAU-YH' | Solexa sequencing (Illumina) | RT-qPCR | Chromium stress | Liu *et al.*, 2015 |
| 48 | *Solanum pimpinellifolium* L3708 | HiSeq 2000 Sequencing System | qRT-PCR | *Phytophthora infestans* (Pathogen resistance) | Luan *et al.*, 2015 |
| 49 | *Triticum aestivum* L. cv. Hanxuan10 and Zhengyin1 w | Genome Analyzer IIx System | Northern blotting, RT-PCR and qPCR | Dehydration stress | Ma *et al.*, 2015 |
| 50 | *Raphanus sativus* L. inbred line 'NAU-LU127' | Illumina HiSeq™ 2000 | RT-qPCR | Bolting and flowering time related | Nie *et al.*, 2015 |
| 51 | *Camelina sativa* | Illumina HiSeq 2000 | RT-PCR | Different developmental stages | Poudel *et al.*, 2015 |
| 52 | *Brassica napus* cultivars Tapidor and Ningyou7 | Illumina HiSeq 2000 | qRT-PCR and RLM-RACE | Double haploid lines | Shen *et al.*, 2015 |
| 53 | *Cicer arietinum* L. genotype IC4958 | Illumina Genome Analyser | Small RNA gel blot and 5'RACE | Phasi-siRNAs discovery | Srivastava *et al.*, 2015 |
| 54 | *Raphanus sativus* L. advanced inbred line 'NAU-YH' | Illumina HiSeq™ 2000 | RT-qPCR | Salinity stress | Sun *et al.*, 2015 |
| 55 | *Brassica rapa* ssp. *pekinensis* | Solexa platform | qRT-PCR and Microscopy | Pollen abortion and Bud development | Wei *et al.*, 2015 |
| 56 | *Gossypium hirsutum* L. cultivar TM-1 | Illumina HiSeq high-throughput sequencing platform | Stem–loop RT-PCR and qRT-PCR | Drought- and salinity-responsive | Xie *et al.*, 2015 |
| 57 | *Solanum lycopersicum* L. cv. 'Ailsa Craig' | Illumina Genome Analyzer I | qRT-PCR | Tomato pedicel abscission | Xu *et al.*, 2015 |
| 58 | *Arachis hypogaea* L. Luhua 14 and *A. glabrata* | Illumina Genome Analyzer | qRT-PCR | Pathogen resistance *Ralstonia solanacearum*, bacterial wilt | Zhao *et al.*, 2015 |
| 59 | *Camellia sinensis* (L.) O. Kuntze cv. lLongjing 43' | Illumina HiSeq 2000 platform | qRT-PCR | Chilling and freezing | Zheng *et al.*, 2015 |
| 60 | *Ipomoea batatas* L.,cv. Ningzishu 1 | Solexa sequencing technology | Computational validation | miRNAome | Bian *et al.*, 2016 |

**Table 1.3** Continued

| Number | Crop | Platform used | Validation method | Experimental conditions | Reference |
|---|---|---|---|---|---|
| 61 | *Brassica napus* | Illumina HiSeq2000 | Computational validation | Synthetic variety | Fu *et al.*, 2016 |
| 62 | *Camellia sinensis* | Illumina HiSeq2500 | qRT-PCR | Drought stress | Liu *et al.*, 2016b |
| 63 | *Moringa oleifera* | Illumina HiSeq | qRT-PCR and Western blotting | Pharmacological potential properties | Pirrò *et al.*, 2016 |
| 64 | *Solanum tuberosum* Zhuangshu No. 3 | Illumina HiSeq2000 | qRT-PCR | Secondary metabolism | Qiao *et al.*, 2016 |
| 65 | *Raphanus sativus* L. advanced inbred line 'NAU-YH' | Illumina HiSeq system | RT-qPCR | Taproot thickness | Yu *et al.*, 2016 |
| 66 | 'Summer Black' grapevine (hybrids of *V. vinifera* and *V. Labrusca* | Solexa sequencing (Illumina) | qRT-PCR and RLM-RACE | In response to exogenous ethylene | Zhao *et al.*, 2016 |
| 67 | *Zea mays* inbredlineB73 | Illumina HiSeq | qRT-PCR | Black-streaked dwarf virus, maize rough dwarf disease (MRDD) | Zhou *et al.*, 2016 |
| 68 | *Solanum lycopersicum* 'Rui Xin' | Illumina HiSeq2500 | Computational validation | Chilling injury | Zuo *et al.*, 2016 |

Hajyzadeh *et al.*, 2015). Similarly novel miRNAs and their targets involved in stress response were also identified in pigeonpea and chickpea (Kompelli *et al*, 2015; Barrera-Figueroa *et al.*, 2011).

Oilseed crops, largely grown for human consumption, are important for Indian agriculture. One of the important oilseed crops, soybean, is grown all over the world. To understand the regulatory network of miRNAs and their functions during seed development as well as to the miRNAome profiling associated with abiotic stress responses in soybean, NGS technology was used to identify the differentially expressed miRNAs (Song *et al.*, 2011; Li *et al.*, 2011b). Besides soybean, members of Brassicaceae family, specifically *Brassica napus*, is a major source of vegetable oil. There are many reports in *Brassica napus*, which used NGS techniques to decipher various molecular mechanisms during different stages of growth, during their interaction with biotic and abiotic stresses, oleic acid content, etc. (Table 1.3). Commercial or cash crops are another important category belonging to agricultural crops. Some of the major crops of India on which miRNA studies have been conducted are cotton (*Gossypium hirsutum*), tea (*Camellia sinensis* L.), *Coffea canephora* and sugarcane. In cotton, using NGS tools, 25 novel miRNAs that were associated with somatic embryogenesis were identified (Yang *et al.*, 2013). Similarly, miRNAs in response to drought and salinity stress were also studied in cotton, and miRNAs which played role in drought and fibre development were identified (Xie *et al.*, 2015). A brief summary of works is given in Table 1.3.

### Horticulture crops

Besides agricultural crops, horticultural crops are also a significant component of farming systems and sources for many dietary supplements. Various studies based on NGS technologies were conducted to identify and characterize miRNAs in these crops (Table 1.3). For instance, miRNAs involved in tomato fruit during ripening, *Phytophthora* infections and cucumber mosaic resistance, and chilling injury were identified (Moxon *et al.*, 2008; Luan *et al.*, 2015; Zou *et al.*, 2016). Similarly, *Verticillium dahlia* responsive miRNAs in *Solanum linnaeanum* (Yang *et al.*, 2013), and those involved in drought stress in *Solanum tuberosum* (Lakhotia *et al.*, 2014) were also identified. Overall, small non-coding

RNAs is a broad class of regulatory RNAs, and behaves as protein counterparts involved in regulating post transcriptional gene silencing and translational repression. Deep sequencing employs efficient, economical, massive parallel sequencing, generating millions of small RNA sequence reads from a given sample. sRNAome by deep sequencing quantifies absolute abundance and allows for the discovery of novel microRNAs that have escaped previous cloning and standard sequencing efforts (Creighton *et al.*, 2009). Some miRNAs can significantly affect plant traits, including virus resistance, nematode resistance, drought and salinity tolerance, heavy metal detoxification, biomass yield, grain yield, fruit development and flower development. Therefore, miRNAs are considered as a newly identified gene resource for the genetic improvement of crop plants (Zheng and Qu, 2015). Therefore, these all studies which analysed and identified miRNAs have the potential to enhance food security by helping breed crop cultivars with improved agronomic traits (Zhou and Luo, 2013). Studies have indicated that small non-coding RNA loci, like protein-coding genes, could be targets of domestication selection and play an important role in crop domestication, and improvement, abiotic stress, plant–pathogen interactions and breeding useful traits (Ruiz-Ferrer and Voinnet, 2009; Wang *et al.*, 2010).

## NGS and metagenomics of plant–microbe interaction

Microbes are the most abundant living organisms on earth. In spite of their abundance, only small fraction of them have been explored and very few of them can be grown in laboratory conditions. Overall, large chunks of the microbial population are out of the reach of scientists largely due to an inability to culture them under laboratory conditions. Quite intriguingly, it has been observed that microorganisms growing on media and those directly obtained from natural samples showed very large differences in their number, displaying so called the great plate count anomaly (Staley *et al.*, 1985). Therefore, lack of standard culture media and growth conditions has largely led to the restricted scientific exploration of these microbial populations. In the context of foregoing observations, it becomes necessary to study the microbial populations in their natural conditions so as to capture the microbial community or metagenome (Woese *et al.*, 1987). Metagenome is the use of advanced techniques to analyse microbial communities taken directly from the natural sources without growing them under laboratory conditions (Chen *et al.*, 2005).

### Approaches used in metagenomics

Broadly two approaches are used to study metagenomics, i.e. sequence-based metagenomics and function-based metagenomics. The sequence-based study depends on identifying the complete genetic sequence of microorganisms present in the sample. In most of the studies targeting microbial communities, 16S rRNA gene sequences were used to identify the species. This method is useful to study the evolutionary conservation and phylogenetic analysis of the microbes present in the samples (Claesson *et al.*, 2010; Creer *et al.*, 2016). On the contrary, function-based metagenomics explores the products of the microbial community to study that specific community (Coughlan *et al.*, 2015). Subsequently it was sequencing techniques, particularly the introduction of NGS techniques, that revolutionized the plant–microbial metagenomic studies. This sequencing technology opened a new way for discovering and analysing these organisms at genome level, which is a culture independent technology that utilizes a combination of various research methods, specifically NGS and bioinformatics tools. Plants and microbes are in continuous interaction with each other in the environment. The association of microbes with the plant can be of endophytic (inside the plant) or epiphytic (attached to the plant). Plant–microbial associations could be positive interactions, neutral associations or negative interactions (pathogens). Almost every part of a plant is inhabited by a microbial community, yet rhizosphere and phyllosphere are the major sites for plant microbe interactions for most of the microbes. Microbiota of these two regions (rhizosphere and phyllosphere) are so closely associated with plants that even they are called as second genome of the plant (Berendsen *et al.*, 2012). Several studies have been carried out to analyse the microbiome of various plants sampled from the phyllosphere, rhizosphere, and various parts, and were found beneficial to the plant (Lugtenberg *et al.*, 2009; Porras-Alfaro *et al.*, 2011; Vorholt *et al.*, 2012). The details of the metagenomics studies aimed at plant–microbial interactions and plant microbiota are given in Tables 1.4 and 1.5. Recently, the 454 NGS technique was used to study the diversity, community structure, and dynamics of endophytic bacteria in different plant species and it found four classes of *Proteobacteria* with *Alphaproteobacteria* as the dominant class and revealed that host plant species had greater influences on type of bacterial communities (Ding and Melcher, 2016).

Besides their application in metagenomics, NGS techniques also have great applications in the study of metatranscriptomics. Metatranscriptomics includes analysis of gene transcripts directly isolated from the entire community of organisms. Metatranscriptome study is also referred to as environmental transcriptomes, microbial community gene expression profiles, microbial community RNAs and whole community transcripts. The metatranscriptome field has opened a door to study various aspects around environmental community, i.e. active community members and metabolic pathways (Urich *et al.*, 2008). The metatranscriptomics approach has been used to identify various genes actively involved in *Eichhornia crassipes* and *Fusarium verticillioidies* association, in strawberry plants for defining the fungal communities associated with different organs of plants (Luo *et al.*, 2015; Abdelfattah *et al.*, 2016). Recently, a study combining NGS and metagenomic analysis was conducted to generate large number of cDNAs using model system tomato pepino mosaic virus. Subsequently, the same approach was used in the study of globe amaranth (*Gomphrena globose*) infected with an unknown pathogen. Therefore, this method hastens the process of development of routine assays for new viral pathogens (Adams *et al.*, 2009).

NGS technique was also used for sequencing whole viral genomes to undertake plant metagenomic studies to discover new viruses. Adams *et al.* (2013) identified the presence of *Maize chlorotic mottle virus* and *Sugarcane mosaic virus*, causal agents

**Table 1.4** Plant–microbial interaction studies using NGS-based metagenomics

| Number | Host | Interacting partner | Environment | Conclusions | Platform | Reference |
|---|---|---|---|---|---|---|
| 1 | Wheat | *Azospirillum brasilense* | Rhizosphere | Up-regulated genes related to nutrient uptake, cell cycle, and nitrogen assimilation that enhance productivity and growth | SOLiD | Camilios-Nato *et al.*, 2014 |
| 2 | Wheat | Microbiome | Rhizosphere | Fertilizers with high nutrient availability and long-term storage leads to less microbe interaction with crop | 454 Pyrosequencing | Ai *et al.*, 2015 |
| 3 | *Eichhornia crassipes* | *Fusarium verticillioidies* | Rhizosphere | Mutualistic action of plant and fungi was efficient for bioremediation | Illumina Hi Seq 2500 | Luo *et al.*, 2015 |
| 4 | Potato | *Burkholderia phytofirmans* PsJN | Rhizosphere | Stress signal perceived by plant are also affects plant endophyte association | Illumina Hi Seq 2000 | Tezerji *et al.*, 2015 |
| 5 | Soybean | Mycovirus | Phyllosphere | Novel mycoviruses were identified | Illumina Hi Seq 2500 | Marzona *et al.*, 2016 |
| 6 | Strawberry | Fungi | Phyllosphere | Diverse fungal organisms inhabits on plants and *Botrytis* and *Cladosporium* were dominant. | 454 GS FLX+System | Abdelfattah *et al.*, 2016 |
| 7 | Rice | *S. epidermidis* | Rhizosphere | *S. epidermidis* of plant and animal origin are diversified at genome level | Illumina MiSeq | Chaudhry *et al.*, 2016 |
| 8 | Non-cultivated plants | Endophytic bacteria | Phyllosphere | Proteobacteria found to be highest phylum (85.42%). Acidobacteria (0.59%) lowest on leaf of all five plants | 454 Pyrosequencing | Ding *et al.*, 2016 |

**Table 1.5** Studies of plant associated microbiota using NGS techniques

| Number | Host | Environment | Conclusions | Platform | References |
|---|---|---|---|---|---|
| 1 | Soybean | Phyllosphere | Composition of microbiota and proteomes remain consistent in different plants species | Roche 454 | Delmotte *et al.*, 2009 |
| 2 | Rice | Phyllosphere and Rhizosphere | Phyllosphere and rhizosphere region analysis to identify bacteria and archea in association with rice | Roche 454 | Knief *et al.*, 2012 |
| 3 | Tamarisk, soybean, *Arabidopsis thaliana* | Phyllosphere | Phyllosphere contains different groups of phototrophic organisms that are phylogenetically diverse | Roche 454 | Atamna-Ismaeel *et al.*, 2012 |
| 4 | Barley | Rhizosphere | Mineral phosphate solubilization genes were identified | Roche 454 | Chhabra *et al.*, 2013 |
| 5 | Lotus | Rhizosphere | No major microbial communities changes with respect to phytic acid utilization. Phytic acid utilization genes were identified | Roche 454 | Unno and Shinano, 2013 |
| 6 | Tomato | Phyllosphere | Different organs of plants have distinct microbial communities | Roche 454 GS Titanium FLX | Ottesen *et al.*, 2013 |
| 7 | Soybean | Rhizosphere | Plants prefer a specific microbial community beneficial to its growth and function | Roche 454 GS-FLX | Mendes *et al.*, 2014 |
| 8 | *Aloe vera* | Rhizosphere | *Firmicutes*, *Actinobacteria* and *Bacteriodetes*, four prominent phyla were identified as beneficial for bioactive compound production | Illumina MiSeq | Akinsanya *et al.*, 2015 |
| 9 | *Genlisea* species | Phyllosphere | Complex food interaction between plant *Genlisea* and its entrapped microbiome | Illumina Hi Seq 2000 | Cao *et al.*, 2015 |

of lethal necrosis in Kenyan maize using NGS (Wangai *et al.*, 2016). A detailed review describing the various roles of NGS in viral diagnostics has already been published by Boonham *et al.* (2014).

NGS technology has also been used successfully to determine the begomoviral genome and their associated satellites from begomovirus-affected tomato and okra plants by using metagenomics (Idris *et al.*, 2014). In another study, NGS and metagenomics approaches together were used to study native bacterial microflora diversity in different anatomical organs of *Solanum lycopersicum* (Ottesen *et al.*, 2013). In tomato, the root microbiome was studied in order to understand the endophytes and their association with root-knot nematodes (*Meloidogyne* spp.) (Tian *et al.*, 2015). Similarly in rice, a metagenomic rice endophyte DNA library was constructed and further 16S rRNA gene sequence information was studied (Sessitsch *et al.*, 2012). Similarly, metagenomic analysis of olive-knot to decipher the role of different bacterial species in the disease establishment has been reported (Passos da Silva *et al.*, 2014). Even the role of the enzymatic repertoire (glycoside hydrolases) of a microbial community in degradation of crude cotton biomass was reported using NGS-based metagenomics approach (Zhang *et al.*, 2016).

## NGS applications in exome and captured sequencing

Selection of genomic regions of interest and enrichment of these regions is called captured sequencing. This technology is a revolutionary process for the selective enrichment of targeted genomic regions from the complex genomic DNA. Targeted-capture is used to enrich the sequences of interest before carrying out NGS, such as repetitive sequences (Syring *et al.*, 2016), exome (Neves *et al.*, 2013), and gene space regions (Zhou and Holliday, 2012). This technology permits the isolation of target loci from the background of the entire genome. In comparison with other sequencing technologies, captured sequencing is inexpensive, quick and simple. The scale of capture can range from several targeted loci to over a million target regions (Agilent 2011; Microarray 2011; NimbleGen 2011), making it adaptable for both small-scale and large scale projects. This technology holds promise for plant genomes because they are large, complex and contain a large amount of repetitive elements. In the present review, we are focusing on captured sequencing and exome sequencing. Exome are comprised of those sequences in DNA that code for a protein. It may also include the functional non-protein-coding sequences such as micro-RNAs, long intergenic non-coding RNA, etc. as well as specific candidate regions. The technique to select and enrich the exomes followed by their sequencing using NGS technology is called exome sequencing (Warr *et al.*, 2015). Initially, PCR was widely used to capture the sequence (Mamanova *et al.*, 2010), followed by circularization of capture sequences using a suitable circularization technique such as molecular inversion probes (Nilsson *et al*, 1994) and spacer multiplex amplification reaction (Krishnakumar *et al.*, 2008), and hybridisation capture methods such as array-based hybrid selection (Albert *et al.*, 2007; Okou *et al.*, 2007) and solution hybridization capture (Gnirke *et al.*, 2009).

## Captured and exome sequencing studies across different crops

Application of this approach has gained momentum in plant systems recently. Targeted sequencing has been done in several crop and tree species using the captured method of sequencing (Table 1.6). Captured sequencing has been used to identify DNA polymorphisms in many polyploidy crop species. It is a powerful tool for studying evolution, population genetics and phylogeographic studies (Carstens *et al.*, 2013; Smith *et al.*, 2014; McCormack *et al.*, 2015). Various targeted studies conducted using NGS techniques have been detailed in Table 1.6.

Exome sequencing is most widely used targeted sequencing method that aims only an informative subset of the genome which varied between 1% and 2% of the genome. Saintenac *et al.* (2011) captured 3.5 Mb of the exonic sequence with coverage of 3.5–7.0% of the exome and studied 3497 genes in durum wheat accessions. In the subsequent year, a novel exome capturing protocol for wheat based on a NimbleGen array was developed by Winfield *et al.* (2012). This protocol was used to sequence exomes of the eight wheat accessions by capturing a 56.5 Mb genomic region to identify SNPs for the genotypic classification of a segregating locus in polyploid wheat (Allen *et al.*, 2013). Other studies in rice and wheat include screening of novel mutations of rice and durum wheat, targeted capture of 107 Mb of non-redundant regions in 62 lines of wheat, exome capture for rapid cloning of *R*-genes in hexaploid bread wheat and exome sequencing in the genomes of rice somaclonal variant (salt-tolerant and drought-tolerant) and parental cultivar (Henry *et al.*, 2014; Udomchalothorn *et al.*, 2014; Jordan *et al.*, 2015; King *et al.*, 2015; Steuernagel *et al.*, 2016).

In maize, exome capture was performed to understand endosperm filling and maturation and to create a population of 1788 lines (Jia *et al.*, 2016). Exome capture kit has also been developed for barley (*Hordeum vulgare* L.) to selectively enrich 61.6 Mb of protein coding sequence (Mascher *et al.*, 2013) which was used for sequencing of exome X-ray mutagenized mutants and wild type genotypes to identify a candidate gene HvMND belonging to the CYP78A family that may affect

**Table 1.6** Plant genome capturing studies performed using NGS platforms

| Number | Plant | Genome coverage | Method | Reference |
|---|---|---|---|---|
| 1 | *Triticum aestivum* | 56.5 Mb | Nimblegen Array technology | Winfield *et al.*, 2012 |
| 2 | *Populus trichocarpa* (black cottonwood) | 20.76 Mb | Agilent technologies | Zhou *et al.*, 2012 |
| 3 | *Saccharum officinarum* | 5.8 Mb | SureSelect Target Enrichment System | Bundock *et al.*, 2012 |
| 4 | *Saccharum hybrid* | 5.8 Mb | SureSelect Target Enrichment System | Bundock *et al.*, 2012 |
| 5 | *Pinus taeda* L. | 21.7 Gbp | Probe based and hybridization capture | Neves *et al.*, 2013 |
| 6 | *Fragaria vesca* | 100× | Mycroarray, Mybait and Illumina | Tennessen *et al.*, 2013 |
| 7 | *Brassica napus* | 5.8 Mb | SureSelectXT | Schiessl *et al.*, 2014 |
| 8 | *Pinus albicaulis* (whitebark pine) | 27 Gb | Hybridization-based target capture | Syring *et al.*, 2016 |

many agricultural traits (Mascher *et al.*, 2014). Kono *et al.* (2016) discovered the distribution of hundreds of SNPs in cultivated and wild accessions of barley and soybean using exome sequencing. *Hordium bulbosum* is the wild relative of cultivated barley that has superior pathogen resistance and stress tolerance which can be crossed to cultivated barley genotype. There was lack of suitable molecular tools to characterize the genetic introgressions from *H. bulbosum* in order to select the beneficial variants and exclude the variants that were not important from breeding point of view. Recently many exome capture studies were reported in this crop for the development of genic markers and genome introgression studies (Wendler *et al.*, 2014, 2015). Russell *et al.* (2016) studied the environmental adaptation in the georeferenced landraces and wild accessions by exome capture.

Besides, exome capturing in switchgrass (*Panicum virgatum*), a potential biofuel feedstock crop, allowed assessment of the genome variation in its two primary ecotypes and identification of variation in *CONSTANS (CO)* and *EARLY HEADING DATE 1 (EHD1)* genes (Evans *et al.*, 2014, 2015). Exome sequencing has also been reported in Eucalypts, black spruce (*Picea mariana*) and black cottonwood (*Populus trichocarpa*) (Zhou and Holliday, 2012; Dasgupta *et al.*, 2015; Pavy *et al.*, 2016) primarily to study the genetic variation.

## Next-genomics sequencing and molecular markers

The discovery and application of molecular markers for the detection and exploitation of DNA polymorphism is one of the most significant achievements in the area of molecular genetics and plant breeding. Applicability of these markers depends on various factors, viz. its physical properties and genomic location, the cost involved, ease of use, and degree of throughput required (Jonah *et al.*, 2011). Molecular markers are generally categorized into two groups, macro-molecules (proteins and deoxyribonucleic acid) and biochemical constituents (secondary metabolites in plants). Secondary metabolites are restricted to the plants and its applicability is not as wide as DNA makers (Joshi *et al.*, 2011). Among the macro-molecules, availability of protein markers is very limited and its analysis is difficult and more tedious than DNA markers. Among all the molecular markers, DNA markers are the most commonly used markers in the field of molecular genetics and plant breeding for various purposes. Single nucleotide polymorphisms (SNPs) have been recognized as potential markers of choice for genome-wide studies due to even distribution throughout the genome (like simple sequence repeats, SSRs), having the advantage over SSRs of being easily typed in large numbers (in high-throughput manner), and signifying variation in both coding and noncoding regions of the genome (Altshuler *et al.*, 2000; Brumfield *et al.*, 2003; Slate *et al.*, 2009). With the advancement of genome sequencing technologies, molecular markers with a known genomic location are becoming more useful and applicable. The existence of various kinds of molecular markers, and differences in their principles, methodological adaptability, and application's suitability need cautious consideration in opting for one or more of such methods for crop improvement programmes. However, these markers are generally based on electrophoretic resolution of DNA fragments, which limits capturing of genetic differences and also this method cannot resolve genetic polymorphisms with less than 5 bp differences (Semagn *et al.*, 2006). Genotyping of considerably large plant populations may take longer duration depending on how to do, what kind of maker system to adopt and how much throughput the adopted system could generate. NGS technology fulfils all the demands of the coming age plant breeding experiments. It is an efficient technology to develop low cost, high-throughput molecular markers for genotyping of such a large plant population in a short period. Using the NGS technologies, several molecular markers were developed to decipher the complex sequences at thousand loci in the genome of all the individuals of a large plant population sample. These NGS technologies include reduced-representation libraries (RRLs; Gore *et al.*, 2009; Hyten *et al.*, 2010), complexity reduction of polymorphic sequences (CRoPS; Mammadov *et al.*, 2010), low coverage multiplexed shotgun genotyping (MSG; Andolfatto *et al.*, 2011), restriction-site associated DNA sequencing (RAD-seq; Pfender *et al.*, 2011), genotyping by sequencing (GBS; Elshire *et al.*, 2011), high-density array (HDR) genotyping (Gunderson, 2009), and sequence-based polymorphic (SBP) marker technology (Sahu *et al.*, 2012).

All of these methods comprise the following basic steps: the digestion of multiple samples of genomic DNA extracted from individuals or set of populations with one or more restriction enzymes; a selection or reduction of the resulting restriction fragments; and NGS of the final set of selected fragments, which should be less than 1 kb in size (avoiding the read-length limits of most of current NGS platforms, except PacBio); bioinformatic analysis to study the association between traits and called variants; and ultimately infer the biological importance from the analysed dataset. Variations in the resulting sequenced fragments can be used as molecular markers in crop breeding programmes (Davey *et al.*, 2011). Though NGS-based markers have enhanced crop breeding programmes, still there are challenges in high-throughput marker generation. These are posed in the form of a way to design the experiment, how many individuals to be screened, which NGS platform and method will be well-suited to minimize the per-sample sequencing cost, etc.

## Major applications of high-throughput markers in crop plants

A wide range of applications and methodologies of genetic markers has been reported in various crop plants (Semagn *et al.*, 2006). Molecular plant breeding aims to improve crop variety in terms of its quantity and quality by applying the latest inventions made in the fields of genetics and genomics. Our understanding about the relationship between genotype and phenotype has been continuously increasing with the help of advanced genomics tools. Some of these applications include (i) surveying allelic diversity in breeding material or natural populations to select the desired genotypes; (ii) marker-assisted selection (MAS) strategies for variety development and germplasm improvement; and (iii) gene pyramiding for gathering multiple agronomically desirable genes within the same cultivar (Jain *et al.*, 2002; Gupta and Varshney, 2004). Of these several applications, one of the major applications of markers are identification of DNA sequences associated with desired traits in crop breeding. This type of application has been described in several crop breeding programmes and greatly benefited the plant breeders in the easy selection of genotypes where phenotypic expressions become difficult to detect individually with utmost breeding value. These expressions may be hindered by many factors such as tissue and age of plant, environmental conditions, expression observing methods and time-frame of expression. These constraints can be easily avoided by the application of molecular markers in selection of genotypes with particular trait(s). Some major applications of high-throughput marker technologies are given with separate subheadings.

### Resequencing, genotyping and diversity analysis

High-throughput genotyping derived from NGS is one of the major applications of molecular markers. Approximately seven million plant accessions, including wild relatives, landraces and human-made advanced varieties/cultivars have been preserved in several, around 1750 national and international, gene banks worldwide (FAO, 2010) and it is a well-known fact that the whole world is dependent on these plants for food, fibre and fuel. Therefore, it is very important for biologists to characterize these accessions and make them available for further crop improvement programmes. In this order, McCouch *et al.* (2012) presented a vision for the potential of genotyping at large-scales such as gene bank collections. In this vision, authors also outlined the constraints in genotyping work at the gene bank level and suggested that applications of NGS may solve many problems related to genetic characterization efforts in gene banks. These major challenges include the need to correctly identify accessions and eliminate duplicate accessions from gene bank collections. Such characterization work has begun with rice genomics, whereby 3000 rice genotypes have been characterized, including identification of SNPs and other structural variations of the genome (3K RGP, 2014). However, almost every county has legal provision to protect the country's genetic resources by not moving or transferring any such materials without approval of the competent institutions or genetic resource management governing body of the country.

### Linkage and association mapping

With the use of NGS and genotyping technologies, it is possible to develop high-throughput molecular markers as well as assume genotyping at large scale in both major and minor genes that can be exercised for generating high-resolution genetic and physical maps. Meanwhile, these approaches can also be exploited to identify genetic variation in germplasm collections of cultivars, landraces and wild species. This genetic variation can be introgressed in elite cultivar or genotype of interest through linkage and association mapping approaches. Moreover, the major genes or superior alleles of the genes or QTLs for the desirable traits can also be identified and introgressed or pyramided in elite cultivars or genotypes of interest using advanced plant breeding approaches such as marker-assisted back crossing (MABC), marker-assisted recurrent selection (MARS), advanced-backcross QTL (AB-QTL), multi-parent advanced generation intercross (MAGIC), or genome-wide selection (GWS) (Varshney and Dubey, 2009).

### Phylogenetic and evolutionary studies

Before the development of NGS technologies, large-scale genome-wide studies were restricted to a few model organisms whose genomes were sequenced. During that time whole genome sequencing was very time-consuming and a laborious project and it was done with collaboration of many countries. Now, the scenario is totally different and even a single laboratory may afford the whole sequencing work that is only possible by the advancement in sequencing technologies. Discovery of molecular markers is directly related to the NGS techniques and the role of markers in the phylogenetic and evolutionary studies is well known. For good quality of phylogeny as well as evolutionary studies, ecologists and evolutionary biologists need data from large numbers of individuals. These studies along with the power of NGS technology have been reported in many non-model organisms for determining the gene flow, population divergence, diversity level at intra- or inter-population, phylogeography, domestication process, and phylogenetic and evolutionary analyses (Grover *et al.*, 2012).

## Conclusion and future prospects of next-generation sequencing

Next-generation sequencing technologies have revolutionized the field of plant genomics. As of today, the large chunk of this success is largely attributed to the second-generation sequencing tools. The areas which were significantly influenced by NGS technologies include genome sequencing, captured sequencing, exome sequencing, metagenomics, and plant transcriptomics; which includes mRNAs and non-coding RNAs and molecular markers and plant breeding. The continuous improvements in the field of sequencing technologies and associated decrease in sequencing costs has opened the door to small laboratories to take up plant genome sequencing projects. In spite of these developments, still there are considerable challenges in sequencing more complex genomes such as cross-pollinated crops and crop with higher polyploids. Nevertheless, researchers have taken interest in sequencing particular genomic regions and exomes, called as captured sequencing. This approach has considerably advanced our knowledge of those crops for which there is no sequence information available or which are only partially sequenced. NGS also has a significant role in metagenomics and enhanced the rate of analysis, especially in those circumstances where it would have been difficult to analyse using traditional tools and approaches. NGS-based transcriptome analysis has contributed significantly in transcriptome profiling of those plant species where no sequence information is available and also improve their ability to immensely contribute for novel genes to plant biology. NGS is playing a vital role in development of high throughput molecular markers as well.

However, the benefits of developments in the field of NGS tools can only be harvested by integrating the different genomics and genetics technologies and also advancements in biometrics and bioinformatics tools and techniques. This is going to be a challenging task in coming years. Finally, though captured sequencing has gained importance in past few years, largely due to sequencing costs and complexity involved in it, the future holds bright for whole genome sequencing, as the cost of sequencing is expected to decrease further and also the developments in the third-generation sequencing techniques have the potential to decrease the complexity by easily assembling of more complex genomes.

## Acknowledgements

TRS is thankful to the Indian council of Agricultural Research for funding under NPTC project and Department of Science and Technology Government of India for JC Bose National Fellowship.

## References


3K RGP (2014). The 3,000 rice genomes project. Giga Science 3, 7.

Abdelfattah, A., Wisniewski, M., Nicosia, M.G.L.D., Cacciola, S.O., and Schena, L. (2016). Metagenomic analysis of fungal diversity on strawberry plants and the effect of management practices on the fungal community structure of aerial organs. PLOS ONE, *11*, e0160470.

Abdel-Ghany, S.E., Hamilton, M., Jacobi, J.L., Ngam, P., Devitt, N., Schilkey, F., Ben-Hur, A., and Reddy, A.S. (2016). A survey of the sorghum transcriptome using single-molecule long reads. Nat. Commun. 7, 11706. http://dx.doi.org/10.1038/ncomms11706

Adams, I.P., Miano, D.W., Kinyua, Z.M., Wangai, A., Kimani, E., Phiri, N., Reeder, R., Harju, V., Glover, R., Hany, U., *et al.* (2013). Use of next-generation sequencing for the identification and characterization of *Maize chlorotic mottle virus* and *Sugarcane mosaic virus* causing maize lethal necrosis in Kenya. Plant Pathol., 62(4), 741–749.

Adams, I.P., Glover, R.H., Monger, W.A., Mumford, R., Jackeviciene, E., Navalinskiene, M., Samuitiene, M., and Boonham, N. (2009). Next-generation sequencing and metagenomic analysis: a universal diagnostic tool in plant virology. Mol. Plant Pathol. *10*, 537–545. http://dx.doi.org/10.1111/j.1364-3703.2009.00545.x

Ai, C., Liang, G., Sun, J., Wang, X., He, P., Zhou, W., and He, X. (2015). Reduced dependence of rhizosphere microbiome on plant-derived carbon in 32-year long-term inorganic and organic fertilized soils. Soil Biol. Biochem. *80*, 70–78.

Akinsanya, M.A., Goh, J.K., Lim, S.P., and Ting, A.S. (2015). Metagenomics study of endophytic bacteria in Aloe vera using next-generation technology. Genom. Data *6*, 159–163. http://dx.doi.org/10.1016/j.gdata.2015.09.004

Albert, T.J., Molla, M.N., Muzny, D.M., Nazareth, L., Wheeler, D., Song, X., Richmond, T.A., Middle, C.M., Rodesch, M.J., Packard, C.J., *et al.* (2007). Direct selection of human genomic loci by microarray hybridization. Nat. Methods *4*, 903–905.

Albert, V.A., Barbazuk, W.B., Der, J.P., Leebens-Mack, J., Ma, H., Palmer, J.D., Rounsley, S., Sankoff, D., Schuster, S.C., Soltis, D.E., *et al.* (2013). The *Amborella* genome and the evolution of flowering plants. Science *342*,1241089.

Allen, A.M., Barker, G.L., Wilkinson, P., Burridge, A., Winfield, M., Coghill, J., Uauy, C., Griffiths, S., Jack, P., Berry, S., *et al.* (2013). Discovery and development of exome-based, co-dominant single nucleotide polymorphism markers in hexaploid wheat (*Triticum aestivum* L.). Plant Biotechnol. J. *11*, 279–295.

Altshuler, D., Pollara, V.J., Cowles, C.R., Van Etten, W.J., Baldwin, J., Linton, L., and Lander, E.S. (2000). An SNP map of the human genome generated by reduced representation shotgun sequencing. Nature *407*, 513–516. http://dx.doi.org/10.1038/35035083

Andolfatto, P., Davison, D., Erezyilmaz, D., Hu, T.T., Mast, J., Sunayama-Morita, T., and Stern, D.L. (2011). Multiplexed shotgun genotyping for rapid and efficient genetic mapping. Genome Res. *21*, 610–617. http://dx.doi.org/10.1101/gr.115402.110

Ansorge, W.J. (2009). Next-generation DNA sequencing techniques. New Biotechnol. *25*, 195–203. http://dx.doi.org/10.1016/j.nbt.2008.12.009

*Arabidopsis* Genome Initiative. (2000). Analysis of the genome sequence of the flowering plant *Arabidopsis thaliana*. Nature *408*, 796–815

Atamna-Ismaeel, N., Finkel, O.M., Glaser, F., Sharon, I., Schneider, R., Post, A.F., Spudich, J.L., von Mering, C., Vorholt, J.A., Iluz, D., *et al.* (2012b). Microbial rhodopsins on leaf surfaces of terrestrial plants. Environ. Microbiol. *14*, 140–146. http://dx.doi.org/10.1111/j.1462-2920.2011.02554.x

Bagnaresi, P., Biselli, C., Orrù, L., Urso, S., Crispino, L., Abbruscato, P., Piffanelli, P., Lupotto, E., Cattivelli, L., and Valè, G. (2012). Comparative transcriptome profiling of the early response to *Magnaporthe oryzae* in durable resistant vs susceptible rice (*Oryza sativa* L.) genotypes. PLOS ONE *7*, e51609. http://dx.doi.org/10.1371/journal.pone.0051609

Banks, J.A., Nishiyama, T., Hasebe, M., Bowman, J., L., Gribskov, M., dePamphilis, C., Albert, V.A., Aono, Naoki., Aoyama, T., Ambrose, B.A., Ashton, N.W., *et al.* (2011). The compact *Selaginella* genome identifies changes in gene content associated with the evolution of vascular plants. Science 332, 960–963.

Barrera-Figueroa, B.E., Gao, L., Diop, N.N., Wu, Z., Ehlers, J.D., Roberts, P.A., Close, T.J., Zhu, J.K., and Liu, R. (2011). Identification and comparative analysis of drought-associated microRNAs in two cowpea genotypes. BMC Plant Biol. *11*, 127. http://dx.doi.org/10.1186/1471-2229-11-127

Bedada, G., Westerbergh, A., Muller, T., Galkin, E., Bdolach, E., Moshelion, M., Fridman, E., and Schmid, K.J. (2014). Transcriptome sequencing of two wild barley (*Hordeum spontaneum* L.) ecotypes differentially adapted to drought stress reveals ecotype-specific transcripts. BMC genomics *15*, 995. http://dx.doi.org/10.1186/1471-2164-15-995.

Bennetzen, J.L., Schmutz, J., Wang, H., Percifield, R., Hawkins, J., Pontaroli, A.C., Estep, M., Feng, L., Vaughn, J.N., Grimwood, J., *et al.* (2012). Reference genome sequence of the model plant Setaria. Nat. Biotechnol. *30*, 555–561. http://dx.doi.org/10.1038/nbt.2196

Berendsen, R.L., Pieterse, C.M., and Bakker, P.A. (2012). The rhizosphere microbiome and plant health. Trends Plant Sci. *17*, 478–486. http://dx.doi.org/10.1016/j.tplants.2012.04.001

Bertioli, D.J., Cannon, S.B., Froenicke, L., Huang, G., Farmer, A.D., Cannon, E.K., Liu, X., Gao, D., Clevenger, J., Dash, S., *et al.* (2016). The genome sequences of *Arachis duranensis* and *Arachis ipaensis*, the diploid ancestors of cultivated peanut. Nat. Genet. *48*, 438–446. http://dx.doi.org/10.1038/ng.3517

Bian, X., Ma, P., Jia, Z., Guo, X., and Xie, Y. (2016). Identification of miRNAs in sweet potato by Solexa sequencing. Russ. J. Plant Physiol. 63, 283–292.

Birol, I., Raymond, A., Jackman, S.D., Pleasance, S., Coope, R., Taylor, G.A., Yuen, M.M., Keeling, C.I., Brand, D., Vandervalk, B.P., *et al.* (2013). Assembling the 20 Gb white spruce (*Picea glauca*) genome from whole-genome shotgun sequencing data. Bioinformatics *29*, 1492–1497. http://dx.doi.org/10.1093/bioinformatics/btt178

Bombarely, A., Moser, M., Amrad, A., Bao, M., Bapaume, L., Barry, C.S., Bliek, M., Boersma, M.R., Borghi, L., Bruggmann, R., *et al.* (2016). Insight into the evolution of the Solanaceae from the parental genomes of *Petunia hybrida*. Nat. Plants 2, 16074. http://dx.doi.org/10.1038/nplants.2016.74

Boonham, N., Kreuze, J., Winter, S., van der Vlugt, R., Bergervoet, J., Tomlinson, J., and Mumford, R. (2014). Methods in virus diagnostics: from ELISA to next generation sequencing. Virus Res *186*, 20–31. http://dx.doi.org/10.1016/j.virusres.2013.12.007

Brumfield, R.T., Beerli, P., Nickerson, D.A., and Edwards, S.V. (2003). The utility of single nucleotide polymorphisms in inferences of population history. Trends Ecol. Evol. *18*, 249–256. http://dx.doi.org/10.1016/S0169-5347(03)00018-1

Bundock, P.C., Casu, R.E., and Henry, R.J. (2012). Enrichment of genomic DNA for polymorphism detection in a non-model highly polyploid crop plant. Plant Biotechnol. J. *10*, 657–667. http://dx.doi.org/10.1111/j.1467-7652.2012.00707.x

Burns, J.A., Paasch, A., Narechania, A., and Kim, E. (2015). Comparative genomics of a bacterivorous green alga reveals evolutionary causalities and consequences of phago-mixotrophic mode of nutrition. Genome Biol. Evol. *7*, 3047–3061. http://dx.doi.org/10.1093/gbe/evv144

Burton, J.N., Adey, A., Patwardhan, R.P., Qiu, R., Kitzman, J.O., and Shendure, J. (2013). Chromosome-scale scaffolding of de novo genome assemblies based on chromatin interactions. Nat. Biotechnol. *31*, 1119–1125. http://dx.doi.org/10.1038/nbt.2727

Butts, C.T., Bierma, J.C., and Martin, R.W. (2016). Novel proteases from the genome of the carnivorous plant *Drosera capensis*: Structural prediction and comparative analysis. Proteins *84*, 1517–33.

Byrne, S.L., Nagy, I., Pfeifer, M., Armstead, I., Swain, S., Studer, B., Mayer, K., Campbell, J.D., Czaban, A., Hentrup, S., *et al.* (2015). A synteny-based draft genome sequence of the forage grass *Lolium perenne*. Plant J. *84*, 816–826. http://dx.doi.org/10.1111/tpj.13037

Camilios-Neto, D., Bonato, P., Wassem, R., Tadra-Sfeir, M.Z., Brusamarello-Santos, L.C., Valdameri, G., Donatti, L., Faoro, H., Weiss, V.A., Chubatsu, L.S., *et al.* (2014). Dual RNA-seq transcriptional analysis of wheat roots colonized by *Azospirillumbrasilense* reveals up-regulation of nutrient acquisition and cell cycle genes. BMC genomics, *15*, 378.

Campo, S., Peris-Peris, C., Siré, C., Moreno, A.B., Donaire, L., Zytnicki, M., Notredame, C., Llave, C., and San Segundo, B. (2013). Identification of a novel microRNA (miRNA) from rice that targets an alternatively spliced transcript of the *Nramp6* (Natural resistance-associated macrophage protein 6) gene involved in pathogen resistance. New Phytol. *199*, 212–227.

Cao, H.X., Schmutzer, T., Scholz, U., Pecinka, A., Schubert, I., and Vu, G.T. (2015). Metatranscriptome analysis reveals host-microbiome interactions in traps of carnivorous Genlisea species. Front. Microbiol. *6*, 526. http://dx.doi.org/10.3389/fmicb.2015.00526

Cao, H., Hastie, A.R., Cao, D., Lam, E.T., Sun, Y., Huang, H., Liu, X., Lin, L., Andrews, W., Chan, S., *et al.* (2014). Rapid detection of structural variation in a human genome using nanochannel-based genome mapping technology. GigaScience 3, 34. http://dx.doi.org/10.1186/2047-217X-3-34

Cao, J.Y., Xu, Y.P., Zhao, L., Li, S.S., and Cai, X.Z. (2016). Tight regulation of the interaction between *Brassica napus* and *Sclerotinia sclerotiorum* at the microRNA level. Plant Mol. Biol. *92*, 39–55. http://dx.doi.org/10.1007/s11103-016-0494-3

Cao, S., Zhu, Q.H., Shen, W., Jiao, X., Zhao, X., Wang, M.B., Liu, L., Singh, S.P., and Liu, Q. (2013). Comparative profiling of miRNA expression in developing seeds of high linoleic and high oleic safflower (*Carthamus tinctorius* L.) plants. Front. Plant Sci. *4*, 489. http://dx.doi.org/10.3389/fpls.2013.00489

Carstens, B.C., Brennan, R.S., Chua, V., Duffie, C.V., Harvey, M.G., Koch, R.A., McMahan, C.D., Nelson, B.J., Newman, C.E., Satler, J.D., *et al.* (2013). Model selection as a tool for phylogeographic inference: an example from the willow *Salix melanopsis*. Mol. Ecol. *22*, 4014–4028. http://dx.doi.org/10.1111/mec.12347

Chao, J., Jin, J., Wang, D., Han, R., Zhu, R., Zhu, Y., and Li, S. (2014). Cytological and transcriptional dynamics analysis of host plant revealed stage-specific biological processes related to compatible rice-*Ustilaginoidea virens* interaction. PLOS ONE, *9*(3), p.e91391. http://dx.doi.org/10.1371/journal.pone.0091391.

Chaudhry, V., and Patil, P.B. (2016). Genomic investigation reveals evolution and lifestyle adaptation of endophytic Staphylococcus epidermidis. Scientific reports, *6*.

Chen, K., and Pachter, L. (2005). Bioinformatics for whole-genome shotgun sequencing of microbial communities. PLoS Comput. Biol. *1*, 106–112. http://dx.doi.org/10.1371/journal.pcbi.0010024

Chen, X. (2009). Small RNAs and their roles in plant development. Annu. Rev. Cell Dev. Biol. *25*, 21–44. http://dx.doi.org/10.1146/annurev.cellbio.042308.113417

Chhabra, S., Brazil, D., Morrissey, J., Burke, J.I., O Gara, F., N Dowling, D. (2013). Characterization of mineral phosphate solubilization traits from a barley rhizosphere soil functional metagenome. Microbiologyopen *2*, 717–724. http://dx.doi.org/10.1002/mbo3.110

Chi, X., Yang, Q., Chen, X., Wang, J., Pan, L., Chen, M., Yang, Z., He, Y., Liang, X., and Yu, S. (2011). Identification and characterization of microRNAs from peanut (*Arachis hypogaea* L.) by high-throughput sequencing. PLoS One *6*, e27530. http://dx.doi.org/10.1371/journal.pone.0027530

Claesson, M.J., Wang, Q., O'Sullivan, O., Greene-Diniz, R., Cole, J.R., Ross, R.P., and O'Toole, P.W. (2010). Comparison of two next-generation sequencing technologies for resolving highly complex microbiota composition using tandem variable 16S rRNA gene regions. Nucleic acids research gkq873.

Cloonan, N., Forrest, A.R., Kolle, G., Gardiner, B.B., Faulkner, G.J., Brown, M.K., Taylor, D.F., Steptoe, A.L., Wani, S., Bethel, G., *et al.* (2008). Stem cell transcriptome profiling via massive-scale mRNA sequencing. Nat. Methods *5*, 613–619. http://dx.doi.org/10.1038/nmeth.1223

Tomato Genome Consortium. (2012). The tomato genome sequence provides insights into fleshy fruit evolution. Nature *485*, 635–641. http://dx.doi.org/10.1038/nature11119

Coughlan, L.M., Cotter, P.D., Hill, C., and Alvarez-Ordóñez, A. (2015). Biotechnological applications of functional metagenomics in the food and pharmaceutical industries. Frontiers in microbiology 6.

Creer, S., Deiner, K., Frey, S., Porazinska, D., Taberlet, P., Thomas, W.K., Potter, C., and Bik, H.M. (2016). The ecologist's field guide to sequence-based identification of biodiversity. Methods in Ecology and Evolution. http://dx.doi.org/10.1111/2041-210X.12574

Creighton, C.J., Reid, J.G., and Gunaratne, P.H. (2009). Expression profiling of microRNAs by deep sequencing. Briefings Bioinf. *10*, 490–497. http://dx.doi.org/10.1093/bib/bbp019

Crespi, M.D., Jurkevitch, E., Poiret, M., d'Aubenton-Carafa, Y., Petrovics, G., Kondorosi, E., and Kondorosi, A. (1994). *enod40*, a gene expressed during nodule organogenesis, codes for a non-translatable RNA involved in plant growth. EMBO J. *13*, 5099–5112.

Dasgupta, M.G., Dharanishanthi, V., Agarwal, I., and Krutovsky, K.V. (2015). Development of genetic markers in *Eucalyptus* species by target enrichment and exome sequencing. PLOS ONE *10*, e0116528. http://dx.doi.org/10.1371/journal.pone.0116528

Dautt-Castro, M., Ochoa-Leyva, A., Contreras-Vergara, C.A., Pacheco-Sanchez, M.A., Casas-Flores, S., Sanchez-Flores, A., Kuhn, D.N., and Islas-Osuna, M.A. (2015). Mango (*Mangifera indica* L.) cv. Kent fruit mesocarp *de novo* transcriptome assembly identifies gene families important for ripening. Frontiers in plant science, 6, 62. http://dx.doi.org/10.3389/fpls.2015.00062.

Davey, J.W., Hohenlohe, P.A., Etter, P.D., Boone, J.Q., Catchen, J.M., and Blaxter, M.L. (2011). Genome-wide genetic marker discovery and genotyping using next-generation sequencing. Nat. Rev. Genet. *12*, 499–510. http://dx.doi.org/10.1038/nrg3012

Delmotte, N., Knief, C., Chaffron, S., Innerebner, G., Roschitzki, B., Schlapbach, R., von Mering, C., and Vorholt, J.A. (2009). Community proteogenomics reveals insights into the physiology of phyllosphere bacteria. Proc. Natl. Acad. Sci. U.S.A. *106*, 16428–16433. http://dx.doi.org/10.1073/pnas.0905240106

De Vega, J.J., Ayling, S., Hegarty, M., Kudrna, D., Goicoechea, J.L., Ergon, Å., Rognli, O.A., Jones, C., Swain, M., Geurts, *et al.* (2015). Red clover (*Trifolium pratense* L.) draft genome provides a platform for trait improvement. Sci. Rep. 5, 17394. http://dx.doi.org/10.1038/srep17394

Ding, T., and Melcher, U. (2016). Influences of Plant species, season and location on leaf endophytic bacterial communities of non-cultivated plants. PLOS ONE *11*, e0150895. http://dx.doi.org/10.1371/journal.pone.0150895

Dorn, K.M., Fankhauser, J.D., Wyse, D.L., and Marks, M.D. (2015). A draft genome of field pennycress (*Thlaspi arvense*) provides tools for the domestication of a new winter biofuel crop. DNA Res. *22*, 121–131. http://dx.doi.org/10.1093/dnares/dsu045

Duangjit, J., Bohanec, B., Chan, A.P., Town, C.D., and Havey, M.J. (2013). Transcriptome sequencing to produce SNP-based genetic maps of onion. Theor. Appl. Genet. *126*, 2093–2101. http://dx.doi.org/10.1007/s00122-013-2121-x

Elshire, R.J., Glaubitz, J.C., Sun, Q., Poland, J.A., Kawamoto, K., Buckler, E.S., and Mitchell, S.E. (2011). A robust, simple genotyping-by-sequencing (GBS) approach for high diversity species. PLOS ONE *6*, e19379. http://dx.doi.org/10.1371/journal.pone.0019379

Evans, J., Crisovan, E., Barry, K., Daum, C., Jenkins, J., Kunde-Ramamoorthy, G., Nandety, A., Ngan, C.Y., Vaillancourt, B., Wei, C.L., *et al.* (2015). Diversity and population structure of northern switchgrass as revealed through exome capture sequencing. Plant J. *84*, 800–815. http://dx.doi.org/10.1111/tpj.13041

Evans, J., Kim, J., Childs, K.L., Vaillancourt, B.,Crisovan, E., Nandety, A., Gerhardt, D.J., Richmond, T.A., Jeddeloh, J.A., Kaeppler, S.M., *et al.* (2014). Nucleotide polymorphism and copy number variant detection using exome capture and next-generation sequencing in the polyploid grass *Panicum virgatum*. Plant J. 79, 993–1008.

Fan, J., Ning, K., Zeng, X., Luo, Y., Wang, D., Hu, J., Li, J., Xu, H., Huang, J., Wan, M., *et al.* (2015). Genomic foundation of starch-to-lipid switch in Oleaginous *Chlorella* spp. Plant. Physiol. *169*, 2444–2461. http://dx.doi.org/10.1104/pp.15.01174

Fang, G., Munera, D., Friedman, D.I., Mandlik, A., Chao, M.C., Banerjee, O., Feng, Z., Losic, B., Mahajan, M.C., and Jabado, O.J., *et al.* (2012). Genome-wide mapping of methylated adenine residues in pathogenic *Escherichia coli* using single-molecule real-time sequencing. Nature biotechnology *30*, 1232–1239.

FAO. (2010). The second report on the state of the world's plant genetic resources for food and agriculture. Food and Agriculture Organization, Rome, Italy.

Feng, J., Liu, S., Wang, M., Lang, Q., and Jin, C. (2014). Identification of microRNAs and their targets in tomato infected with *Cucumber mosaic virus* based on deep sequencing. Planta *240*, 1335–1352. http://dx.doi.org/10.1007/s00425-014-2158-3

Feng, J., Wang, J., Fan, P., Jia, W., Nie, L., Jiang, P., Chen, X., Lv, S., Wan, L., Chang, S., *et al.* (2015). High-throughput deep sequencing reveals that microRNAs play important roles in salt tolerance of euhalophyte *Salicornia europaea*. BMC Plant Biol. *15*, 63. http://dx.doi.org/10.1186/s12870-015-0451-3

Flórez-Zapata, M.V.N., Reyes-Valdés. M.H., and Martínez, O. (2016). Long non-coding RNAs are major contributors to transcriptome changes in sunflower meiocytes with different recombination rates. BMC Genomics *17*, 490 http://dx.doi.org/10.1186/s12864-016-2776-1

França, L.T., Carrilho, E., and Kist, T.B. (2002). A review of DNA sequencing techniques. Q. Rev. Biophys. *35*, 169–200.

Fu, Y., Xiao, M., Yu, H., Mason, A.S., Yin, J., Li, J., Zhang, D., and Fu, D. (2016). Small RNA changes in synthetic *Brassica napus*. Planta *244*, 607–622. http://dx.doi.org/10.1007/s00425-016-2529-z

Gao, R., Liu, P., Irwanto, N., Loh, R., and Wong, S.M. (2016). Upregulation of LINC-AP2 is negatively correlated with AP2 gene expression with Turnip crinkle virus infection in *Arabidopsis thaliana*. Plant Cell Rep. 35, 2257–2267. http://dx.doi.org/10.1007/s00299-016-2032-9

Gardner, E.M., Johnson, M.G., Ragone, D., Wickett, N.J., and Zerega, N.J. (2016). Low-coverage, whole-genome sequencing of *Artocarpus camansi* (Moraceae) for phylogenetic marker development and gene discovery. Appl. Plant Sci. *4*, apps.1600017.

Garg, R., Shankar, R., Thakkar, B., Kudapa, H., Krishnamurthy, L., Mantri, N., Varshney, R.K., Bhatia,

S., and Jain, M. (2016). Transcriptome analyses reveal genotype-and developmental stage-specific molecular responses to drought and salinity stresses in chickpea. Scientific Reports, *6*. http://dx.doi.org/10.1038/srep19228.

Gentile, A., Ferreira, T.H., Mattos, R.S., Dias, L.I., Hoshino, A.A., Carneiro, M.S., Souza, G.M., Calsa, T., Nogueira, R.M., Endres, L., *et al.* (2013). Effects of drought on the microtranscriptome of field-grown sugarcane plants. Planta *237*, 783–798. http://dx.doi.org/10.1007/s00425-012-1795-7

Gnirke, A., Melnikov, A., Maguire, J., Rogov, P., LeProust, E.M., Brockman, W., Fennell, T., Giannoukos, G., Fisher, S., Russ, C., *et al.* (2009). Solution hybrid selection with ultra-long oligonucleotides for massively parallel targeted sequencing. Nat. Biotechnol. *27*, 182–189. http://dx.doi.org/10.1038/nbt.1523

Goff, S.A., Ricke, D., Lan, T.H., Presting, G., Wang, R., Dunn, M., Glazebrook, J., Sessions, A., Oeller, P., Varma, H., *et al.* (2002). A draft sequence of the rice genome (*Oryza sativa* L. ssp. *japonica*). Science *296*, 92–100. http://dx.doi.org/10.1126/science.1068275

Gore, M.A., Chia, J.M., Elshire, R.J., Sun, Q., Ersoz, E.S., Hurwitz, B.L., Peiffer, J.A., McMullen, M.D., Grills, G.S., Ross-Ibarra, J., *et al.* (2009). A first-generation haplotype map of maize. Science *326*, 1115–1117. http://dx.doi.org/10.1126/science.1177837

Grover, C.E., Salmon, A., and Wendel, J.F. (2012). Targeted sequence capture as a powerful tool for evolutionary analysis. Am. J. Bot. *99*, 312–319. http://dx.doi.org/10.3732/ajb.1100323

Gunderson, K.L. (2009). Whole-genome genotyping on bead arrays. Methods Mol. Biol. *529*, 197–213. http://dx.doi.org/10.1007/978-1-59745-538-1_13

Guo, L., Qiu, J., Han, Z., Ye, Z., Chen, C., Liu, C., Xin, X., Ye, C.Y., Wang, Y.Y., Xie, H., *et al.* (2015). A host plant genome (*Zizania latifolia*) after a century-long endophyte infection. Plant J. *83*, 600–609. http://dx.doi.org/10.1111/tpj.12912

Guo, W., Wu, G., Yan, F., Lu, Y., Zheng, H., Lin, L., Chen, H., and Chen, J. (2012). Identification of novel *Oryza sativa* miRNAs in deep sequencing-based small RNA libraries of rice infected with *Rice stripe virus*. PLOS ONE *7*, e46443. http://dx.doi.org/10.1371/journal.pone.0046443

Gupta, P.K., and Varshney, R.K. (2004). Cereal genomics: an overview. In P.K. Gupta, and R.K. Varshney, eds. (Kluwer Academic Publishers, Dordrecht, The Netherlands), pp 639–643.

Gupta, V., Estrada, A.D., Blakley, I., Reid, R., Patel, K., Meyer, M.D., Andersen, S.U., Brown, A.F., Lila, M.A., and Loraine, A.E. (2015). RNASeq analysis and annotation of a draft blueberry genome assembly identifies candidate genes involved in fruit ripening, biosynthesis of bioactive compounds, and stage-specific alternative splicing. GigaScience *4*, 5.

Hajyzadeh, M., Turktas, M., Khawar, K.M., and Unver, T. (2015). miR408 overexpression causes increased drought tolerance in chickpea. Gene *555*, 186–193. http://dx.doi.org/10.1016/j.gene.2014.11.002

Hanschen, E.R., Marriage, T.N., Ferris, P.J., Hamaji, T., Toyoda, A., Fujiyama, A., Neme, R., Noguchi, H., Minakuchi, Y., Suzuki, M., *et al.* (2016). The Gonium pectorale genome demonstrates co-option of cell cycle regulation during the evolution of multicellularity. Nat. Commun. *7*, 11370. http://dx.doi.org/10.1038/ncomms11370

Harris, T.D., Buzby, P.R., Babcock, H., Beer, E., Bowers, J., Braslavsky, I., Causey, M., Colonell, J., Dimeo, J., Efcavitch, J.W., *et al.* (2008). Single-molecule DNA sequencing of a viral genome. Science *320*, 106–109. http://dx.doi.org/10.1126/science.1150427

He, Y., Xiao, H., Deng, C., Xiong, L., Nie, H., and Peng, C. (2016). Survey of the genome of Pogostemon cablin provides insights into its evolutionary history and sesquiterpenoid biosynthesis. Sci. Rep. *6*, 26405. http://dx.doi.org/10.1038/srep26405

He, Y., Yang, Z., Hong, N., Wang, G., Ning, G., and Xu, W. (2015). Deep sequencing reveals a novel closterovirus associated with wild rose leaf rosette disease. Mol. Plant Pathol. *16*, 449–458. http://dx.doi.org/10.1111/mpp.12202

Heisel, S.E., Zhang, Y., Allen, E., Guo, L., Reynolds, T.L., Yang, X., Kovalic, D., and Roberts, J.K. (2008). Characterization of unique small RNA populations from rice grain. PLOS ONE 3(8), e2871.

Henry, I.M., Nagalakshmi, U., Lieberman, M.C., Ngo, K.J., Krasileva, K.V., Vasquez-Gross, H., Akhunova, A., Akhunov, E., Dubcovsky, J., Tai, T.H., *et al.* (2014). Efficient genome-wide detection and cataloging of EMS-induced mutations using exome capture and next-generation sequencing. Plant Cell *26*, 1382–1397.

Heo, J.B., and Sung, S. (2011). Vernalization-mediated epigenetic silencing by a long intronic noncoding RNA. Science *331*, 76–79. http://dx.doi.org/10.1126/science.1197349

Hirakawa, H., Okada, Y., Tabuchi, H., Shirasawa, K., Watanabe, A., Tsuruoka, H., Minami, C., Nakayama, S., Sasamoto, S., Kohara, M., *et al.* (2015). Survey of genome sequences in a wild sweet potato, *Ipomoea trifida* (H. B. K.) G. Don. DNA Res. 171–9. http://dx.doi.org/10.1093/dnares/dsv002

Hirakawa, H., Kaur, P., Shirasawa, K., Nichols, P., Nagano, S., Appels, R., Erskine, W., and Isobe, S.N. (2016). Draft genome sequence of subterranean clover, a reference for genus *Trifolium*. Sci. Rep. *6*, 30358. http://dx.doi.org/10.1038/srep30358

Hirota, K., Miyoshi, T., Kugou, K., Hoffman, C.S., Shibata, T., and Ohta, K. (2008). Stepwise chromatin remodelling by a cascade of transcription initiation of non-coding RNAs. Nature *456*, 130–134. http://dx.doi.org/10.1038/nature07348

Hsieh, L.C., Lin, S.I., Shih, A.C., Chen, J.W., Lin, W.Y., Tseng, C.Y., Li, W.H., and Chiou, T.J. (2009). Uncovering small RNA-mediated responses to phosphate deficiency in *Arabidopsis* by deep sequencing. Plant Physiol. *151*, 2120–2132. http://dx.doi.org/10.1104/pp.109.147280

Huang, D., Koh, C., Feurtado, J.A., Tsang, E.W., and Cutler, A.J. (2013). MicroRNAs and their putative targets in *Brassica napus* seed maturation. BMC Genomics *14*, 140. http://dx.doi.org/10.1186/1471-2164-14-140

Hussain, S., Yin, H., Peng, S., Khan, F.A., Khan, F., Sameeullah, M., Hussain, H.A., Huang, J., Cui, K., and Nie, L. (2016). Comparative transcriptional profiling of primed and non-primed rice seedlings under submergence Stress. Front. Plant Sci. *7*, 1125. http://dx.doi.org/10.3389/fpls.2016.01125.

Hyten, D.L., Cannon, S.B., Song, Q., Weeks, N., Fickus, E.W., Shoemaker, R.C., Specht, J.E., Farmer, A.D., May, G.D., and Cregan, P.B. (2010). High-throughput SNP discovery through deep resequencing of a reduced representation library to anchor and orient scaffolds in the soybean whole genome sequence. BMC Genomics *11*, 38. http://dx.doi.org/10.1186/1471-2164-11-38

Ibarra-Laclette, E., Lyons, E., Hernández-Guzmán, G., Pérez-Torres, C.A., Carretero-Paulet, L., Chang, T.H., Lan, T., Welch, A.J., Juárez, M.J., Simpson, J., *et al.* (2013). Architecture and evolution of a minute plant genome. Nature *498*, 94–98. http://dx.doi.org/10.1038/nature12132

Idris, A., Al-Saleh, M., Piatek, M.J., Al-Shahwan, I., Ali, S., and Brown, J.K. (2014). Viral metagenomics: analysis of begomoviruses by illumina high-throughput sequencing. Viruses, *6*(3), 1219–1236.

Imelfort, M., and Edwards, D. (2009). *De novo* sequencing of plant genomes using second-generation technologies. Briefings Bioinf. *10*, 609–618. http://dx.doi.org/10.1093/bib/bbp039

Iorizzo, M., Ellison, S., Senalik, D., Zeng, P., Satapoomin, P., Huang, J., Bowman, M., Iovene, M., Sanseverino, W., Cavagnaro, P., *et al.* (2016). A high-quality carrot genome assembly provides new insights into carotenoid accumulation and asterid genome evolution. Nat. Genet. *48*, 657–666. http://dx.doi.org/10.1038/ng.3565

Iorizzo, M., Senalik, D.A., Grzebelus, D., Bowman, M., Cavagnaro, P.F., Matvienko, M., Ashrafi, H., Van Deynze, A., and Simon, P.W. (2011). De novo assembly and characterization of the carrot transcriptome reveals novel genes, new markers, and genetic diversity. BMC Genomics *12*, 389. http://dx.doi.org/10.1186/1471-2164-12-389

Jagadeeswaran, G., Zheng, Y., Li, Y.F., Shukla, L.I., Matts, J., Hoyt, P., Macmil, S.L., Wiley, G.B., Roe, B.A., Zhang, W., *et al.* (2009). Cloning and characterization of small RNAs from *Medicago truncatula* reveals four novel legume-specific microRNA families. New Phytol. *184*, 85–98. http://dx.doi.org/10.1111/j.1469-8137.2009.02915.x

Jain, M. (2012). Next-generation sequencing technologies for gene expression profiling in plants. Brief. Funct. Genomics. *11*, 63–70. http://dx.doi.org/10.1093/bfgp/elr038

Jain, M., Chevala, V.V., and Garg, R. (2014). Genome-wide discovery and differential regulation of conserved and novel microRNAs in chickpea via deep sequencing. J. Exp. Bot. *65*, 5945–5958. http://dx.doi.org/10.1093/jxb/eru333

Jain, S.M., Brar, D.S., and Ahloowalia, B.S. (2002). Molecular techniques in crop improvement. (Kluwer Academic Publishers, Dordrecht, The Netherlands), pp 433–444.

Jarvis, D.E., Ho, Y.S., Lightfoot, D.J., Schmöckel, S.M., Li, B., Borm, T.J., Ohyanagi, H., Mineta, K., Michell, C.T., Saber, N., *et al.* (2016). The genome of *Chenopodium quinoa*. Nature *542*, 307–312.

Jia, S., Li, A., Morton, K., Avoles-Kianian, P., Kianian, S.F., Zhang, C., and Holding, D. (2016). A population of deletion mutants and an integrated mapping and exome-seq pipeline for gene discovery in maize. G3 *6*, 2385–2395. http://dx.doi.org/10.1534/g3.116.030528

Jiao, Y., Wickett, N.J., Ayyampalayam, S., Chanderbali, A.S., Landherr, L., *et al.* (2011). Ancestral polyploidy in seed plants and angiosperms. Nature. *473*, 97–100.

Jonah, P.M., Bello, L.L., Lucky, O., Midau, A., and Moruppa, S.M. (2011). Review: the importance of molecular markers in plant breeding programmes. GJSFR 11, 5–12.

Jordan, K.W., Wang, S., Lun, Y., Gardiner, L.J., MacLachlan, R., Hucl, P., Wiebe, K., Wong, D., Forrest, K.L., Sharpe, A.G., *et al.* (2015). A haplotype map of allohexaploid wheat reveals distinct patterns of selection on homoeologous genomes. Genome Biol. *16*, 48. http://dx.doi.org/10.1186/s13059-015-0606-4

Joshi, R.K., Megha, S., Basu, U., Rahman, M.H., and Kav, N.V. (2016). Genome wide identification and functional prediction of long non-coding RNAs responsive to *Sclerotinia sclerotiorum* infection in *Brassica napus*. PLOS ONE *11*. http://dx.doi.org/10.1371/journal.pone.0158784.

Joshi, S.P., Prabhakar, K., Ranjekar, P.K., and Gupta, V.S. (1999). Molecular markers in plant genome analysis. http://www.ias.ac.in/currsci/jul25/articles15.htm. pp 1–19.

Kang, Y.J., Satyawan, D., Shim, S., Lee, T., Lee, J., Hwang, W.J., Kim, S.K., Lestari, P., Laosatit, K., Kim, K.H., *et al.* (2015). Draft genome sequence of adzuki bean, *Vigna angularis*. Sci. Rep. *5*, 8069. http://dx.doi.org/10.1038/srep08069

Kawahara, Y., Oono, Y., Kanamori, H., Matsumoto, T., Itoh, T., and Minami, E. (2012). Simultaneous RNA-seq analysis of a mixed transcriptome of rice and blast fungus interaction. PLOS ONE *7*, e49423. http://dx.doi.org/10.1371/journal.pone.0049423

Kehoe, M.A., Coutts, B.A., Buirchell, B.J., and Jones, R.A. (2014). Plant virology and next generation sequencing: experiences with a Potyvirus. PLOS ONE *9*, e104580. http://dx.doi.org/10.1371/journal.pone.0104580

Kellner, F., Kim, J., Clavijo, B.J., Hamilton, J.P., Childs, K.L., Vaillancourt, B., Cepela, J., Habermann, M., Steuernagel, B., Clissold, L., *et al.* (2015). Genome-guided investigation of plant natural product biosynthesis. Plant J. *82*, 680–692. http://dx.doi.org/10.1111/tpj.12827

Kim, E.D., and Sung, S. (2012). Long noncoding RNA: unveiling hidden layer of gene regulatory networks. Trends Plant Sci. *17*, 16–21. http://dx.doi.org/10.1016/j.tplants.2011.10.008

King, R., Bird, N., Ramirez-Gonzalez, R., Coghill, J.A., Patil, A., Hassani-Pak, K., Uauy, C., and Phillips, A.L. (2015). Mutation scanning in wheat by exon capture and next-generation sequencing. PLOS ONE *10*, e0137549. http://dx.doi.org/10.1371/journal.pone.0137549

Knief, C., Delmotte, N., Chaffron, S., Stark, M., Innerebner, G., Wassmann, R., von Mering, C., and Vorholt, J.A. (2012). Metaproteogenomic analysis of microbial communities in the phyllosphere and rhizosphere of rice. ISME J. *6*, 1378–1390. http://dx.doi.org/10.1038/ismej.2011.192

Kompelli, S.K., Kompelli, V.S.P., Enjala, C., and Suravajhala, P. (2015.) Genome-wide identification of miRNAs in pigeonpea (*Cajanus cajan* L.). Aust. J. Crop Sci *9*, 215–222.

Kono, T.J., Fu, F., Mohammadi, M., Hoffman, P.J., Liu, C., Stupar, R.M., Smith, K.P., Tiffin, P., Fay, J.C., and Morrell, P.L. (2016). The role of deleterious substitutions in crop genomes. Mol. Biol. Evol. *33*, 2307–2317. http://dx.doi.org/10.1093/molbev/msw102

Krishnakumar, S., Zheng, J., Wilhelmy, J., Faham, M., Mindrinos, M., and Davis, R. (2008). A comprehensive assay for targeted multiplex amplification of human DNA sequences. Proc. Natl. Acad. Sci. U.S.A. *105*, 9296–9301. http://dx.doi.org/10.1073/pnas.0803240105

Lakhotia, N., Joshi, G., Bhardwaj, A.R., Katiyar-Agarwal, S., Agarwal, M., Jagannath, A., Goel, S., and Kumar, A. (2014). Identification and characterization of miRNAome in root, stem, leaf and tuber developmental stages of potato (*Solanum tuberosum* L.) by high-throughput sequencing. BMC Plant Biol. *14*, 6. http://dx.doi.org/10.1186/1471-2229-14-6

Lee, H., Golicz, A.A., Bayer, P.E., Jiao, Y., Tang, H., Paterson, A.H., Sablok, G., Krishnaraj, R.R., Chan, C.K., Batley, J., *et al.* (2016). The genome of a southern hemisphere seagrass species (*Zostera muelleri*). Plant Physiol. *172*, 272–83.

Lee, H., Gurtowski, j., Yoo, S., Nattestad, M., Marcus, S., Goodwin, S., McCombie, W.R., and Schatz, M. (2016). Third-generation sequencing and the future of genomics. http://dx.doi.org/10.1101/048603

Leushkin, E.V., Sutormin, R.A., Nabieva, E.R., Penin, A.A., Kondrashov, A.S., and Logacheva, M.D. (2013). The miniature genome of a carnivorous plant *Genlisea aurea* contains a low number of genes and short noncoding sequences. BMC Genomics *14*, 476.17.

Levene, M.J., Korlach, J., Turner, S.W., Foquet, M., Craighead, H.G., and Webb, W.W. (2003). Zero-mode waveguides for single-molecule analysis at high concentrations. Science *299*, 682–686. http://dx.doi.org/10.1126/science.1079700

Li, C., Qiao, Z., Qi, W., Wang, Q., Yuan, Y., Yang, X., Tang, Y., Mei, B., Lv, Y., Zhao, H., *et al.* (2015). Genome-wide characterization of cis-acting DNA targets reveals the transcriptional regulatory framework of *opaque2* in maize. Plant. Cell *27*, 532–545. http://dx.doi.org/10.1105/tpc.114.134858

Li, F., Fan, G., Wang, K., Sun, F., Yuan, Y., Song, G., Li, Q., Ma, Z., Lu, C., Zou, C., *et al.* (2014). Genome sequence of the cultivated cotton *Gossypium arboreum*. Nat. Genet. *46*, 567–572. http://dx.doi.org/10.1038/ng.2987

Li, H., Dong, Y., Sun, Y., Zhu, E., Yang, J., Liu, X., Xue, P., Xiao, Y., Yang, S., Wu, J., *et al.* (2011a). Investigation of the microRNAs in safflower seed, leaf, and petal by high-throughput sequencing. Planta *233*, 611–619. http://dx.doi.org/10.1007/s00425-010-1327-2

Li, H., Dong, Y., Yin, H., Wang, N., Yang, J., Liu, X., Wang, Y., Wu, J., and Li, X. (2011b). Characterization of the stress associated microRNAs in *Glycine max* by deep sequencing. BMC Plant Biol. *11*, 170. http://dx.doi.org/10.1186/1471-2229-11-170

Li, J., Ding, Q., Wang, F., Zhang, Y., Li, H., and Gao, J. (2015a). Integrative analysis of mRNA and miRNA expression profiles of the tuberous root development at seedling stages in turnips. PLOS ONE *10*, e0137983. http://dx.doi.org/10.1371/journal.pone.0137983

Li, J., Wu, L.Q., Zheng, W.Y., Wang, R.F., and Yang, L.X. (2015b). Genome-wide identification of microRNAs responsive to high temperature in rice (*Oryza sativa*) by high-throughput deep sequencing. J. Agron. Crop Sci. *201*, 379–388.

Li, R., Fan, W., Tian, G., Zhu, H., He, L., Cai, J., Huang, Q., Cai, Q., Li, B., Bai, Y., *et al.* (2010). The sequence and *de novo* assembly of the giant panda genome. Nature *463*, 311–317. http://dx.doi.org/10.1038/nature08696

Li, X., Jin, F., Jin, L., Jackson, A., Ma, X., Shu, X., Wu, D., and Jin, G. (2015). Characterization and comparative profiling of the small RNA transcriptomes in two phases of flowering in *Cymbidium ensifolium*. BMC Genomics *16*, 622. http://dx.doi.org/10.1186/s12864-015-1764-1

Li, X., Sun, H., Pei, J., Dong, Y., Wang, F., Chen, H., Sun, Y., Wang, N., Li, H., and Li, Y. (2012). *De novo* sequencing and comparative analysis of the blueberry transcriptome to discover putative genes related to antioxidants. Gene *511*, 54–61. http://dx.doi.org/10.1016/j.gene.2012.09.021

Li, Y., Lu, Y.G., Shi, Y., Wu, L., Xu, Y.J., Huang, F., Guo, X.Y., Zhang, Y., Fan, J., Zhao, J.Q., *et al.* (2014). Multiple rice microRNAs are involved in immunity against the blast fungus *Magnaporthe oryzae*. Plant Physiol. *164*, 1077–1092. http://dx.doi.org/10.1104/pp.113.230052

Li, Y.F., Zheng, Y., Jagadeeswaran, G., and Sunkar, R. (2013). Characterization of small RNAs and their target genes in wheat seedlings using sequencing-based approaches. Plant Sci. *203*, 17–24.

Lipshitz, H.D., Peattie, D.A., and Hogness, D.S. (1987). Novel transcripts from the Ultrabithorax domain of the bithorax complex. Genes. Dev. *1*, 307–322.

Liu, H., Qin, C., Chen, Z., Zuo, T., Yang, X., Zhou, H., Xu, M., Cao, S., Shen, Y., Lin, H., *et al.* (2014). Identification of miRNAs and their target genes in developing maize ears by combined small RNA and degradome sequencing. BMC Genomics *15*, 25. http://dx.doi.org/10.1186/1471-2164-15-25

Liu, Q., Lan, Y., Wen, C., Zhao, H., Wang, J., and Wang, Y. (2016a). Transcriptome sequencing analyses between the cytoplasmic male sterile line and its maintainer line in Welsh onion (*Allium fistulosum* L.). Int. J. Mol. Sci. 17(7). http://dx.doi.org/10.3390/ijms17071058

Liu, S.C., Xu, Y.X., Ma, J.Q., Wang, W.W., Chen, W., Huang, D.J., Fang, J., Li, X.J., and Chen, L. (2016b). Small RNA and degradome profiling reveals important roles for microRNAs and their targets in tea plant response to drought stress. Physiol. Plant http://dx.doi.org/10.1111/ppl.12477.

Liu, W., Xu, L., Wang, Y., Shen, H., Zhu, X., Zhang, K., Chen, Y., Yu, R., Limera, C., and Liu, L. (2015). Transcriptome-wide analysis of chromium-stress responsive microRNAs to explore miRNA-mediated regulatory networks in radish (*Raphanus sativus* L.). Sci. Rep. *11*, 5. http://dx.doi.org/10.1038/srep14024.

Liu, X., Zhao, B., Zheng, H.J., Hu, Y., Lu, G., Yang, C.Q., Chen, J.D., Chen, J.J., Chen, D.Y., Zhang, L., *et al.* (2015). *Gossypium barbadense* genome sequence provides insight into the evolution of extra-long staple fiber and specialized metabolites. Sci. Rep. *5*, 14139. http://dx.doi.org/10.1038/srep14139

Loss-Morais, G., Ferreira, D.C., Margis, R., Alves-Ferreira, M., and Corrêa, R.L. (2014). Identification of novel and conserved microRNAs in *Coffea canephora* and *Coffea arabica*. Genet. Mol. Biol. 37, 671–682. http://dx.doi.org/10.1590/S1415-47572014005000020

Lu, M., An, H., and Li, L. (2016). Genome survey sequencing for the characterization of the genetic background of *Rosa roxburghii* tratt and leaf ascorbate metabolism genes. PLOS ONE *11*, e0147530. http://dx.doi.org/10.1371/journal.pone.0147530

Lu, X., Chen, X., Mu, M., Wang, J., Wang, X., Wang, D., Yin, Z., Fan, W., Wang, S., Guo, L., *et al.* (2016). Genome-wide analysis of long noncoding RNAs and their responses to drought stress in cotton (*Gossypium hirsutum* L.). PLOS ONE *11*, e0156723. http://dx.doi.org/10.1371/journal.pone.0156723

Luan, Y., Cui, J., Zhai, J., Li, J., Han, L., and Meng, J. (2015). High-throughput sequencing reveals differential expression of miRNAs in tomato inoculated with *Phytophthora infestans*. Planta *241*, 1405–1416.

Lugtenberg, B., and Kamilova, F. (2009). Plant-growth-promoting rhizobacteria. Annu. Rev. Microbiol. *63*, 541–556. http://dx.doi.org/10.1146/annurev.micro.62.081307.162918

Luo, B., Gu, W., Zhong, J., Wang, Y., and Zhang, G. (2015). Revealing crosstalk of plant and fungi in the symbiotic roots of sewage-cleaning *Eichhornia crassipes* using direct *de novo* metatranscriptomic analysis. Scientific Reports, 5. http://dx.doi.org/10.1038/srep15407.

Luria, N., Sela, N., Yaari, M., Feygenberg, O., Kobiler, I., Lers, A., and Prusky, D. (2014). De-novo assembly of mango fruit peel transcriptome reveals mechanisms of mango response to hot water treatment. BMC Genomics *15*, 957. http://dx.doi.org/10.1186/1471-2164-15-957

Ma, X., Xin, Z., Wang, Z., Yang, Q., Guo, S., Guo, X., Cao, L., and Lin, T. (2015). Identification and comparative analysis of differentially expressed miRNAs in leaves of two wheat (*Triticum aestivum* L.) genotypes during dehydration stress. BMC Plant Biol. *15*, 1. http://dx.doi.org/10.1186/s12870-015-0413-9.

Mahuku, G., Wangai, A., Sadessa, K., Teklewold, A., Wegary, D., Adams, I., Smith, J., Bottomley, E., Bryce, S., Braidwood, L., *et al.* (2015). First report of *Maize chlorotic mottle virus* and Maize lethal necrosis on maize in Ethiopia. *Plant Disease*, *99*(12), 1870.

Mamanova, L., Coffey, A.J., Scott, C.E., Kozarewa, I., Turner, E.H., Kumar, A., Howard, E., Shendure, J., and Turner, D.J. (2010). Target-enrichment strategies for next-generation sequencing. Nat. Methods *7*, 111–118. http://dx.doi.org/10.1038/nmeth.1419

Mammadov, J.A., Chen, W., Ren, R., Pai, R., Marchione, W., Yalçin, F., Witsenboer, H., Greene, T.W., Thompson, S.A., and Kumpatla, S.P. (2010). Development of highly polymorphic SNP markers from the complexity reduced portion of maize [*Zea mays* L.] genome for use in marker-assisted breeding. Theor. Appl. Genet. *121*, 577–588. http://dx.doi.org/10.1007/s00122-010-1331-8

Mardis, E.R. (2008). The impact of next-generation sequencing technology on genetics. Trends Genet. *24*, 133–141. http://dx.doi.org/10.1016/j.tig.2007.12.007

Mardis, E.R. (2013). Next-generation sequencing platforms. Annu. Rev. Anal. Chem. *6*, 287–303. http://dx.doi.org/10.1146/annurev-anchem-062012-092628

Margulies, M., Egholm, M., Altman, W.E., Attiya, S., Bader, J.S., Bemben, L.A., Berka, J., Braverman, M.S., Chen, Y.J., Chen, Z., *et al.* (2005). Genome sequencing in microfabricated high-density picolitre reactors. Nature *437*, 376–380.

Mariner, P.D., Walters, R.D., Espinoza, C.A., Drullinger, L.F., Wagner, S.D., Kugel, J.F., and Goodrich, J.A. (2008). Human Alu RNA is a modular transacting repressor of mRNA transcription during heat shock. Mol. Cell *29*, 499–509. http://dx.doi.org/10.1016/j.molcel.2007.12.013

Martianov, I., Ramadass, A., Serra Barros, A., Chow, N., and Akoulitchev, A. (2007). Repression of the human dihydrofolate reductase gene by a non-coding interfering transcript. Nature *445*, 666–670.

Martínez-García, P.J., Crepeau, M.W., Puiu, D., Gonzalez-Ibeas, D., Whalen, J., Stevens, K.A., Paul, R., Butterfield, T.S., Britton, M.T., Reagan, R.,L., *et al.* (2016). The walnut (*Juglans regia*) genome sequence reveals diversity in genes coding for the biosynthesis of non-structural polyphenols. Plant J. *87*, 507–32.

Marzano, S.Y.L., andDomier, L.L. (2016). Reprint of 'Novel mycoviruses discovered from metatranscriptomics survey of soybean phyllospherephytobiomes'. Virus Res. *219*, 11–21.

Mascher, M., Jost, M., Kuon, J.E., Himmelbach, A., Aßfalg, A., Beier, S., Scholz, U., Graner, A., and Stein, N. (2014). Mapping-by-sequencing accelerates forward genetics in barley. Genome Biol. *15*, R78. http://dx.doi.org/10.1186/gb-2014-15-6-r78

Mascher, M., Richmond, T.A., Gerhardt, D.J., Himmelbach, A., Clissold, L., Sampath, D., Ayling, S., Steuernagel, B., Pfeifer, M., D'Ascenzo, M., *et al.* (2013). Barley whole exome capture: a tool for genomic research in the genus *Hordeum* and beyond. Plant J. *76*, 494–505. http://dx.doi.org/10.1111/tpj.12294

Maxam, A.M., and Gilbert, W. (1977). A new method for sequencing DNA. Proc. Natl. Acad. Sci. U.S.A. *74*, 560–564.

McCormack, J.E., Tsai, W.L., and Faircloth, B.C. (2016). Sequence capture of ultraconserved elements from bird museum specimens. Mol. Ecol. Resour. *16*, 1189–1203. http://dx.doi.org/10.1111/1755-0998.12466

McCouch, S.R., McNally, K.L., Wang, W., and Sackville Hamilton, R. (2012). Genomics of gene banks: A case study in rice. Am. J. Bot. *99*, 407–423. http://dx.doi.org/10.3732/ajb.1100385

Mendes, L.W., Kuramae, E.E., Navarrete, A.A., van Veen, J.A., and Tsai, S.M. (2014). Taxonomical and functional microbial community selection in soybean rhizosphere. ISME J. *8*, 1577–1587. http://dx.doi.org/10.1038/ismej.2014.17

Merchant, S.S., Prochnik, S.E., Vallon, O., Harris, E.H., Karpowicz, S.J., Witman, G.B., Terry, A., Salamov, A., Fritz-Laylin, L.K., Maréchal-Drouard, L., *et al.* (2007). The *Chlamydomonas* genome reveals the evolution of key animal and plant functions. Science *318*, 245–250.

Michael, T.P., and VanBuren, R. (2015). Progress, challenges and the future of crop genomes. Curr. Opin. Plant Biol. *24*, 71–81. http://dx.doi.org/10.1016/j.pbi.2015.02.002

Ming, R., VanBuren, R., Wai, C.M., Tang, H., Schatz, M.C., Bowers, J.E., Lyons, E., Wang, M.L., Chen, J., Biggers, E., *et al.* (2015). The pineapple genome and the evolution of CAM photosynthesis. Nat. Genet. *47*, 1435–1442. http://dx.doi.org/10.1038/ng.3435

Mortazavi, A., Williams, B.A., McCue, K., Schaeffer, L., and Wold, B. (2008). Mapping and quantifying mammalian transcriptomes by RNA-Seq. Nat. Methods *5*, 621–628. http://dx.doi.org/10.1038/nmeth.1226

Moxon, S., Jing, R., Szittya, G., Schwach, F., Rusholme Pilcher, R.L., Moulton, V., and Dalmay, T. (2008). Deep sequencing of tomato short RNAs identifies microRNAs targeting genes involved in fruit ripening. Genome Res. *18*, 1602–1609. http://dx.doi.org/10.1101/gr.080127.108

Neale, D.B., Wegrzyn, J.L., Stevens, K.A., Zimin, A.V., Puiu, D., Crepeau, M.W., Cardeno, C., Koriabine, M., Holtz-Morris, A.E., Liechty, J.D., *et al.* (2014). Decoding the massive genome of loblolly pine using haploid DNA and novel assembly strategies. Genome Biol. *15*, R59. http://dx.doi.org/10.1186/gb-2014-15-3-r59

Neves, L.G., Davis, J.M., Barbazuk, W.B., and Kirst, M. (2013). Whole-exome targeted sequencing of the uncharacterized pine genome. Plant. J. *75*, 146–156. http://dx.doi.org/10.1111/tpj.12193

Nguyen, V.T., Kiss, T., Michels, A.A., and Bensaude, O. (2001). 7SK small nuclear RNA binds to and inhibits the activity of CDK9/cyclin T complexes. Nature *414*, 322–325. http://dx.doi.org/10.1038/35104581

Nie, S., Li, C., Xu, L., Wang, Y., Huang, D., Muleke, E.M., Sun, X., Xie, Y., and Liu, L. (2016). *De novo* transcriptome analysis in radish (*Raphanus sativus* L.) and identification of critical genes involved in bolting and flowering. BMC Genomics *17*, 389. http://dx.doi.org/10.1186/s12864-016-2633-2

Nie, S., Xu, L., Wang, Y., Huang, D., Muleke, E.M., Sun, X., Wang, R., Xie, Y., Gong, Y., and Liu, L. (2015). Identification of bolting-related microRNAs and their targets reveals complex miRNA-mediated flowering-time regulatory networks in radish (*Raphanus sativus* L.). Sci. Rep. 5, http://dx.doi.org/10.1038/srep14034.

Nilsson, M., Malmgren, H., Samiotaki, M., Kwiatkowski, M., Chowdhary, B.P., and Landegren, U. (1994). Padlock probes: circularizing oligonucleotides for localized DNA detection. Science *265*, 2085–2088.

Nowak, M.D., Russo, G., Schlapbach, R., Huu, C.N., Lenhard, M., and Conti, E. (2015). The draft genome of *Primula veris* yields insights into the molecular basis of heterostyly. Genome Biol. *16*, 12. http://dx.doi.org/10.1186/s13059-014-0567-z

Nystedt, B., Street, N.R., Wetterbom, A., Zuccolo, A., Lin, Y.C., Scofield, D.G., Vezzi, F., Delhomme, N., Giacomello, S., Alexeyenko, A., *et al.* (2013). The Norway spruce genome sequence and conifer genome evolution. Nature *497*, 579–584. http://dx.doi.org/10.1038/nature12211

Okou, D.T., Steinberg, K.M., Middle, C., Cutler, D.J., Albert, T.J., and Zwick, M.E. (2007). Microarray-based genomic selection for high-throughput resequencing. Nat. Methods *4*, 907–909.

Olsen, J.L., Rouzé, P., Verhelst, B., Lin, Y.C., Bayer, T., Collen, J., Dattolo, E., De Paoli, E., Dittami, S., Maumus, *et al.* (2016). The genome of the seagrass *Zostera marina* reveals angiosperm adaptation to the sea. Nature *530*, 331–335. http://dx.doi.org/10.1038/nature16548

Ota, S., Oshima, K., Yamazaki, T., Kim, S., Yu, Z., Yoshihara, M., Takeda, K., Takeshita, T., Hirata, A., Bišová, K., *et al.* (2016). Highly efficient lipid production in the green alga *Parachlorella kessleri*: draft genome and transcriptome endorsed by whole-cell 3D ultrastructure. Biotechnol. Biofuels 9, 13.

Ottesen, A.R., González Peña, A., White, J.R., Pettengill, J.B., Li, C., Allard, S., Rideout, S., Allard, M., Hill, T., Evans, P., *et al.* (2013b). Baseline survey of the anatomical microbial ecology of an important food plant: *Solanum lycopersicum* (tomato). BMC Microbiol. *13*, 114. http://dx.doi.org/10.1186/1471-2180-13-114

Pankin, A., Campoli, C., Dong, X., Kilian, B., Sharma, R., Himmelbach, A., Saini, R., Davis, S.J., Stein, N., Schneeberger, K. *et al.* (2014). Mapping-by-sequencing identifies HvPHYTOCHROME C as a candidate gene for the early maturity 5 locus modulating the circadian clock and photoperiodic flowering in barley. Genetics *198*, 383–396.

Passos da Silva, D., Castañeda-Ojeda, M.P., Moretti, C., Buonaurio, R., Ramos, C., and Venturi, V. (2014). Bacterial multispecies studies and microbiome analysis of a plant disease. Microbiology *160*, 556–566. http://dx.doi.org/10.1099/mic.0.074468-0

Pavy, N., Gagnon, F., Deschênes, A., Boyle, B., Beaulieu, J., and Bousquet, J. (2016). Development of highly reliable in silico SNP resource and genotyping assay from exome capture and sequencing: an example from black spruce (*Picea mariana*). Mol. Ecol. Resour. *16*, 588–598. http://dx.doi.org/10.1111/1755-0998.12468

Pendleton, M., Sebra, R., Pang, A.W., Ummat, A., Franzen, O., Rausch, T., Stütz, A.M., Stedman, W., Anantharaman, T., Hastie, A., *et al.* (2015). Assembly and diploid architecture of an individual human genome via single-molecule technologies. Nat. Methods *12*, 780–786. http://dx.doi.org/10.1038/nmeth.3454

Pfender, W.F., Saha, M.C., Johnson, E.A., and Slabaugh, M.B. (2011). Mapping with RAD (restriction-site associated DNA) markers to rapidly identify QTL for stem rust resistance in *Lolium perenne*. Theor. Appl. Genet. *122*, 1467–1480. http://dx.doi.org/10.1007/s00122-011-1546-3

Pingault, L., Choulet, F., Alberti, A., Glover, N., Wincker, P., Feuillet, C., and Paux, E. (2015). Deep transcriptome sequencing provides new insights into the structural and functional organization of the wheat genome. Genome Biol. *16*, 29. http://dx.doi.org/10.1186/s13059-015-0601-9

Pirrò, S., Zanella, L., Kenzo, M., Montesano, C., Minutolo, A., Potestà, M., Sobze, M.S., Canini, A., Cirilli, M., Muleo, R., *et al.* (2016). MicroRNA from *Moringa oleifera*: identification by high throughput sequencing and their potential contribution to plant medicinal value. PLOS ONE *11*, e0149495. http://dx.doi.org/10.1371/journal.pone.0149495

Porras-Alfaro, A., and Bayman, P. (2011). Hidden fungi, emergent properties: endophytes and microbiomes. Annu. Rev. Phytopathol. *49*, 291–315. http://dx.doi.org/10.1146/annurev-phyto-080508-081831

Poudel, S., Aryal, N., and Lu, C. (2015). Identification of microRNAs and transcript targets in *Camelina sativa* by deep sequencing and computational methods. PLOS ONE *10*, e0121542. http://dx.doi.org/10.1371/journal.pone.0121542

Prochnik, S.E., Umen, J., Nedelcu, A.M., Hallmann, A., Miller, S.M., Nishii, I., Ferris, P., Kuo, A., Mitros, T., Fritz-Laylin, L.K., *et al.* (2010). Genomic analysis of organismal complexity in the multicellular green alga

*Volvox carteri*. Science *329*, 223–226. http://dx.doi.org/10.1126/science.1188800

Qiao, Y., Zhang, J., Zhang, J., Wang, Z., Ran, A., Guo, H., Wang, D., and Zhang, J. (2016). Integrated RNA-seq and sRNA-seq analysis reveals miRNA effects on secondary metabolism in *Solanum tuberosum* L. Mol. Genet. Genomics 1–16, http://dx.doi.org/10.1007/s00438-016-1253-5

Quan, M., Chen, J., and Zhang, D. (2015). Exploring the secrets of long noncoding RNAs. Int. J. Mol. Sci. *16*, 5467–5496. http://dx.doi.org/10.3390/ijms16035467

Que, Y., Su, Y., Guo, J., Wu, Q., and Xu, L., 2014. A global view of transcriptome dynamics during *Sporisorium scitamineum* challenge in sugarcane by RNA-Seq. PLOS ONE, 9(8), p.e106476, http://dx.doi.org/10.1371/journal.pone.0106476.

Rahman, H., Jagadeeshselvam, N., Valarmathi, R., Sachin, B., Sasikala, R., Senthil, N., Sudhakar, D., Robin, S., and Muthurajan, R. (2014). Transcriptome analysis of salinity responsiveness in contrasting genotypes of finger millet (*Eleusine coracana* L.) through RNA-sequencing. Plant molecular biology *85*, 485–503.

Rastogi, S., Kalra, A., Gupta, V., Khan, F., Lal, R.K., Tripathi, A.K., Parameswaran, S., Gopalakrishnan, C., Ramaswamy, G., and Shasany, A.K. (2015). Unravelling the genome of Holy basil: an 'incomparable' 'elixir of life' of traditional Indian medicine. BMC Genomics *16*, 413. http://dx.doi.org/10.1186/s12864-015-1640-z

Rensing, S.A., Lang, D., Zimmer, A.D., Terry, A., Salamov, A., Shapiro, H., Nishiyama, T., Perroud, P.F., Lindquist, E.A., Kamisugi, Y., *et al.* (2008). The Physcomitrella genome reveals evolutionary insights into the conquest of land by plants. Science *319*, 64–69.

Rinn, J.L., and Chang, H.Y. (2012). Genome regulation by long noncoding RNAs. Annu. Rev. Biochem. *81*, 145–166. http://dx.doi.org/10.1146/annurev-biochem-051410-092902

Roberts, R.J., Carneiro, M.O., and Schatz, M.C. (2013). The advantages of SMRT sequencing. Genome Biol. *14*, 405. http://dx.doi.org/10.1186/gb-2013-14-6-405

Ross, M.G., Russ, C., Costello, M., Hollinger, A., Lennon, N.J., Hegarty, R., Nusbaum, C., and Jaffe, D.B. (2013). Characterizing and measuring bias in sequence data. Genome Biol. *14*, R51. http://dx.doi.org/10.1186/gb-2013-14-5-r51

Rowland, L.J., Alkharouf, N., Darwish, O., Ogden, E.L., Polashock, J.J., Bassil, N.V., and Main, D. (2012). Generation and analysis of blueberry transcriptome sequences from leaves, developing fruit, and flower buds from cold acclimation through deacclimation. BMC plant biology, *12*, 1. http://dx.doi.org/10.1186/1471-2229-12-46.

Ruiz-Ferrer, V., and Voinnet, O. (2009). Roles of plant small RNAs in biotic stress responses. Annu. Rev. Plant Biol. *60*, 485–510. http://dx.doi.org/10.1146/annurev.arplant.043008.092111

Russell, J., Mascher, M., Dawson, I.K., Kyriakidis, S., Calixto, C., Freund, F., Bayer, M., Milne, I., Marshall-Griffiths, T., Heinen, S., *et al.* (2016). Exome sequencing of geographically diverse barley landraces and wild relatives gives insights into environmental adaptation. Nature Genet. *48*, 1024–1030. http://dx.doi.org/10.1038/ng.3612.

Sahu, B.B., Sumit, R., Srivastava, S.K., and Bhattacharyya, M.K. (2012). Sequence based polymorphic (SBP) marker technology for targeted genomic regions: its application in generating a molecular map of the *Arabidopsis thaliana* genome. BMC Genomics *13*, 20. http://dx.doi.org/10.1186/1471-2164-13-20

Saintenac, C., Jiang, D., and Akhunov, E.D. (2011). Targeted analysis of nucleotide and copy number variation by exon capture in allotetraploid wheat genome. Genome Biol. *12*, R88. http://dx.doi.org/10.1186/gb-2011-12-9-r88

Sanger, F., Nicklen, S., and Coulson, A.R. (1977). DNA sequencing with chain-terminating inhibitors. Proc. Natl. Acad. Sci. U.S.A. *74*, 5463–5467.

Scaglione, D., Reyes-Chin-Wo, S., Acquadro, A., Froenicke, L., Portis, E., Beitel, C., Tirone, M., Mauro, R., Lo Monaco, A., Mauromicale, G., *et al.* (2016). The genome sequence of the outbreeding globe artichoke constructed *de novo* incorporating a phase-aware low-pass sequencing strategy of F1 progeny. Sci. Rep. *6*, 19427.

Schatz, M.C., Delcher, A.L., and Salzberg, S.L. (2010). Assembly of large genomes using second-generation sequencing. Genome Res. *20*, 1165–1173. http://dx.doi.org/10.1101/gr.101360.109

Schiessl, S., Samans, B., Hüttel, B., Reinhard, R., and Snowdon, R.J. (2014). Capturing sequence variation among flowering-time regulatory gene homologs in the allopolyploid crop species Brassica napus. Front. Plant Sci. 5, 404. http://dx.doi.org/10.3389/fpls.2014.00404

Schmutz, J., Cannon, S.B., Schlueter, J., Ma, J., Mitros, T., Nelson, W., Hyten, D.L., Song, Q., Thelen, J.J., Cheng, J., *et al.* (2010). Genome sequence of the palaeopolyploid soybean. Nature *463*, 178–183. http://dx.doi.org/10.1038/nature08670

Schnable, P.S., Ware, D., Fulton, R.S., Stein, J.C., Wei, F., Pasternak, S., Liang, C., Zhang, J., Fulton, L., Graves, T.A., *et al.* (2009). The B73 maize genome: complexity, diversity, and dynamics. Science *326*, 1112–1115. http://dx.doi.org/10.1126/science.1178534

Schreiber, A.W., Hayden, M.J., Forrest, K.L., Kong, S.L., Langridge, P., and Baumann, U. (2012). Transcriptome-scale homoeolog-specific transcript assemblies of bread wheat. BMC Genomics *13*, 492. http://dx.doi.org/10.1186/1471-2164-13-492

Semagn, K., Bjornstad, A., and Ndjiondjop, M.N. (2006). An overview of molecular marker methods for plants. Afr. J. Biotech. *5*, 2540–2568.

Sessitsch, A., Hardoim,P., Döring,J., Weilharter, A., Krause, A. *et al.* (2012). Functional characteristics of an endophyte community colonizing rice roots as revealed by metagenomic analysis. Mol. Plant Microbe Interact., http://dx.doi.org/10.1094/MPMI-08-11-0204

Severin, A.J., Woody, J.L., Bolon, Y.T., Joseph, B., Diers, B.W., Farmer, A.D., Muehlbauer, G.J., Nelson, R.T., Grant, D., Specht, J.E., *et al.* (2010). RNA-Seq Atlas of *Glycine max*: a guide to the soybean transcriptome. BMC Plant Biol. *10*, 160. http://dx.doi.org/10.1186/1471-2229-10-160

Sheibani-Tezerji, R., Rattei, T., Sessitsch, A., Trognitz, F., and Mitter, B. (2015). Transcriptome profiling of the endophyte *Burkholderia phytofirmans* PsJN indicates sensing of the plant environment and drought stress.

mBio 6, e00621-15. http://dx.doi.org/10.1128/mBio.00621-15
Shen, D., Suhrkamp, I., Wang, Y., Liu, S., Menkhaus, J., Verreet, J.A., Fan, L., and Cai, D. (2014). Identification and characterization of microRNAs in oilseed rape (*Brassica napus*) responsive to infection with the pathogenic fungus *Verticillium longisporum* using Brassica AA (*Brassica rapa*) and CC (*Brassica oleracea*) as reference genomes. New Phytol. *204*, 577–594.
Shen, E., Zou, J., Hubertus Behrens, F., Chen, L., Ye, C., Dai, S., Li, R., Ni, M., Jiang, X., Qiu, J., *et al.* (2015). Identification, evolution, and expression partitioning of miRNAs in allopolyploid *Brassica napus*. J. Exp. Bot. *66*, 7241–7253. http://dx.doi.org/10.1093/jxb/erv420
Singh, D., Singh, P.K., Chaudhary, S., Mehla, K., and Kumar, S. (2012). Exome sequencing and advances in crop improvement. Adv. Genet. *79*, 87–121. http://dx.doi.org/10.1016/B978-0-12-394395-8.00003-7
Singh, V.K., and Jain, M. (2014). Transcriptome profiling for discovery of genes involved in shoot apical meristem and flower development. Genom. Data *2*, 135–138. http://dx.doi.org/10.1016/j.gdata.2014.06.004
Slate, J., Gratten, J., Beraldi, D., Stapley, J., Hale, M., and Pemberton, J.M. (2009). Gene mapping in the wild with SNPs: guidelines and future directions. Genetica *136*, 97–107. http://dx.doi.org/10.1007/s10709-008-9317-z
Smith, B.T., Harvey, M.G., Faircloth, B.C., Glenn, T.C., and Brumfield, R.T. (2014). Target capture and massively parallel sequencing of ultraconserved elements for comparative studies at shallow evolutionary time scales. Systematic biology 63, 83–95.
Song, Q.X., Liu, Y.F., Hu, X.Y., Zhang, W.K., Ma, B., Chen, S.Y., and Zhang, J.S. (2011). Identification of miRNAs and their target genes in developing soybean seeds by deep sequencing. BMC Plant Biol. *11*, 5. http://dx.doi.org/10.1186/1471-2229-11-5
Srivastava, S., Singh, R.K., Pathak, G., Goel, R., Asif, M.H., Sane, A.P., and Sane, V.A. (2016). Comparative transcriptome analysis of unripe and mid-ripe fruit of *Mangifera indica* (var. 'Dashehari') unravels ripening associated genes. Scientific Reports, *6*, http://dx.doi.org/10.1038/srep32557.
Srivastava, S., Zheng, Y., Kudapa, H., Jagadeeswaran, G., Hivrale, V., Varshney, R.K., and Sunkar, R. (2015). High throughput sequencing of small RNA component of leaves and inflorescence revealed conserved and novel miRNAs as well as phasiRNA loci in chickpea. Plant. Sci. *235*, 46–57. http://dx.doi.org/10.1016/j.plantsci.2015.03.002
Staley, J.T., and Konopka, A. (1985). Measurement of in situ activities of nonphotosynthetic microorganisms in aquatic and terrestrial habitats. Annu. Rev. Microbiol. *39*, 321–346. http://dx.doi.org/10.1146/annurev.mi.39.100185.001541
Steuernagel, B., Periyannan, S.K., Hernández-Pinzón, I., Witek, K., Rouse, M.N., Yu, G., Hatta, A., Ayliffe, M., Bariana, H., Jones, J.D., *et al.* (2016). Rapid cloning of disease-resistance genes in plants using mutagenesis and sequence capture. Nat. Biotechnol. *34*, 652–655. http://dx.doi.org/10.1038/nbt.3543
Sun, X., Xu, L., Wang, Y., Luo, X., Zhu, X., Kinuthia, K.B., Nie, S., Feng, H., Li, C., and Liu, L. (2016). Transcriptome-based gene expression profiling identifies differentially expressed genes critical for salt stress response in radish (*Raphanus sativus* L.). Plant cell reports, 35(2), 329–346.
Sun, X., Xu, L., Wang, Y., Yu, R., Zhu, X., Luo, X., Gong, Y., Wang, R., Limera, C., Zhang, K. *et al.* (2015). Identification of novel and salt-responsive miRNAs to explore miRNA-mediated regulatory network of salt stress response in radish (*Raphanus sativus* L.). BMC genomics 16(1), 1. http://dx.doi.org/10.1186/s12864-015-1416-5
Syring, J.V., Tennessen, J.A., Jennings, T.N., Wegrzyn, J., Scelfo-Dalbey, C., and Cronn, R. (2016). Targeted capture sequencing in whitebark pine reveals range-wide demographic and adaptive patterns despite challenges of a large, repetitive genome. Front. Plant Sci. 7, 484.
Tanaka, H., Hirakawa, H., Kosugi, S., Nakayama, S., Ono, A., Watanabe, A., Hashiguchi, M., Gondo, T., Ishigaki, G., Muguerza, M., *et al.* (2016). Sequencing and comparative analyses of the genomes of zoysiagrasses. DNA Res. *23*, 171–180. http://dx.doi.org/10.1093/dnares/dsw006
Tanase, K., Nishitani, C., Hirakawa, H., Isobe, S., Tabata, S., Ohmiya, A., and Onozaki, T. (2012). Transcriptome analysis of carnation (*Dianthus caryophyllus* L.) based on next-generation sequencing technology. BMC Genomics *13*, 292. http://dx.doi.org/10.1186/1471-2164-13-292
Tang, W., Zheng, Y., Dong, J., Yu, J., Yue, J., Liu, F., Guo, X., Huang, S., Wisniewski, M., Sun, J., *et al.* (2016). Comprehensive transcriptome profiling reveals long noncoding RNA expression and alternative splicing regulation during fruit development and ripening in kiwifruit (*Actinidia chinensis*). Front. Plant Sci. 7, 335, http://dx.doi.org/10.3389/fpls.2016.00335.
Tang, Z., Zhang, L., Xu, C., Yuan, S., Zhang, F., Zheng, Y., and Zhao, C. (2012). Uncovering small RNA-mediated responses to cold stress in a wheat thermosensitive genic male-sterile line by deep sequencing. Plant Physiol. *159*, 721–738. http://dx.doi.org/10.1104/pp.112.196048
Tennessen, J.A., Govindarajulu, R., Liston, A., and Ashman, T.L. (2013). Targeted sequence capture provides insight into genome structure and genetics of male sterility in a gynodioecious diploid strawberry, *Fragaria vesca* ssp. *bracteata* (Rosaceae). G3 3, 1341–1351.
Tian, B.Y., Cao, Y., and Zhang, K.Q. (2015). Metagenomic insights into communities, functions of endophytes, and their associates with infection by root-knot nematode, *Meloidogyne incognita,* in tomato roots. Scientific Reports, 5.
Tian, Y., Tian, Y., Luo, X., Zhou, T., Huang, Z., Liu, Y., Qiu, Y., Hou, B., Sun, D., Deng, H., *et al.* (2014). Identification and characterization of microRNAs related to salt stress in broccoli, using high-throughput sequencing and bioinformatics analysis. BMC Plant Biol. *14*, 226. http://dx.doi.org/10.1186/s12870-014-0226-2
Tian, Y., Zeng, Y., Zhang, J., Yang, C., Yan, L., Wang, X., Shi, C., Xie, J., Dai, T., Peng, L., *et al.* (2015). High quality reference genome of drumstick tree (*Moringa oleifera* Lam.), a potential perennial crop. Sci. China Life Sci. *58*, 627–638. http://dx.doi.org/10.1007/s11427-015-4872-x

Tombuloglu, G., Tombuloglu, H., Sakcali, M.S., and Unver, T. (2015). High-throughput transcriptome analysis of barley (*Hordeum vulgare*) exposed to excessive boron. Gene 557, 71–81. http://dx.doi.org/10.1016/j.gene.2014.12.012

Turktas, M., Kurtoglu, K.Y., Dorado, G., Zhang, B., and Hernandez, P. (2015). Sequencing of plant genomes – a review. Turk J. Agric. For. *39*, 361–376. http://dx.doi.org/10.3906/tar-1409-93

Tuskan, G.A., Difazio, S., Jansson, S., Bohlmann, J., Grigoriev, I., Hellsten, U., Putnam, N., Ralph, S., Rombauts, S., Salamov, A., *et al.* (2006). The genome of black cottonwood, *Populus trichocarpa* (Torr. & Gray). Science *313*, 1596–1604.

Udomchalothorn, T., Plaimas, K., Comai, L., Buaboocha, T., and Chadchawan,S. (2014). Molecular karyotyping and exome analysis of salt-tolerant rice mutant from somaclonal variation. The Plant Genome 7, http://dx.doi.org/10.3835/plantgenome2014.04.0016.

Ulitsky, I., and Bartel, D.P. (2013). lincRNAs: genomics, evolution, and mechanisms. Cell *154*, 26–46. http://dx.doi.org/10.1016/j.cell.2013.06.020

Unno, Y., and Shinano, T. (2013). Metagenomic analysis of the rhizosphere soil microbiome with respect to phytic acid utilization. Microbes. Environ. *28*, 120–127.

Urich, T., Lanzén, A., Qi, J., Huson, D.H., Schleper, C., and Schuster, S.C. (2008). Simultaneous assessment of soil microbial community structure and function through analysis of the meta-transcriptome. PLOS ONE *3*, e2527. http://dx.doi.org/10.1371/journal.pone.0002527

VanBuren, R., Bryant, D., Bushakra, J.M., Vining, K.J., Edger, P.P., Rowley, E.R., Priest, H.D., Michael, T.P., Lyons, E., Filichkin, S.A., *et al.* (2016). The genome of black raspberry (*Rubus occidentalis*). Plant J. *87*, 535–547. http://dx.doi.org/10.1111/tpj.13215

VanBuren, R., Bryant, D., Edger, P.P., Tang, H., Burgess, D., Challabathula, D., Spittle, K., Hall, R., Gu, J., Lyons, E., *et al.* (2015). Single-molecule sequencing of the desiccation-tolerant grass *Oropetium thomaeum*. Nature *527*, 508–511. http://dx.doi.org/10.1038/nature15714

Van Hoeck, A., Horemans, N., Monsieurs, P., Cao, H.X., Vandenhove, H., and Blust, R. (2015). The first draft genome of the aquatic model plant Lemna minor opens the route for future stress physiology research and biotechnological applications, Biotechnol. Biofuels *8*, 1.

Varshney, R.K., and Dubey, A. (2009). Novel genomic tools and modern genetic and breeding approaches for crop improvement. J. Pant Biochem. Biot. *18*, 127–138.

Vicentini, R., Bem, D., Van Sluys, M.A., Nogueira, F.T.S., and Vincentz, M. (2012). Gene content analysis of sugarcane Public ESTs reveals thousands of missing coding-genes and an unexpected pool of grasses conserved ncRNAs. Tropical Plant Biol. *5*, 199–205, http://dx.doi.org/10.1007/s12042-012-9103-z

Vogel, J.P., Garvin, D.F., Mockler, T.C., Schmutz, J., Rokhsar, D., Bevan, M.W., Barry, K., Lucas, S., Harmon-Smith, Miranda., Lail, K., *et al.* (2010). Genome sequencing and analysis of the model grass *Brachypodium distachyon*. Nature 463,763–768.

Vorholt, J.A. (2012). Microbial life in the phyllosphere. Nat. Rev. Microbiol. *10*, 828–840. http://dx.doi.org/10.1038/nrmicro2910

Wang, H., Jiang, J., Chen, S., Qi, X., Peng, H., Li, P., Song, A., Guan, Z., Fang, W., Liao, Y., *et al.* (2013). Next-generation sequencing of the *Chrysanthemum nankingense* (Asteraceae) transcriptome permits large-scale unigene assembly and SSR marker discovery. PLOS ONE, 8(4), 62293, http://dx.doi.org/10.1371/journal.pone.0062293.

Wang, H., Niu, Q.W., Wu, H.W., Liu, J., Ye, J., Yu, N., and Chua, N.H. (2015a). Analysis of non-coding transcriptome in rice and maize uncovers roles of conserved lncRNAs associated with agriculture traits. Plant. J. *84*, 404–416. http://dx.doi.org/10.1111/tpj.13018

Wang, L., Yu, X., Wang, H., Lu, Y.Z., de Ruiter, M., Prins, M., and He, Y.K. (2011). A novel class of heat-responsive small RNAs derived from the chloroplast genome of Chinese cabbage (*Brassica rapa*). BMC Genomics *12*, 289. http://dx.doi.org/10.1186/1471-2164-12-289

Wang, M., Yuan, D., Tu, L., Gao, W., He, Y., Hu, H., Wang, P., Liu, N., Lindsey, K., and Zhang, X. (2015c). Long noncoding RNAs and their proposed functions in fibre development of cotton (*Gossypium* spp.). New Phytol. *207*, 1181–1197. http://dx.doi.org/10.1111/nph.13429

Wang, T.Z., Liu, M., Zhao, M.G., Chen, R., and Zhang, W.H. (2015b). Identification and characterization of long non-coding RNAs involved in osmotic and salt stress in *Medicago truncatula* using genome-wide high-throughput sequencing. BMC Plant Biol. *15*, 131. http://dx.doi.org/10.1186/s12870-015-0530-5

Wang, W., Haberer, G., Gundlach, H., Gläßer, C., Nussbaumer, T., Luo, M.C., Lomsadze, A., Borodovsky, M., Kerstetter, R.A., Shanklin, J., *et al.* (2014). The *Spirodela polyrhiza* genome reveals insights into its neotenous reduction fast growth and aquatic lifestyle. Nat. Commun. *5*, 3311. http://dx.doi.org/10.1038/ncomms4311

Wang, X., Elling, A.A., Li, X., Li, N., Peng, Z., He, G., Sun, H., Qi, Y., Liu, X.S., and Deng, X.W. (2009). Genome-wide and organ-specific landscapes of epigenetic modifications and their relationships to mRNA and small RNA transcriptomes in maize. Plant Cell *21*, 1053–1069. http://dx.doi.org/10.1105/tpc.109.065714

Wang, Y., Shen, D., Bo, S., Chen, H., Zheng, J., Zhu, Q.H., Cai, D., Helliwell, C., and Fan, L. (2010). Sequence variation and selection of small RNAs in domesticated rice. BMC Evol. Biol. *10*, 119. http://dx.doi.org/10.1186/1471-2148-10-119

Wangai, A.W., Redinbaugh, M.G., Kinyua, Z.M., Miano, D.W., Leley, P.K., Kasina, M., Mahuku, G., Scheets, K., and Jeffers, D. (2016). First report of *maize chlorotic mottle virus* and maize lethal necrosis in Kenya. Virus Research.

Warr, A., Robert, C., Hume, D., Archibald, A., Deeb, N., and Watson, M. (2015). Exome sequencing: current and future perspectives. G3 *5*, 1543–1550. http://dx.doi.org/10.1534/g3.115.018564

Watson, J.D., and Crick, F.H. (1953). Molecular structure of nucleic acids; a structure for deoxyribose nucleic acid. Nature *171*, 737–738.

Wei, B., Cai, T., Zhang, R., Li, A., Huo, N., Li, S., Gu, Y.Q., Vogel, J., Jia, J., Qi, Y., *et al.* (2009). Novel microRNAs uncovered by deep sequencing of small RNA transcriptomes in bread wheat (*Triticum aestivum* L.)

and *Brachypodium distachyon* (L.) Beauv. Funct. Integr. Genomics 9, 499–511.

Wei, X., Zhang, X., Yao, Q., Yuan, Y., Li, X., Wei, F., Zhao, Y., Zhang, Q., Wang, Z., Jiang, W., *et al.* (2015). The miRNAs and their regulatory networks responsible for pollen abortion in Ogura-CMS Chinese cabbage revealed by high-throughput sequencing of miRNAs, degradomes, and transcriptomes. Front. Plant Sci. *6*, 894, http://dx.doi.org/10.3389/fpls.2015.00894.

Wendler, N., Mascher, M., Himmelbach, A., Johnston, P., Pickering, R., and Stein, N. (2015). Bulbosum to go: A toolbox to utilize *Hordeum vulgare/bulbosum* introgressions for breeding and beyond. Mol. Plant *8*, 1507–1519. http://dx.doi.org/10.1016/j.molp.2015.05.004

Wendler, N., Mascher, M., Nöh, C., Himmelbach, A., Scholz, U., Ruge-Wehling, B., and Stein, N. (2014). Unlocking the secondary gene-pool of barley with next-generation sequencing. Plant Biotechnol. J. *12*, 1122–1131. http://dx.doi.org/10.1111/pbi.12219

Wilhelm, B.T., Marguerat, S., Watt, S., Schubert, F., Wood, V., Goodhead, I., Penkett, C.J., Rogers, J., and Bahler, J. (2008). Dynamic repertoire of a eukaryotic transcriptome surveyed at single-nucleotide resolution. Nature *453*, 1239–1243.

Willing, E.M., Rawat, V., Mandáková, T., Maumus, F., James, G.V., Nordström, K.J., Becker, C., Warthmann, N., Chica, C., Szarzynska, B., *et al.* (2015). Genome expansion of Arabis alpina linked with retrotransposition and reduced symmetric DNA methylation. Nat. Plants *1*, 14023. http://dx.doi.org/10.1038/nplants.2014.23

Willingham, A.T., Orth, A.P., Batalov, S., Peters, E.C., Wen, B.G., Aza-Blanc, P., Hogenesch, J.B., and Schultz, P.G. (2005). A strategy for probing the function of noncoding RNAs finds a repressor of NFAT. Science *309*, 1570–1573.

Winfield, M.O., Wilkinson, P.A., Allen, A.M., Barker, G.L., Coghill, J.A., Burridge, A., Hall, A., Brenchley, R.C., D'Amore, R., Hall, N., *et al.* (2012). Targeted re-sequencing of the allohexaploid wheat exome. Plant Biotechnol. J. *10*, 733–742. http://dx.doi.org/10.1111/j.1467-7652.2012.00713.x

Woese, C.R. (1987). Bacterial evolution. Microbiol. Rev. *51*, 221–271.

Wu, W., Yang, Y.L., He, W.M., Rouard, M., Li, W.M., Xu, M., Roux, N., and Ge, X.J. (2016). Whole genome sequencing of a banana wild relative *Musa itinerans* provides insights into lineage-specific diversification of the *Musa* genus. Sci. Rep. *6*, 31586. http://dx.doi.org/10.1038/srep31586

Xiao, L., Yang, G., Zhang, L., Yang, X., Zhao, S., Ji, Z., Zhou, Q., Hu, M., Wang, Y., Chen, M., *et al.* (2015). The resurrection genome of *Boea hygrometrica*: A blueprint for survival of dehydration. Proc. Natl. Acad. Sci. U.S.A. *112*, 5833–5837. http://dx.doi.org/10.1073/pnas.1505811112

Xiao, M., Zhang, Y., Chen, X., Lee, E. J., Barber, C. J., Chakrabarty, R., Desgagné-Penix, I., Haslam, T.M., Kim, Y.B., Liu, E., et al. (2013). Transcriptome analysis based on next-generation sequencing of non-model plants producing specialized metabolites of biotechnological interest. J. Biotechnol. *166*, 122–134.

Xie, F., Wang, Q., Sun, R., and Zhang, B. (2015). Deep sequencing reveals important roles of microRNAs in response to drought and salinity stress in cotton. J. Exp. Bot. *66*, 789–804. http://dx.doi.org/10.1093/jxb/eru437

Xin, M., Wang, Y., Yao, Y., Xie, C., Peng, H., Ni, Z., and Sun, Q. (2010). Diverse set of microRNAs are responsive to powdery mildew infection and heat stress in wheat (*Triticum aestivum* L.). BMC Plant Biol. *10*, 123. http://dx.doi.org/10.1186/1471-2229-10-123

Xu, M.Y., Dong, Y., Zhang, Q.X., Zhang, L., Luo, Y.Z., Sun, J., Fan, Y.L., and Wang, L. (2012). Identification of miRNAs and their targets from *Brassica napus* by high-throughput sequencing and degradome analysis. BMC Genomics *13*, 421. http://dx.doi.org/10.1186/1471-2164-13-421

Xu, T., Wang, Y., Liu, X., Lv, S., Feng, C., Qi, M., and Li, T. (2015). Small RNA and degradome sequencing reveals microRNAs and their targets involved in tomato pedicel abscission. Planta *242*, 963–984. http://dx.doi.org/10.1007/s00425-015-2318-0

Xu, X., Passey, T., Wei,F., Saville, R., and Harrison, R.J. (2015). Amplicon-based metagenomics identified candidate organisms in soils that caused yield decline in strawberry. Hortic. Res. 2:15022. http://dx.doi.org/10.1038/hortres.2015.22..

Xu, Y., Gao, S., Yang, Y., Huang, M., Cheng, L., Wei, Q., Fei, Z., Gao, J., and Hong, B. (2013). Transcriptome sequencing and whole genome expression profiling of chrysanthemum under dehydration stress. BMC Genomics *14*, 662. http://dx.doi.org/10.1186/1471-2164-14-662

Yang, L., Jue, D., Li, W., Zhang, R., Chen, M., and Yang, Q. (2013). Identification of MiRNA from eggplant (*Solanum melongena* L.) by small RNA deep sequencing and their response to *Verticillium dahliae* infection. PLOS ONE *8*, e72840. http://dx.doi.org/10.1371/journal.pone.0072840

Yang, X., and Li, L. (2012). Analyzing the microRNA transcriptome in plants using deep sequencing data. Biology *1*, 297–310. http://dx.doi.org/10.3390/biology1020297

Yang, X., Wang, L., Yuan, D., Lindsey, K., and Zhang, X. (2013). Small RNA and degradome sequencing reveal complex miRNA regulation during cotton somatic embryogenesis. J. Exp. Bot. *64*, 1521–1536. http://dx.doi.org/10.1093/jxb/ert013

Yao, S., Jiang, C., Huang, Z., Torres-Jerez, I., Chang, J., Zhang, H., Udvardi, M., Liu, R., and Verdier, J. (2016). The *Vigna unguiculata* Gene Expression Atlas (VuGEA) from *de novo* assembly and quantification of RNA-seq data provides insights into seed maturation mechanisms. Plant J. *88*, 318–327. http://dx.doi.org/10.1111/tpj.13279

Yao, Y., Guo, G., Ni, Z., Sunkar, R., Du, J., Zhu, J.K., and Sun, Q. (2007). Cloning and characterization of microRNAs from wheat (*Triticum aestivum* L.). Genome Biol. *8*, R96.

Yao, Y., Ni, Z., Peng, H., Sun, F., Xin, M., Sunkar, R., Zhu, J.K., and Sun, Q. (2010). Non-coding small RNAs responsive to abiotic stress in wheat (*Triticum aestivum* L.). Funct. Integr. Genomics. *10*, 187–190. http://dx.doi.org/10.1007/s10142-010-0163-6

Yasui, Y., Hirakawa, H., Ueno, M., Matsui, K., Katsube-Tanaka, T., Yang, S.J., Aii, J., Sato, S., and Mori, M. (2016). Assembly of the draft genome of buckwheat and its applications in identifying agronomically useful genes. DNA Res. *23*, 215–224. http://dx.doi.org/10.1093/dnares/dsw012

Yu, J., Hu, S., Wang, J., Wong, G.K., Li, S., Liu, B., Deng, Y., Dai, L., Zhou, Y., Zhang, X., *et al.* (2002). A draft sequence of the rice genome (*Oryza sativa* L. ssp. indica). Science *296*, 79–92. http://dx.doi.org/10.1126/science.1068037

Yu, R., Wang, J., Xu, L., Wang, Y., Wang, R., Zhu, X., Sun, X., Luo, X., Xie, Y., *et al.* (2016). Transcriptome profiling of taproot reveals complex regulatory networks during taproot thickening in radish (*Raphanus sativus* L.). Front. Plant Sci. *7*, 1210, http://dx.doi.org/10.3389/fpls.2016.01210.

Yu, X., Wang, H., Lu, Y., de Ruiter, M., Cariaso, M., Prins, M., van Tunen, A., and He, Y. (2012). Identification of conserved and novel microRNAs that are responsive to heat stress in *Brassica rapa*. J. Exp. Bot. *63*, 1025–1038. http://dx.doi.org/10.1093/jxb/err337

Zeng, X., Long, H., Wang, Z., Zhao, S., Tang, Y., Huang, Z., Wang, Y., Xu, Q., Mao, L., Deng, G., *et al.* (2015). The draft genome of Tibetan hulless barley reveals adaptive patterns to the high stressful Tibetan Plateau. Proc. Natl. Acad. Sci. U.S.A. *112*, 1095–1100. http://dx.doi.org/10.1073/pnas.1423628112

Zhang, G., Guo, G., Hu, X., Zhang, Y., Li, Q., Li, R., Zhuang, R., Lu, Z., He, Z., Fang, X., *et al.* (2010). Deep RNA sequencing at single base-pair resolution reveals high complexity of the rice transcriptome. Genome Res. *20*, 646–654. http://dx.doi.org/10.1101/gr.100677.109

Zhang, G., Liu, P., Zhang, L., Wei, W., Wang, X., Wei, D., and Wang, W. (2016). Bioprospecting metagenomics of a microbial community on cotton degradation: Mining for new glycoside hydrolases. J. Biotechnol. *234*, 35–42.

Zhang, G., Liu, X., Quan, Z., Cheng, S., Xu, X., Pan, S., Xie, M., Zeng, P., Yue, Z., Wang, W., *et al.* (2012). Genome sequence of foxtail millet (*Setaria italica*) provides insights into grass evolution and biofuel potential. Nat. Biotechnol. *30*, 549–554. http://dx.doi.org/10.1038/nbt.2195

Zhang, G., Tian, Y., Zhang, J., Shu, L., Yang, S., Wang, W., Sheng, J., Dong, Y., and Chen, W. (2015). Hybrid de novo genome assembly of the Chinese herbal plant danshen (*Salvia miltiorrhiza* Bunge). Gigascience *4*, 62. http://dx.doi.org/10.1186/s13742-015-0104-3

Zhang, G.Q., Xu, Q., Bian, C., Tsai, W.C., Yeh, C.M., Liu, K.W., Yoshida, K., Zhang, L.S., Chang, S.B., Chen, F., *et al.* (2015). The *Dendrobium catenatum* Lindl. genome sequence provides insights into polysaccharide synthase, floral development and adaptive evolution. Sci. Rep. *5*, 19029.

Zhang, J., Liang, S., Duan, J., Wang, J., Chen, S., Cheng, Z., Zhang, Q., Liang, X., and Li, Y. (2012). *De novo* assembly and characterisation of the transcriptome during seed development, and generation of genic-SSR markers in peanut (*Arachis hypogaea* L.). BMC Genomics *13*, 90. http://dx.doi.org/10.1186/1471-2164-13-90

Zhang, J., Tian, Y., Yan, L., Zhang, G., Wang, X., Zeng, Y., Zhang, J., Ma, X., Tan, Y., Long, N., *et al.* (2016). Genome of plant maca (*Lepidium meyenii*) illuminates genomic basis for high-altitude adaptation in the central Andes. Mol. Plant. *9*, 1066–1077. http://dx.doi.org/10.1016/j.molp.2016.04.016

Zhang, J., Xu, Y., Huan, Q., and Chong, K. (2009). Deep sequencing of *Brachypodium* small RNAs at the global genome level identifies microRNAs involved in cold stress response. BMC Genomics *10*, 449. http://dx.doi.org/10.1186/1471-2164-10-449

Zhang, N., Yang, J., Wang, Z., Wen, Y., Wang, J., He, W., Liu, B., Si, H., and Wang, D. (2014). Identification of novel and conserved microRNAs related to drought stress in potato by deep sequencing. PLOS ONE *9*, e95489. http://dx.doi.org/10.1371/journal.pone.0095489

Zhang, R., Marshall, D., Bryan, G.J., and Hornyik, C. (2013). Identification and characterization of miRNA transcriptome in potato by high-throughput sequencing. PLOS ONE *8*, e57233. http://dx.doi.org/10.1371/journal.pone.0057233

Zhang, X., Zou, Z., Zhang, J., Zhang, Y., Han, Q., Hu, T., Xu, X., Liu, H., Li, H., and Ye, Z. (2011). Over-expression of sly-miR156a in tomato results in multiple vegetative and reproductive trait alterations and partial phenocopy of the *sft* mutant. FEBS Lett. *585*, 435–439. http://dx.doi.org/10.1016/j.febslet.2010.12.036

Zhang, X.N., Li, X., and Liu, J.H. (2014). Identification of conserved and novel cold-responsive microRNAs in trifoliate orange (*Poncirus trifoliata* (L.) Raf.) using high-throughput sequencing. Plant Mol. Biol. Rep. *32*, 328–341.

Zhang, Y.C., Liao, J.Y., Li, Z.Y., Yu, Y., Zhang, J.P., Li, Q.F., Qu, L.H., Shu, W.S., and Chen, Y.Q. (2014). Genome-wide screening and functional analysis identify a large number of long noncoding RNAs involved in the sexual reproduction of rice. Genome Biol. *15*, 512. http://dx.doi.org/10.1186/s13059-014-0512-1

Zhang, Y.C., Yu, Y., Wang, C.Y., Li, Z.Y., Liu, Q., Xu, J., Liao, J.Y., Wang, X.J., Qu, L.H., Chen, F., *et al.* (2013). Overexpression of microRNA OsmiR397 improves rice yield by increasing grain size and promoting panicle branching. Nat. Biotechnol. *31*, 848–852. http://dx.doi.org/10.1038/nbt.2646

Zhao, C., Xia, H., Cao, T., Yang, Y., Zhao, S., Hou, L., Zhang, L., Li, C., and Zhang, X. (2015). Small RNA and degradome deep sequencing reveals peanut microRNA roles in response to pathogen infection. Plant Mol. Biol. Rep. 33, 1013–1029.

Zhao, F., Wang, C., Han, J., Zhu, X., Li, X., Wang, X., and Fang, J. (2016). Characterization of miRNAs responsive to exogenous ethylene in grapevine berries at whole genome level. Funct. Integr. Genomics 1–23. http://dx.doi.org/10.1007/s10142-016-0514-z

Zheng, C., Zhao, L., Wang, Y., Shen, J., Zhang, Y., Jia, S., Li, Y., and Ding, Z. (2015). Integrated RNA-Seq and sRNA-Seq analysis identifies chilling and freezing responsive key molecular players and pathways in tea plant (*Camellia sinensis*). PLOS ONE *10*, e0125031. http://dx.doi.org/10.1371/journal.pone.0125031

Zheng, L.L., and Qu, L.H. (2015). Application of microRNA gene resources in the improvement of agronomic traits in rice. Plant Biotechnol. J. *13*, 329–336. http://dx.doi.org/10.1111/pbi.12321

Zhou, L., and Holliday, J.A. (2012). Targeted enrichment of the black cottonwood (*Populus trichocarpa*) gene space using sequence capture. BMC Genomics *13*, 703. http://dx.doi.org/10.1186/1471-2164-13-703

Zhou, M., and Luo, H. (2013). MicroRNA-mediated gene regulation: Potential applications for plant genetic engineering. Plant Mol. Biol. *83*, 59–75.

Zhou, Y., Xu, Z., Duan, C., Chen, Y., Meng, Q., Wu, J., Hao, Z., Wang, Z., Li, M., Yong, H., *et al.* (2016). Dual transcriptome analysis reveals insights into the response to *Rice black-streaked dwarf virus* in maize. J. Exp. Bot. *67*, 4593–4609. http://dx.doi.org/10.1093/jxb/erw244

Zhu, Q.H., and Wang, M.B. (2012). Molecular Functions of Long Non-Coding RNAs in Plants. Genes 3, 176–190. http://dx.doi.org/10.3390/genes3010176

Zhuang, Y., Zhou, X.H., and Liu, J. (2014). Conserved miRNAs and their response to salt stress in wild eggplant *Solanum linnaeanum* roots. Int. J. Mol. Sci. *15*, 839–849. http://dx.doi.org/10.3390/ijms15010839

Zimin, A., Stevens, K.A., Crepeau, M.W., Holtz-Morris, A., Koriabine, M., Marçais, G., Puiu, D., Roberts, M., Wegrzyn, J.L., de Jong, P.J., *et al.* (2014). Sequencing and assembly of the 22-gb loblolly pine genome. Genetics *196*, 875–890. http://dx.doi.org/10.1534/genetics.113.159715

Zou, C., Wang, Q., Lu, C., Yang, W., Zhang, Y., Cheng, H., Feng, X., Prosper, M.A., and Song, G. (2016). Transcriptome analysis reveals long noncoding RNAs involved in fiber development in cotton (*Gossypium arboreum*). Sci. China. Life. Sci. *59*, 164–171. http://dx.doi.org/10.1007/s11427-016-5000-2

Zuo, J., Wang, Q., Han, C., Ju, Z., Cao, D., Zhu, B., Luo, Y., and Gao, L. (2016). SRNAome and degradome sequencing analysis reveals specific regulation of sRNA in response to chilling injury in tomato fruit. Physiol. Plant 6, http://dx.doi.org/10.1111/ppl.12509.

# Next-generation Sequencing Promoted the Release of Reference Genomes and Discovered Genome Evolution in Cereal Crops

2

Yong Huang†, Haiyang Liu† and Yongzhong Xing*

National Key Lab of Crop Genetic Improvement, Huazhong Agricultural University, Wuhan, China.

*Correspondence: yzxing@mail.hzau.edu.cn
†These authors contributed equally

https://doi.org/10.21775/9781910190654.03

## Abstract

In recent decades, next-generation sequencing (NGS) was developed and brought biology into a new era. Rice, maize, wheat, sorghum and barley are the most important cereal crops and feed most of the world's population. Great progress in the study of cereal genomes has been made with the help of NGS. Reference genome sequence assembly and re-sequencing have grown exponentially. Thus, evolution and comparative genomics are renewed, including origin verification, evolution tracking and so on. In this review, we briefly record the development of sequencing technology, the comparison of NGS methods and platforms and summarize the bioinformatics tools used for NGS data analysis. We describe how NGS accelerates reference genome assembly and new evolutionary findings. We finally discuss how to discover more valuable resources and improve cereal breeding in the future.

## Introduction

Over the three decades since Frederick Sanger and Walter Gilbert received the Nobel Prize for Chemistry in 1980, extraordinary progress has been made in genome sequencing technologies. Particularly since the new century, genome sequencing has made breakthroughs that have greatly increased the understanding of plant genomes and biology. To date, the most widely used first-generation sequencer is a typical Applied Biosystems' (ABI) 3730XL capillary electrophoresis sequencer, which has a longer read length (more than 1000 bp) and high accuracy rate (99.999%); the cost per 1000 bases is $0.5, and a many as 600,000 bases are generated per day (Shendure and Ji, 2008). However, due to the limitations of throughput and the relatively high cost of sequencing, the first-generation sequencer is unable to meet the needs of large-scale genome sequencing, such as deep sequencing and repeated sequencing, which prompted the birth of second-generation sequencing technologies, also known as next-generation sequencing (NGS).

## Theories and platforms of NGS technologies

Over the last decade, NGS technologies have formed gradually and developed maturely. Roche/454 pyrosequencer, Illumina sequencer and ABI SOLiD sequencer are the main representatives. The approaches of NGS used in these three sequencers fall into two categories: sequencing by synthesis (SBS) and sequencing by ligation (SBL) (Fig. 2.1). No matter what kind of methods are used for sequencing, a library needs to be prepared

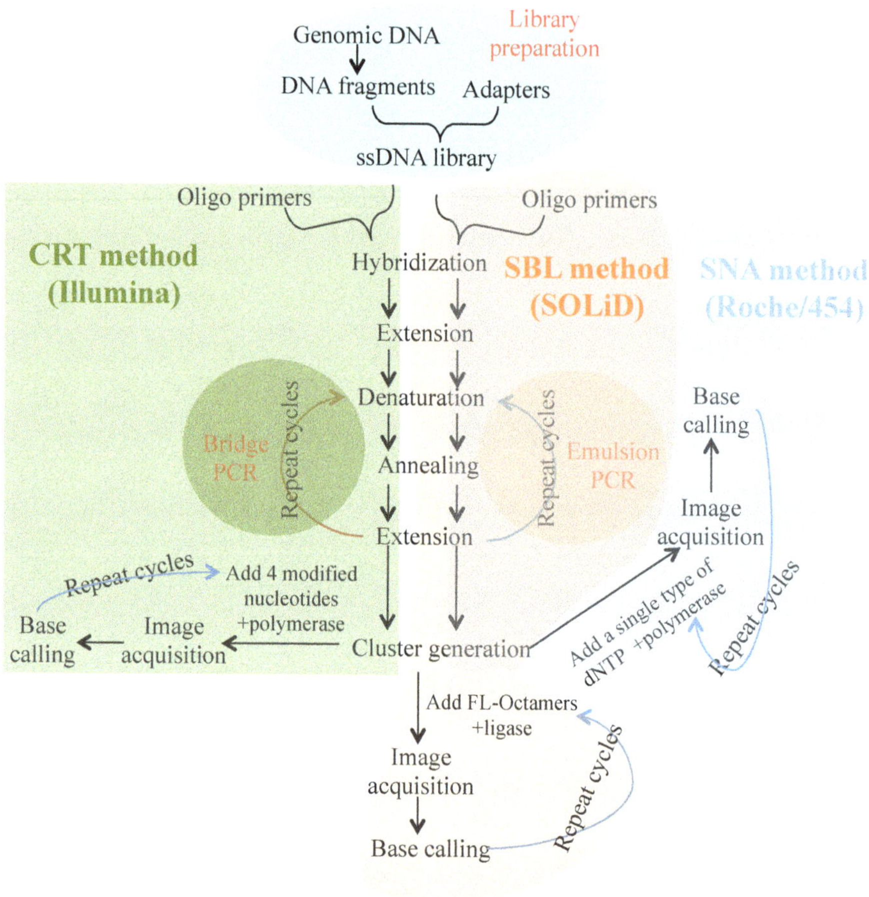


**Figure 2.1** The workflow of the approaches of next-generation sequencing. Aqua green region represents library-construction process. Olive area, light blue region and light orange region represent the process of CRT, SNA and SBL methods, respectively. In the CRT method, modified nucleotide is that fluorophore-labelled, terminally blocked nucleotides. ssDNA, single-strand DNA; CRT, cyclic reversible termination; SNA, single-nucleotide addition; SBL, sequencing by ligation; FL-Octamers, fluorescence labelling octamers.

by randomly breaking the genomic DNA into small fragments and adding common adapters to the fragment ends. Subsequently, a DNA template is primed by a sequence that is complementary to the adapter sequence. In order to achieve the required signal for sequencing, different types of polymerase chain reactions (PCRs) are performed at different positions. After several rounds of amplification, respective DNA clusters are generated.

In SBS approaches, PCR and fluorescence sequencing are repeatedly performed on DNA samples filled with chips. SBS approaches are divided into cyclic reversible termination (CRT) and single-nucleotide addition (SNA). The uniqueness of the CRT approach is that the ribose 3′-OH group of the nucleotide is blocked (Fig. 2.1). During each cycle, a fragment of DNA template combined with an adapter will incorporate just one nucleotide as the blocked 3′ group prevents additional incorporates. After nucleotide incorporation, unincorporated nucleotides are washed away, and the image is obtained to determine which base was incorporated in each cluster; then, the cycle begins again. The Illumina platform, currently the most cost-effective and most widely used, is a representative of the CRT approach (Goodwin *et al.*, 2016; Mardis, 2008; Metzker, 2010). Unlike CRT approaches, SNA approaches do not require the dNTPs to be blocked because only one type of dNTP is flowed across the PicoTiterPlate (PTP) wells in each round of the sequencing reaction. SNA approaches are applied on a Roche/454 pyrosequencing platform, which is the founder of and greatly promotes the development of NGS technologies (Rothberg and Leamon, 2008). In the CRT approaches of Illumina, bridge PCR and sequencing are carried out on the surface of the flow cell, while in the SNA approaches of Roche/454, emulsion PCR and sequencing are carried out in two different places: microbeads and PTP wells, respectively.

The unique feature of SBL approaches is that they replace the polymerization reaction during sequencing by synthesis with a ligation reaction (Fig. 2.1). ABI SOLiD sequencing technology is a typical SBL method. With SOLiD, each sequencing cycle introduces a degenerate population of fluorophore-labelled octamers composed of known nucleotides in the first and second positions. After ligation and imaging, the octamer is chemically cleaved between positions 5 and 6, removing the fluorescent label and leaving a free end for the next cycle of ligation. Single sequencing includes five rounds of sequencing reactions and ultimately can complete all positions of the nucleotide sequence; each position of the base is queried twice (Metzker, 2010).

In order to better explore DNA sequence information, a new generation of sequencing methods has been developed as single-molecule sequencing technology, also known as third-generation sequencing technology. It performs faster data reading and has great application potential. The read length of third-generation sequencing technology can reach several or even dozens of kilobases, thereby allowing the resolution of highly complex genomes with many long repetitive elements, copy number variations (CNVs) and structural variations. The most popular of the third-generation sequencing platforms is the single-molecule real-time (SMRT) sequencing method used by Pacific Biosciences (PacBio). In SMRT approaches, the polymerase is fixed at the bottom of the well and allows the DNA template to go through the zero-mode waveguides (ZMW) where sequencing can take place rather than polymerase binding and travelling along the DNA template as in NGS methods. All four dNTPs are labelled and available for incorporation. When the polymerase incorporates a fluorophore-labelled nucleotide, the emitted light is recorded by a camera, and the corresponding base sequence information then is transformed by the computer analyser (Goodwin *et al.*, 2016).

## Comparison of NGS platforms

As new technologies emerge, existing problems are exacerbated or new problems arise. Due to double detection, the original sequencing accuracy of SOLiD is as high as 99.94%, and the accuracy of 15× coverage is even up to 99.999%. Therefore, SOLiD should provide the highest sequencing accuracy in the current NGS technology. Despite its accuracy, the SOLiD platform still has a number of limitations, such as displaying AT bias (Rieber *et al.*, 2013). The maximum read length of the Illumina, 454, SOLiD and PacBio platforms is 600 bp, 1000 bp, 75 bp and ~20,000 bp, respectively (Table 2.1). The very short read length may be the greatest limitation of the SOLiD platform preventing its widespread adoption. In NGS technologies, due to dependence on the template amplification phase prior to sequencing, the assembly of a genome with a high GC content is particularly limited (Aird *et al.*, 2011; Dohm *et al.*, 2008; Niu *et al.*, 2010). The unbiased and longer read length of SMRT sequencing markedly improved genome assembly with a high GC content through gap filling and repeat resolution (Shin *et al.*, 2013).

Although providing huge amounts of data, NGS platforms have higher error rates (~0.1 to 15%) and generally shorter (35 to 700 bp) read lengths than those of traditional Sanger sequencing platforms (Liu *et al.*, 2012). Compared with the huge genome, short read length makes sequence assembly extremely difficult. Although having vast quantities of sequence data and a depth of sequencing coverage reaching dozens or even hundreds of times, many users are still unable to complete the genome sequence assembly. Although third-generation sequencing technology overcomes the read length limitation of the NGS approaches, there remains a series of problems, such as greater cost, lower throughput and higher error rate of ~14% with insertion/deletion (InDel) errors (Carneiro *et al.*, 2012). Therefore, third-generation sequencing is always not used independently. Combinations of first-, second- and third-generation sequencing technologies play to their respective advantages and complement each other, leading to a high-quality genome sequence. For example, the high-quality genomes of ZS97 and MH63 were sequenced using a bacterial artificial chromosome (BAC)-by-BAC approach and PacBio SMRT technology, supplemented with

**Table 2.1** Property summary of NGS platforms

| Platform | MRL (bp) | Throughput | Run time | RPFC | Data quality | PE | ER (%) |
|---|---|---|---|---|---|---|---|
| Illumina Miniseq | 2×150 | 0.6–7.5 Gb | 4–24 hours | 25 million | ≥80% of base above Q30 at 2×150 bp | Substitution | <1 |
| Illumina MiSeq | 2×300 | 0.3–15 Gb | 5–55 hours | 25 million | ≥80% of base above Q30 at 2×150 bp | Substitution | <0.1 |
| Illumina Nextseq500 | 2×150 | 20–39 Gb | 15–26 hours | 130 million | ≥75% of base above Q30 at 2×150 bp | Substitution | <1 |
| | 2×150 | 30–120 Gb | 12–30 hours | 400 million | ≥75% of base above Q30 at 2×150 bp | Substitution | <1 |
| Illumina HiSeq2500 | 2×250 | 10–300 Gb | 7–60 hours | 300 million | ≥80% of base above Q30 at 2×150 bp | Substitution | 0.1 |
| | 2×125 | 50–1000 Gb | <1–6 days | 2 billion | ≥80% of base above Q30 at 2×100 bp | Substitution | 0.1 |
| Illumina HiSeq3000 | 2×150 | 125–750 Gb | <1–3.5 days | 2.5 billion | ≥75% of base above Q30 at 2×150 bp | Substitution | 0.1 |
| Illumina HiSeq4000 | 2×150 | 125–1500 Gb | <1–3.5 days | 2.5 billion | ≥75% of base above Q30 at 2×150 bp | Substitution | 0.1 |
| Illumina HiSeq X | 2×150 | 900–1800 Gb | <3 days | 3 billion | ≥75% of base above Q30 at 2×150 bp | Substitution | 0.1 |
| SOLiD5500xl | 1×75 | 80–320 Gb | 6–10 days | 1.4 billion | ~99.99% | A-T bias | >0.01 |
| 454 GS FLX XL+ | 1000 | 700 Mb | 23 hours | 1 million | Q20 read length of 400 bp | Indel | 1 |
| 454 GS FLX XLR70 | 600 | 450 Mb | 10 hours | 1 million | Q20 read length of 400 bp | Indel | 1 |
| PacBio RS II | ~20,000 | 500 Mb–1 Gb | 4 hours | 55,000 | Q20-Q24 for bases without indels | Indel | ~14 |

ER, error ratio; MRL, maximum read length (bp); PE, primary errors; RPFC, reads per flow cell.
Q20 and Q30 mean the probability that the base was miscalled is 1% and 0.1%, respectively.

Illumina whole-genome shotgun (WGS) data (Zhang *et al.*, 2016).

## Bioinformatic tools for analysing NGS data

The emergence of NGS technology makes it possible to re-sequence entire genomes more efficiently and economically and in greater depth than ever before. When the reference genome is available, NGS data can be assembled into a new genome by alignment. However, the assembly of *de novo* sequencing is a greater challenge without a reference genome. Compared with traditional sequencing technology, the NGS raw data are more prodigious because of short sequences with a higher error rate. Therefore, further approaches for short-read assembly need to be proposed and implemented (Imelfort and Edwards, 2009). In addition, the identification of DNA sequence polymorphism within a genome is of utmost importance for dissecting genetic variation. The process of polymorphic screening includes the assembly and alignment of raw data, single nucleotide polymorphism (SNP) discovery and haplotype analysis.

### Assembly and alignment of raw data

Different NGS platforms have different types of errors in sequencing. The Illumina, 454 and SOLiD platforms primarily produce errors of substitutions, InDel and A-T bias, respectively. The per-base quality score (Q) measures the error possibility of the base in the read. When Q is higher, the base error probability is smaller. The formula is $Q = -10\log P(\text{error})$. For example, Q20 and Q30 represent the probabilities of base error of 1% and 0.1%, respectively. In general, Q30 is the evaluation criteria for the Illumina platform. Illumina guarantees that the Illumina HiSeqX, the most widespread used high-throughput sequencing platform, has a greater than 75% base quality above Q30 at 2 × 150 bp.

Correct sequence alignment is the key to polymorphism detection. Alignment errors may result in SNP and genotype detection errors. Therefore, alignment accuracy is vital for variation detection. Separating the true difference between point mutations and InDels requires improved alignment algorithms. There is a major issue in efficiently aligning short reads to a reference genome and handling ambiguity or lack of accuracy in this alignment. Most alignment algorithms for NGS data are based on either 'Hashing' or 'Burrows–Wheeler transform' (BWT). 'Hashing' includes the software MAQ, SOAP, Novoalign and Stampy. MAQ is used to map and assemble short reads. It can also report SNPs and InDels using a simple assembly visualizer (Li *et al.*, 2008a). SOAP efficiently aligns gapped and ungapped short reads onto reference sequences (Li *et al.*, 2008b). Stampy uses a hybrid mapping algorithm and a detailed statistical model to achieve both speed and sensitivity. It is particularly suitable for the reads that include sequence variation. Both Novoalign and Stampy show good performance for longer paired-end reads (Lunter and Goodson, 2011). BWT-based aligners, for example, Bowtie (Langmead *et al.*, 2009), SOAP2 (Li *et al.*, 2009) and BWA24 (Li and Durbin, 2009), which are less sensitive than the best hash-based mappers, are fast and particularly useful for aligning repetitive reads. The greater is the difference between the sequencing and reference genomes, the more difficult is the alignment; thus, the alignment can be solved by analysing the longer reads and pair-end reads.

## SNP discovery and haplotype analysis

SNP is DNA sequence variation that occurs when a single nucleotide of the genome sequence is altered. SNP is the most abundant form of genetic variation existing between any diverse genotypes. Many high-throughput technologies, such as the Illumina platform, have been developed to efficiently genotype SNPs. The main aims underlying SNP calling is to obtain polymorphic loci or different base sites with reference sequences. With low-coverage sequencing data, it is difficult to detect SNPs and assemble the haplotypes with low non-reference allele frequency, as there is often considerable uncertainty associated with the results. A very high coverage is needed because certain regions may not be covered by the reads. According to the quality value of Q20, analyses filter and retain high-confidence bases. The bioinformatics tools that are used for SNP discovery and haplotype analysis include BreakDancer, VarScan, MAQ, INTERSNP, and Atlas-SNP, as well as HapHunt, HapCUT, BEAGLE 3.0, and Hap-seqX, respectively. BreakDancer can sensitively and accurately detect InDels ranging from 10 base pairs to 1 megabase pair, which are difficult to detect via a single conventional approach (Chen *et al.*, 2009). VarScan can detect SNPs and indels with high sensitivity and specificity in both Roche/454 sequencing and Illumina sequencing (Koboldt *et al.*, 2009). MAQ is accurate and efficient, rapidly aligns short reads to the reference genome, and discovers variants, including SNPs and short indels (Li *et al.*, 2008a). INTERSNP examines each position in the genome; a SNP is called whenever two samples differ at the same position (Herold *et al.*, 2009). Atlas-SNP is used for SNP and InDel discovery from genome resequencing using NGS technologies (Wheeler *et al.*, 2008). HapHunt uses K-means clustering to solve the haplotype-phasing problem and consists of identifying all haplotypes in an individual or population (Page *et al.*, 2014). HapCUT, which is based on computing max-cuts, is an efficient and accurate algorithm for the haplotype assembly problem (Bansal and Bafna, 2008). BEAGLE 3.0, which uses a haplotype frequency model, produces most likely haplotypes and sampled haplotypes for each individual (Browning and Browning, 2009). Hap-seqX can save all the intermediate results to memory, which not only solves the memory issue but also significantly improves the running time, and predicts haplotypes from whole-genome sequencing data (He and Eskin, 2013).

## The pathways to assemble a reference genome in cereal crops

The reference genome sequence is the foundation of functional genomics and comparative genomics. With a high-quality reference genome, it is easy to develop millions of polymorphism markers and construct a super high-density genetic linkage map in crops, which can be used to discover more favourable genes for crop breeding. Since first-generation sequencing technology was invented in 1973, scientists have attempted to sequence the genomes of microbes, plants, animals and humans. There are two main strategies to sequence the whole genome. One is WGS, which requires constructing a sequencing library using whole-genome DNA and sequencing enough folds to cover the whole genome. The other is clone-to-clone strategy, which requires constructing a library; selecting the minimum tiling path clones, which cover the whole genome sequence; and then sequencing the clones directly or sequencing in shotgun.

## Reference genomes in the Sanger sequence era

The first reference genome sequence in crops came from rice, the model monocot, in 2002 (Goff *et al.*, 2002; Yu *et al.*, 2002). After 6-fold WGS with Sanger sequencing, Beijing Genomics Institute (BGI) and Syngenta completed the draft genome sequence of *indica* cultivar 93–11 and *japonica* cultivar Nipponbare separately in 2 years. Both draft genomes coved more than 90% per cent of the genome (Goff *et al.*, 2002; Yu *et al.*, 2002). However, for scientists, a draft map for model plants was not enough. In 2005, the IRGSP (International Rice Genome Sequencing Project) released the map-based genome sequence of Nipponbare (including the Syngenta sequencing data), which took seven years and coved 95% of the genome in the first version. When comparing the draft and map-based genome sequences, nearly 80% match well (International Rice Genome Sequencing, 2005). The rice reference genome has greatly promoted rice molecular genetics and biology. The achievement of the rice reference genome inspired other crop consortiums to form genome-sequencing programs. In 2009, the maize genome B73 was completed by BAC-to-BAC in Sanger and covered 89% of the whole 2.3 Gb genome. The maize genome is abundant with transposons, which occupy more than 85% of whole genome (Schnable *et al.*, 2009). In the same year, sorghum reference genome was assembled with WGS data and BAC pair-end sequences (Paterson *et al.*, 2009). In 2010, the soybean Williams 82 genome was finished by 8-fold WGS in Sanger and covered 80% of the whole 1115 Mb (Schmutz *et al.*, 2010). In the Sanger sequencing era, the WGS was timesaving but had difficulty covering the whole genome; the genome coverage was not as high as that from clone-by-clone, especially for complex genomes. Clone-by-clone is standard at first, but it's more time consuming and, more importantly, requires longer scaffolds in complex genome assembly.

## Reference genomes assembled from next-generation sequencing

As mentioned above, second-generation sequencing includes massively parallel sequencing. The length and quantity are not as good as those of Sanger sequencing, but second-generation sequencing is cheaper and higher throughput; thus, genome sequencing has entered a new stage. The human 1000 Genomes Project, 1000 Plants Genome Project and the 3000 Rice Genome Project, among others, have been initiated (1000 Genomes Project Consortium, 2015; Li *et al.*, 2014; Matasci *et al.*, 2014). Crop reference genomes have been made breakthroughs, and several cereal crops have had reference genome sequences released in recent decades (Table 2.2).

It is unknown whether there is a significant difference in reference sequence quality between NGS and Sanger sequencing. Huang *et al.* (2009) reported that with 68× WGS in Illumina and 4× WGS in Sanger, the assembly genome coved 73% of the cucumber whole genome, which demonstrated that massively parallel sequencing was powerful for genome sequencing. Schatz *et al.* (2014) reported *de novo* assemblies of three strains of rice with 110-fold WGS in Illumina HiSeq 2000. When comparing the *de novo* assembled Nipponbare sequence with the IRGSP genome sequence, 91.2% was covered and 99.94% was identical, on average. The error rate of sequencing and assembly is at most 0.06%. That means that regarding accuracy, the *de novo* assembly genome was nearly as high quality as the reference genome from clone-to-clone assembly (Schatz *et al.*, 2014). Then, some complex genome underwent *de novo* assembly of the reference genome with high-fold WGS in Illumina HiSeq 2000.

Using wheat-A genome progenitor *Triticum urartu* var. G1812 as an example, 100-fold WGS in Illumina HiSeq 2000 obtained 79% coverage of the whole 4.94 Gb (Ling *et al.*, 2013). Compared with the Illumina HiSeq platform, the Roche/454 platform has longer reads of 600 to 1000 bp, which is a similar length as that of Sanger sequencing.

**Table 2.2** Reference genome sequence obtained by NGS in cereal crops

| Species | Chromosome no. | Genome size | Coverage | Gene no. | TEs (%) | Strategies | Reference |
|---|---|---|---|---|---|---|---|
| *Oryza sativa* ssp. *japonica* cv. Nipponbare | $2n=2x=24$ | 389 Mb | 82% | 39,083 | - | WGS, 110× | Schatz *et al.*, 2014 |
| *Oryza sativa* ssp. *indica* cv. IR64 | $2n=2x=24$ | 389 Mb | 81% | 37,758 | - | WGS, 110× | |
| *Oryza sativa* ssp. *Indica* cv. DJ123 | $2n=2x=24$ | 389 Mb | 83% | 37,812 | - | WGS, 110× | |
| *Oryza sativa* ssp. *indica* cv. Zhenshan 97 | $2n=2x=24$ | 384 Mb | 91% | 54,831 | 41 | Clone-by-clone, and WGS, 200× | Zhang *et al.*, 2016 |
| *Oryza sativa* ssp. *indica* cv. Minghui 63 | $2n=2x=24$ | 386 Mb | 93% | 57,174 | 42 | | |
| *Triticum aestivum* var. Chinese spring | $2n=6x=42$ | 17 Gb | 61% | 124,201 | 77 | WGS | IWGSC, 2014 |
| *Sorghum bicolor* var. BTx623 | $2n=2x=20$ | 730 Mb | 96% | 34,496 | 55 | WGS, 8.5× +clone libraries | Paterson *et al.*, 2009 |
| *Setaria italica* var. Zhang gu | $2n=2x=18$ | 490 Mb | 86% | 38,801 | 46 | WGS, 82× | Zhang *et al.*, 2012 |
| *Hordeum vulgare* cv. Morex | $2n=2x=14$ | 5.1 Gb | 90% | 24,287 | 84 | Clone-by-clone and WGS, 50× | International Barley Genome Sequencing *et al.*, 2012 |
| *Hordeum vulgare* var. Nudum | $2n=2x=14$ | 4.48 Gb | 89% | 36,151 | 81 | WGS, 178× | Zeng *et al.*, 2015 |

Tes, transposon elements; WGS, whole-genome shotgun; IWGSC, International Wheat Genome Sequencing Consortium.

Would the Roche/454 platform perform better than Illumina with shorter reads? Chalhoub *et al.* first assembled the allopolyploid *Brassica napus* oilseed genome using Illumina HiSeq data and some Roche/454 and Sanger data (Chalhoub *et al.*, 2014). According to Diguistini *et al.*'s study in fungus, Roche/454 has more advantages than Illumina, and Illumina, Roche/454 and Sanger combined together are powerful (Diguistini *et al.*, 2009). With 94-fold WGS in Roche/454, the wheat D-genome progenitor *Aegilops tauschii* var. AL8/78 assembly genome covered 97% of the 4.36 Gb genome (Zeng *et al.*, 2015). Even with the complex hexaploid bread wheat *Triticum aestivum* var. Chinese spring, a 17 Gb draft genome assembly was obtained by WGS on the Roche/454 platform based on chromosome isolation methods, covering 61% of the total genome (Consortium, 2014). As a special advantage of long reads and high throughput, Roche/454 combined with other platforms, such as Illumina HiSeq 2000 and Sanger sequencing, is a powerful strategy.

Zhang *et al.* completed five *Oryza* AA genome assemblies with 88–95% coverage using Roche/454 and Illumina HiSeq 2000 (Zhang *et al.*, 2014). In addition, after sequencing BAC clones by Roche/454 and assembly with 50-fold WGS sequence data, 90% of the 5.1 Gb *Hordeum vulgare* cv. Morex genome was obtained (International Barley Genome Sequencing *et al.*, 2012). Combined with the third-generation sequencing of long read sequences by 20 kb, it is possible to assemble a more complex genome and obtain a more accurate genome. The two cases of cereal crop genome assembly with third-generation sequencing are Zhenshan 97 and Minghui 63. With 110× Minimum tiling paths (MTPs) of the BAC clone sequence, the assembled genome with 200× WGS Illumina sequence data covered 90.6% and 93.2%, respectively (Zhang *et al.*, 2016).

In addition to cereal crops, many other crops have also been recently sequenced (URL: https://genomevolution.org/wiki/index.php/Sequenced_plant_genomes). The past decade was the genome sequencing explosion decade, as massively parallel sequencing became popular. A combination of cheaper and high-throughput NGS with increasing computational capabilities would create an age in which sequencing could flourish.

## Challenges for cereals with complex genomes

In addition to sequencing technology, other technology also accelerates genome assembly. During scaffold assembly, a physical map is useful. The fingerprint of the BAC end is mostly used in the past, but Hi-C, Optical Nano-mapping and nano-channel array technologies were established independently of BAC clone libraries, which are time-consuming to construct (Hastie *et al.*, 2013; Kaplan and Dekker, 2013; Korbel and Lee, 2013; Lam *et al.*, 2012). In spite of the advanced sequencing methods and mapping technology, some crops with complex genome structures still did not achieve a high-quality reference genome, such as sugar cane, which is challenged by high polyploidy and aneuploidy and a large size of 10 Gb. Sequenced wheat also has a draft genome sequence, which only covered 61% of the genome, still a long way from the 'reference genome'. In the future, to quickly obtain higher quality reference genomes for such crops, especially for genomes with complex structures, there are several pathways for improvement: (1) longer and high-fidelity sequence technology – a longer read will significantly improve the genome assembly and the highly repetitive region; and (2) new technology to sample chromosome-specific DNA, such as isolating every chromosome specifically. With high-quality and less complex samples, it will be easier to sequence and assemble genomes.

## The genome evolution and germplasm resources in cereal crops

Maize, rice and wheat provide the most important foods in the world, and they all are cereals. However, it is not clear why and how they evolved into completely different crops. The reference genome provides us the chance to answer this evolution question. Evolution has always been presented with a phylogenetic tree. However, most phylogenetic trees are built on a single gene or several genes and are most likely incorrect (Degnan and Rosenberg, 2006). A genome sequence comparison would provide a more accurate phylogenetic tree. When building phylogenetic trees with oxtail millet (*Setaria italica*) and other *Poaceae* species, all the *Panicoideae* millet, sorghum and maize genome sequences cluster together, and millet appears to have split from sorghum and maize approximately 27 Mya. When looking at phylogenetic trees in detail, it was found that a whole genome duplication (WGD) in *Poaceae* occurred before the split millet (Jiao *et al.*, 2014; Zhang *et al.*, 2012).

Wild rice has different types of genomes that exhibit different characters in genome size and plant architecture (Table 2.3); even wild rice in the same AA genome type has diversified to adapt to different environments (Ge *et al.*, 1999; Jacquemin *et al.*, 2013; Zhang *et al.*, 2014). Domestication from wild species to cultivars sometimes is complex due to the interference from artificial selection. Scientists have debated the mechanism of rice domestication for many years before NGS. When building phylogenetic trees with a few different domestication genes, two completely opposite hypotheses are generated. One that *indica* and *japonica* cultivars originate from the same progenitor, while the other hypothesis is that these cultivars have independent origins from two different progenitors (Sang and Ge, 2007). Both hypotheses are supported by casual domestication related genes; using the no shattering trait as an example, *shattering 4* mutated by a functional SNP was found in all cultivar subpopulations, while the dominant *SH4* existed only in wild rice (Li *et al.*, 2006; Lin *et al.*, 2007). The domestication of *sh4* indicated that all the cultivars originate from a single progenitor. However, the result was opposite in *qSH1*, in which the shattering alleles were found mainly in *indica* subpopulations, indicating that *indica* and *japonica* originate from two different progenitors (Konishi *et al.*, 2006). Finally, this problem was solved by a Chinese group who re-sequenced 446 wild rice and 1083 cultivars. Then, a phylogenetic tree based on whole-genome sequences indicated that the wild rice was divided into three subgroups (Or-I, Or-II, Or-III), the ancient *japonica* subgroup was first domesticated from Or-III, and the ancient *japonica* subgroup evolved the *aromatic*, *temperate japonica* and *tropical japonica* subspecies. Then, the *indica* subgroup generated from the cross of ancient *japonica* and Or-I subgroup wild rice eventually resulted in the

**Table 2.3** Reference genomes of rice wild relatives revealed by NGS

| Species | Genome type | Genome size | Coverage | Gene no. | TE (%) | Strategies | Reference |
|---|---|---|---|---|---|---|---|
| *Oryza rufipogon* var. W1943 | AA | – | 406 Mb | – | – | WGS, 100× | Huang *et al.*, 2012 |
| *Oryza brachyantha* var. IRGC101232 | FF | 297 Mb | 88% | 32,038 | 29 | WGS 104×, BAC clones | Chen *et al.*, 2013 |
| *Oryza glaberrima* cv. CG14 | AA | 358 Mb | 88% | 33,164 | 34 | Clone-by-clone and WGS | Wang *et al.*, 2014 |
| *Oryza glumaepatula* | AA | 366 Mb | 91% | 41,605 | 30 | WGS, 86× | Zhang *et al.*, 2014 |
| *Oryza meridionalis* | AA | 388 Mb | 88% | 39,106 | 30 | WGS, 60× | |
| *Oryza glaberrima* | AA | 370 Mb | 93% | 41,476 | 29 | WGS, 56× | |
| *Oryza barthii* | AA | 376 Mb | 89% | 42,283 | 30 | WGS, 51× | |
| *Oryza nivara* | AA | 395 Mb | 95% | 41,490 | 28 | WGS, 73× | |
| *Oryza barthii* | AA | – | 308 Mb | – | – | WGS, 110× | http://ensembl.gramene.org/species.html |
| *Oryza longistaminata* | AA | – | 326 Mb | – | – | WGS | |
| *Oryza glumaepatula* | AA | – | 373 Mb | – | – | WGS | |
| *Oryza meridionalis* | AA | – | 336 Mb | – | – | WGS | |
| *Oryza nivara* | AA | – | 338 Mb | – | – | WGS | |
| *Oryza punctata* | BB | – | 394 Mb | – | – | WGS | |

TE, transposon element; WGS, whole-genome shotgun.

*indica* and *aus* subspecies (Huang *et al.*, 2012). A similar method was used to analyse the origin of African rice, cotton, wheat, fleshy fruit, etc. (Meyer *et al.*, 2016).

A comparison of genome sequences told us not only about the domestication but also the improvement process. With 5.4× coverage on average, several teams from China re-sequenced six elite maize inbred lines that were the parents of the most productive hybrids in China and identified the complementation of presence/absence and other deletion variation contributing to heterosis together, which means these variations were selected during modern breeding (Lai *et al.*, 2010). Furthermore, they increased the population to 278 temperate maize inbred lines from different stages of breeding history and found that after domestication, there was some introgression from wild relatives to increase the diversity of cultivars, and the favourable alleles of key loci for agronomy traits were selected for past breeding. They also noted that some relative fraction of rare alleles would be helpful in future breeding programmes (Jiao *et al.*, 2012). In a word, NGS also helped to identify variations, favourable alleles and breeding processes during past artificial selection.

At the same time, to capture all the QTLs from whole germplasm resources, only studying the natural variations that exist in reference genome sequence is not enough, as some of the cultivar- or subpopulation-specific DNA sequences or genes would be missed. The pan-genome containing a core genome and variable genome has been assembled in many crops using NGS, and much more accurate QTLs were identified in the variable genome when studied in association with the pan-genome. In a study by Yao *et al.* in rice, when 840 metabolic traits associated with the pan-genome from 1483 cultivars were compared with the reference genome, 23.5% of the traits had higher association signals and 41.6% had associated SNP locations concordant with associated variable genome sequences (Schatz *et al.*, 2014; Yao *et al.*, 2015).

Crops always suffer from artificial selection. Empirical selection could enhance cultivars' yield potential and stress tolerance. However, it is not enough to directly or efficiently improve a target trait. If more genes were identified in nature, they could be used for marker-assisted selection for breeding. Therefore, mining favourable alleles from genetic resources becomes more important. Genome-wide association studies (GWAS) are a robust way to establish the relationship between genome variation and phenotype. Hundreds of QTLs/genes have been detected based on natural variation in cereals (Bai *et al.*, 2016; Han *et al.*, 2016; Huang *et al.*, 2010; Magwa *et al.*, 2016; Tian *et al.*, 2011). It is believed that more QTLs will be discovered from nature by GWAS for the future breeding programmes. Therefore, NGS is encouraged to mine the favourable alleles recently mutated in cultivars. From what has been discussed above, to trace all the natural variation and artificial selection and to more sufficiently explore and apply the potential of germplasm resources in the era of NGS, we should make the following efforts.

Build higher quality genome sequences. Genome sequences are the foundation of the study of genome evolution. With a pan-genome, we could gather complete information for a species, mainly in association studies now but later in other genome study areas, such as comparative genomics (Yao *et al.*, 2015).

Develop more accurate sequencing technologies. Now in population genetic and evolution studies, most studies pay more attention to SNPs but little attention to InDels due to the low confidence of InDels with NGS (Huang *et al.*, 2012; Huang *et al.*, 2010). We could cover both SNP and InDel variations if the InDel information could be obtained as easily and accurately as SNPs.

Increase the germplasm resources, such as adding more wild species and neighbour species. More germplasm resources would provide more natural variations to explore, study and apply (Zhang *et al.*, 2014).

Identify genetic interaction within populations. The natural variation was contributed not only by loci but also by loci–loci interactions and loci–environment interactions. It has been studied in rice yield heterosis by sequencing an immortalized F2 population and a sufficiently large F2 population (Huang *et al.*, 2016; Zhou *et al.*, 2012). Genetic improvement would be pushed forwards by using favourable alleles from wild relatives and the latest mutations in cultivars identified by NGS.

## Acknowledgements

This work was partially supported by natural science foundation of China (91535301) and Agriculture Public Welfare Scientific Research Project of China (201303008).

## References


1000 Genomes Project Consortium (2015). A global reference for human genetic variation. Nature *526*, 68–74. http://dx.doi.org/10.1038/nature15393

Aird, D., Ross, M.G., Chen, W.S., Danielsson, M., Fennell, T., Russ, C., Jaffe, D.B., Nusbaum, C., and Gnirke, A. (2011). Analyzing and minimizing PCR amplification bias in Illumina sequencing libraries. Genome Biol. *12*, R18. http://dx.doi.org/10.1186/gb-2011-12-2-r18

Bai, X., Zhao, H., Huang, Y., Xie, W., Han, Z., Zhang, B., Guo, Z., Yang, L., Dong, H., Xue, W., *et al.* (2016). Genome-wide association analysis reveals different genetic control in panicle architecture between and rice. Plant Genome *9*, 02. http://dx.doi.org/10.3835/plantgenome2015.11.0115

Bansal, V., and Bafna, V. (2008). HapCUT: an efficient and accurate algorithm for the haplotype assembly problem. Bioinformatics *24*, i153–9. http://dx.doi.org/10.1093/bioinformatics/btn298

Browning, B.L., and Browning, S.R. (2009). A unified approach to genotype imputation and haplotype-phase inference for large data sets of trios and unrelated individuals. Am. J. Hum. Genet. *84*, 210–223. http://dx.doi.org/10.1016/j.ajhg.2009.01.005

Carneiro, M.O., Russ, C., Ross, M.G., Gabriel, S.B., Nusbaum, C., and DePristo, M.A. (2012). Pacific biosciences sequencing technology for genotyping and variation discovery in human data. BMC Genomics *13*, 375. http://dx.doi.org/10.1186/1471-2164-13-375

Chalhoub, B., Denoeud, F., Liu, S., Parkin, I.A., Tang, H., Wang, X., Chiquet, J., Belcram, H., Tong, C., Samans, B., *et al.* (2014). Plant genetics. Early allopolyploid evolution in the post-Neolithic *Brassica napus* oilseed genome. Science *345*, 950–953. http://dx.doi.org/10.1126/science.1253435

Chen, J., Huang, Q., Gao, D., Wang, J., Lang, Y., Liu, T., Li, B., Bai, Z., Luis Goicoechea, J., Liang, C., *et al.* (2013). Whole-genome sequencing of Oryza brachyantha reveals mechanisms underlying *Oryza* genome evolution. Nat. Commun. *4*, 1595. http://dx.doi.org/10.1038/ncomms2596

Chen, K., Wallis, J.W., McLellan, M.D., Larson, D.E., Kalicki, J.M., Pohl, C.S., McGrath, S.D., Wendl, M.C., Zhang, Q., Locke, D.P., *et al.* (2009). BreakDancer: an algorithm for high-resolution mapping of genomic structural variation. Nat. Methods. *6*, 677–681. http://dx.doi.org/10.1038/nmeth.1363

Consortium, I.W.G.S. (2014). A chromosome-based draft sequence of the hexaploid bread wheat (Triticum aestivum) genome. Science *345*, 1251788. http://dx.doi.org/10.1126/science.1251788

Degnan, J.H., and Rosenberg, N.A. (2006). Discordance of species trees with their most likely gene trees. PLOS Genet. *2*, e68. http://dx.doi.org/10.1371/journal.pgen.0020068

Diguistini, S., Liao, N.Y., Platt, D., Robertson, G., Seidel, M., Chan, S.K., Docking, T.R., Birol, I., Holt, R.A., Hirst, M., *et al.* (2009). De novo genome sequence assembly of a filamentous fungus using Sanger, 454 and Illumina sequence data. Genome Biol. *10*, R94. http://dx.doi.org/10.1186/gb-2009-10-9-r94

Dohm, J.C., Lottaz, C., Borodina, T., and Himmelbauer, H. (2008). Substantial biases in ultra-short read data sets from high-throughput DNA sequencing. Nucleic Acids Res. *36*, e105. http://dx.doi.org/10.1093/nar/gkn425

Ge, S., Sang, T., Lu, B.R., and Hong, D.Y. (1999). Phylogeny of rice genomes with emphasis on origins of allotetraploid species. Proc. Natl. Acad. Sci. U.S.A. *96*, 14400–14405.

Goff, S.A., Ricke, D., Lan, T.H., Presting, G., Wang, R., Dunn, M., Glazebrook, J., Sessions, A., Oeller, P., Varma, H., *et al.* (2002). A draft sequence of the rice genome (Oryza sativa L. ssp. japonica). Science *296*, 92–100. http://dx.doi.org/10.1126/science.1068275

Goodwin, S., McPherson, J.D., and McCombie, W.R. (2016). Coming of age: ten years of next-generation sequencing technologies. Nat. Rev. Genet. *17*, 333–351. http://dx.doi.org/10.1038/nrg.2016.49

Han, Z., Zhang, B., Zhao, H., Ayaad, M., and Xing, Y. (2016). Genome-wide association studies reveal that diverse heading date genes respond to short and long day lengths between Indica and Japonica Rice. Front. Plant Sci. *7*, 1270. http://dx.doi.org/10.3389/fpls.2016.01270

Hastie, A., Lam, E., Chan, T., Requa, M., Austin, T.A.M., Trintchouk, F., Saghbini, M., Lai, Y., Mak, A., and Kwok, P. Structural Variation Discovery by De Novo Genome mapping of the human genome at the single molecule level using nanochannel linearization. http://bionanogenomics.com/wp-content/uploads/2013/10/ESHG2013_poster.pdf

He, D., and Eskin, E. (2013). Hap-seqX: expedite algorithm for haplotype phasing with imputation using sequence data. Gene *518*, 2–6. http://dx.doi.org/10.1016/j.gene.2012.11.093

Herold, C., Steffens, M., Brockschmidt, F.F., Baur, M.P., and Becker, T. (2009). INTERSNP: genome-wide interaction analysis guided by a priori information. Bioinformatics *25*, 3275–3281. http://dx.doi.org/10.1093/bioinformatics/btp596

Huang, S., Li, R., Zhang, Z., Li, L., Gu, X., Fan, W., Lucas, W.J., Wang, X., Xie, B., Ni, P., *et al.* (2009). The genome of the cucumber, Cucumis sativus L. Nat. Genet. *41*, 1275–1281. http://dx.doi.org/10.1038/ng.475

Huang, X., Kurata, N., Wei, X., Wang, Z.X., Wang, A., Zhao, Q., Zhao, Y., Liu, K., Lu, H., Li, W., *et al.* (2012). A map of rice genome variation reveals the origin of cultivated rice. Nature *490*, 497–501. http://dx.doi.org/10.1038/nature11532

Huang, X., Wei, X., Sang, T., Zhao, Q., Feng, Q., Zhao, Y., Li, C., Zhu, C., Lu, T., Zhang, Z., *et al.* (2010). Genome-wide association studies of 14 agronomic traits in rice landraces. Nat. Genet. *42*, 961–967. http://dx.doi.org/10.1038/ng.695

Huang, X., Yang, S., Gong, J., Zhao, Q., Feng, Q., Zhan, Q., Zhao, Y., Li, W., Cheng, B., Xia, J., *et al.* (2016). Genomic architecture of heterosis for yield traits in rice. Nature *537*, 629–633. http://dx.doi.org/10.1038/nature19760

Imelfort, M., and Edwards, D. (2009). De novo sequencing of plant genomes using second-generation technologies. Briefings Bioinf. *10*, 609–618. http://dx.doi.org/10.1093/bib/bbp039

Mayer, K.F., Waugh, R., Brown, J.W., Schulman, A., Langridge, P., Platzer, M., Fincher, G.B., Muehlbauer, G.J., Sato, K., Close, T.J., *et al.* (2012). A physical, genetic and functional sequence assembly of the barley genome. Nature *491*, 711–716. http://dx.doi.org/10.1038/nature11543

International Rice Genome Sequencing Project. (2005). The map-based sequence of the rice genome. Nature *436*, 793–800. http://dx.doi.org/10.1038/nature03895

Jacquemin, J., Bhatia, D., Singh, K., and Wing, R.A. (2013). The International Oryza Map Alignment Project: development of a genus-wide comparative genomics platform to help solve the 9 billion-people question. Curr. Opin. Plant Biol. *16*, 147–156. http://dx.doi.org/10.1016/j.pbi.2013.02.014

Jiao, Y., Li, J., Tang, H., and Paterson, A.H. (2014). Integrated syntenic and phylogenomic analyses reveal an ancient genome duplication in monocots. Plant Cell *26*, 2792–2802. http://dx.doi.org/10.1105/tpc.114.127597

Jiao, Y., Zhao, H., Ren, L., Song, W., Zeng, B., Guo, J., Wang, B., Liu, Z., Chen, J., Li, W., *et al.* (2012). Genome-wide genetic changes during modern breeding of maize. Nat. Genet. *44*, 812–815. http://dx.doi.org/10.1038/ng.2312

Kaplan, N., and Dekker, J. (2013). High-throughput genome scaffolding from in vivo DNA interaction frequency. Nat. Biotechnol. *31*, 1143–1147. http://dx.doi.org/10.1038/nbt.2768

Koboldt, D.C., Chen, K., Wylie, T., Larson, D.E., McLellan, M.D., Mardis, E.R., Weinstock, G.M., Wilson, R.K., and Ding, L. (2009). VarScan: variant detection in massively parallel sequencing of individual and pooled samples. Bioinformatics *25*, 2283–2285. http://dx.doi.org/10.1093/bioinformatics/btp373

Konishi, S., Izawa, T., Lin, S.Y., Ebana, K., Fukuta, Y., Sasaki, T., and Yano, M. (2006). An SNP caused loss of seed shattering during rice domestication. Science *312*, 1392–1396. http://dx.doi.org/10.1126/science.1126410

Korbel, J.O., and Lee, C. (2013). Genome assembly and haplotyping with Hi-C. Nat. Biotechnol. *31*, 1099–1101. http://dx.doi.org/10.1038/nbt.2764

Lai, J., Li, R., Xu, X., Jin, W., Xu, M., Zhao, H., Xiang, Z., Song, W., Ying, K., Zhang, M., *et al.* (2010). Genome-wide patterns of genetic variation among elite maize inbred lines. Nat. Genet. *42*, 1027–1030. http://dx.doi.org/10.1038/ng.684

Lam, E.T., Hastie, A., Lin, C., Ehrlich, D., Das, S.K., Austin, M.D., Deshpande, P., Cao, H., Nagarajan, N., Xiao, M., *et al.* (2012). Genome mapping on nanochannel arrays for structural variation analysis and sequence assembly. Nat. Biotechnol. *30*, 771–776. http://dx.doi.org/10.1038/nbt.2303

Langmead, B., Trapnell, C., Pop, M., and Salzberg, S.L. (2009). Ultrafast and memory-efficient alignment of short DNA sequences to the human genome. Genome Biol. *10*, R25. http://dx.doi.org/10.1186/gb-2009-10-3-r25

Li, C., Zhou, A., and Sang, T. (2006). Rice domestication by reducing shattering. Science *311*, 1936–1939. http://dx.doi.org/10.1126/science.1123604

Li, H., and Durbin, R. (2009). Fast and accurate short read alignment with Burrows-Wheeler transform. Bioinformatics *25*, 1754–1760. http://dx.doi.org/10.1093/bioinformatics/btp324

Li, H., Ruan, J., and Durbin, R. (2008a). Mapping short DNA sequencing reads and calling variants using mapping quality scores. Genome Res. *18*, 1851–1858. http://dx.doi.org/10.1101/gr.078212.108

Li, J.Y., Wang, J., and Zeigler, R.S. (2014). The 3,000 rice genomes project: new opportunities and challenges for future rice research. GigaScience *3*, 8. http://dx.doi.org/10.1186/2047-217X-3-8

Li, R., Li, Y., Kristiansen, K., and Wang, J. (2008b). SOAP: short oligonucleotide alignment program. Bioinformatics *24*, 713–714. http://dx.doi.org/10.1093/bioinformatics/btn025

Li, R., Yu, C., Li, Y., Lam, T.W., Yiu, S.M., Kristiansen, K., and Wang, J. (2009). SOAP2: an improved ultrafast tool for short read alignment. Bioinformatics 25, 1966–1967. http://dx.doi.org/10.1093/bioinformatics/btp336

Lin, Z., Griffith, M.E., Li, X., Zhu, Z., Tan, L., Fu, Y., Zhang, W., Wang, X., Xie, D., and Sun, C. (2007). Origin of seed shattering in rice (Oryza sativa L.). Planta *226*, 11–20. http://dx.doi.org/10.1007/s00425-006-0460-4

Ling, H.Q., Zhao, S., Liu, D., Wang, J., Sun, H., Zhang, C., Fan, H., Li, D., Dong, L., Tao, Y., *et al.* (2013). Draft genome of the wheat A-genome progenitor Triticum urartu. Nature *496*, 87–90. http://dx.doi.org/10.1038/nature11997

Liu, L., Li, Y., Li, S., Hu, N., He, Y., Pong, R., Lin, D., Lu, L., and Law, M. (2012). Comparison of next-generation sequencing systems. J. Biomed. Biotechnol. *2012*, 251364. http://dx.doi.org/10.1155/2012/251364

Lunter, G., and Goodson, M. (2011). Stampy: a statistical algorithm for sensitive and fast mapping of Illumina sequence reads. Genome Res. *21*, 936–939. http://dx.doi.org/10.1101/gr.111120.110

Magwa, R.A., Zhao, H., and Xing, Y. (2016). Genome-wide association mapping revealed a diverse genetic basis of seed dormancy across subpopulations in rice (Oryza sativa L.). BMC Genet. *17*, 28. http://dx.doi.org/10.1186/s12863-016-0340-2

Mardis, E.R. (2008). Next-generation DNA sequencing methods. Annu. Rev. Genomics. Hum. Genet. *9*, 387–402. http://dx.doi.org/10.1146/annurev.genom.9.081307.164359

Matasci, N., Hung, L.H., Yan, Z., Carpenter, E.J., Wickett, N.J., Mirarab, S., Nguyen, N., Warnow, T., Ayyampalayam, S., Barker, M., *et al.* (2014). Data access for the 1,000 Plants (1KP) project. GigaScience 3, 17. http://dx.doi.org/10.1186/2047-217X-3-17

Metzker, M.L. (2010). Sequencing technologies – the next generation. Nat. Rev. Genet. *11*, 31–46. http://dx.doi.org/10.1038/nrg2626

Meyer, R.S., Choi, J.Y., Sanches, M., Plessis, A., Flowers, J.M., Amas, J., Dorph, K., Barretto, A., Gross, B., Fuller, D.Q., *et al.* (2016). Domestication history and geographical adaptation inferred from a SNP map of African rice. Nat. Genet. *48*, 1083–1088. http://dx.doi.org/10.1038/ng.3633

Niu, B., Fu, L., Sun, S., and Li, W. (2010). Artificial and natural duplicates in pyrosequencing reads of metagenomic data. BMC Bioinf. *11*, 187. http://dx.doi.org/10.1186/1471-2105-11-187

Page, J.T., Liechty, Z.S., Huynh, M.D., and Udall, J.A. (2014). BamBam: genome sequence analysis tools for biologists. BMC Res. Notes. 7, 829. http://dx.doi.org/10.1186/1756-0500-7-829

Paterson, A.H., Bowers, J.E., Bruggmann, R., Dubchak, I., Grimwood, J., Gundlach, H., Haberer, G., Hellsten, U., Mitros, T., Poliakov, A., *et al.* (2009). The Sorghum bicolor genome and the diversification of grasses. Nature *457*, 551–556. http://dx.doi.org/10.1038/nature07723

Rieber, N., Zapatka, M., Lasitschka, B., Jones, D., Northcott, P., Hutter, B., Jäger, N., Kool, M., Taylor, M., Lichter, P., *et al.* (2013). Coverage bias and sensitivity of variant calling for four whole-genome sequencing technologies. PLOS ONE *8*, e66621. http://dx.doi.org/10.1371/journal.pone.0066621

Rothberg, J.M., and Leamon, J.H. (2008). The development and impact of 454 sequencing. Nat. Biotechnol. *26*, 1117–1124. http://dx.doi.org/10.1038/nbt1485

Sang, T., and Ge, S. (2007). The puzzle of rice domestication. J. Integr. Plant Biol. *49*, 760–768. http://dx.doi.org/10.1111/j.1744-7909.2007.00510.x

Schatz, M.C., Maron, L.G., Stein, J.C., Hernandez Wences, A., Gurtowski, J., Biggers, E., Lee, H., Kramer, M., Antoniou, E., Ghiban, E., *et al.* (2014). Whole genome de novo assemblies of three divergent strains of rice, *Oryza sativa*, document novel gene space of aus and indica. Genome Biol. *15*, 506. http://dx.doi.org/10.1186/PREACCEPT-2784872521277375

Schmutz, J., Cannon, S.B., Schlueter, J., Ma, J., Mitros, T., Nelson, W., Hyten, D.L., Song, Q., Thelen, J.J., Cheng, J., *et al.* (2010). Genome sequence of the palaeopolyploid soybean. Nature *463*, 178–183. http://dx.doi.org/10.1038/nature08670

Schnable, P.S., Ware, D., Fulton, R.S., Stein, J.C., Wei, F., Pasternak, S., Liang, C., Zhang, J., Fulton, L., Graves, T.A., *et al.* (2009). The B73 maize genome: complexity, diversity, and dynamics. Science *326*, 1112–1115. http://dx.doi.org/10.1126/science.1178534

Shendure, J., and Ji, H. (2008). Next-generation DNA sequencing. Nat. Biotechnol. *26*, 1135–1145. http://dx.doi.org/10.1038/nbt1486

Shin, S.C., Ahn, D.H., Kim, S.J., Lee, H., Oh, T.J., Lee, J.E., and Park, H. (2013). Advantages of single-molecule real-time sequencing in high-GC content genomes. PLOS ONE *8*, e68824. http://dx.doi.org/10.1371/journal.pone.0068824

Tian, F., Bradbury, P.J., Brown, P.J., Hung, H., Sun, Q., Flint-Garcia, S., Rocheford, T.R., McMullen, M.D., Holland, J.B., and Buckler, E.S. (2011). Genome-wide association study of leaf architecture in the maize nested association mapping population. Nat. Genet. *43*, 159–162. http://dx.doi.org/10.1038/ng.746

Wang, M., Yu, Y., Haberer, G., Marri, P.R., Fan, C., Goicoechea, J.L., Zuccolo, A., Song, X., Kudrna, D., Ammiraju, J.S., *et al.* (2014). The genome sequence of African rice (*Oryza glaberrima*) and evidence for independent domestication. Nat. Genet. *46*, 982–988. http://dx.doi.org/10.1038/ng.3044

Wheeler, D.A., Srinivasan, M., Egholm, M., Shen, Y., Chen, L., McGuire, A., He, W., Chen, Y.J., Makhijani, V., Roth, G.T., *et al.* (2008). The complete genome of an individual by massively parallel DNA sequencing. Nature *452*, 872–876. http://dx.doi.org/10.1038/nature06884

Yao, W., Li, G., Zhao, H., Wang, G., Lian, X., and Xie, W. (2015). Exploring the rice dispensable genome using a metagenome-like assembly strategy. Genome Biol. *16*, 187. http://dx.doi.org/10.1186/s13059-015-0757-3

Yu, J., Hu, S., Wang, J., Wong, G.K., Li, S., Liu, B., Deng, Y., Dai, L., Zhou, Y., Zhang, X., *et al.* (2002). A draft sequence of the rice genome (Oryza sativa L. ssp. indica). Science *296,* 79–92. http://dx.doi.org/10.1126/science.1068037

Zeng, X., Long, H., Wang, Z., Zhao, S., Tang, Y., Huang, Z., Wang, Y., Xu, Q., Mao, L., Deng, G., *et al.* (2015). The draft genome of Tibetan hulless barley reveals adaptive patterns to the high stressful Tibetan Plateau. Proc. Natl. Acad. Sci. U.S.A. *112,* 1095–1100. http://dx.doi.org/10.1073/pnas.1423628112

Zhang, G., Liu, X., Quan, Z., Cheng, S., Xu, X., Pan, S., Xie, M., Zeng, P., Yue, Z., Wang, W., *et al.* (2012). Genome sequence of foxtail millet (Setaria italica) provides insights into grass evolution and biofuel potential. Nat. Biotechnol. *30,* 549–554. http://dx.doi.org/10.1038/nbt.2195

Zhang, J., Chen, L.L., Xing, F., Kudrna, D.A., Yao, W., Copetti, D., Mu, T., Li, W., Song, J.M., Xie, W., *et al.* (2016). Extensive sequence divergence between the reference genomes of two elite indica rice varieties Zhenshan 97 and Minghui 63. Proc. Natl. Acad. Sci. U.S.A. *113,* E5163–5171. http://dx.doi.org/10.1073/pnas.1611012113

Zhang, Q.J., Zhu, T., Xia, E.H., Shi, C., Liu, Y.L., Zhang, Y., Liu, Y., Jiang, W.K., Zhao, Y.J., Mao, S.Y., *et al.* (2014). Rapid diversification of five Oryza AA genomes associated with rice adaptation. Proc. Natl. Acad. Sci. U.S.A. *111,* E4954–62. http://dx.doi.org/10.1073/pnas.1418307111

Zhou, G., Chen, Y., Yao, W., Zhang, C., Xie, W., Hua, J., Xing, Y., Xiao, J., and Zhang, Q. (2012). Genetic composition of yield heterosis in an elite rice hybrid. Proc. Natl. Acad. Sci. U.S.A. *109,* 15847–15852. http://dx.doi.org/10.1073/pnas.1214141109

# Advanced Applications of Next-generation Sequencing Technologies to Orchid Biology

3

Chuan-Ming Yeh[1†], Zhong-Jian Liu[2,3,4†] and Wen-Chieh Tsai[5,6,7*]

[1]Division of Strategic Research and Development, Graduate School of Science and Engineering, Satitama University, Saitama, Japan.
[2]Shenzhen Key Laboratory for Orchid Conservation and Utilization, The National Orchid Conservation Center of China and The Orchid Conservation and Research Center of Shenzhen, Shenzhen, China.
[3]The Center for Biotechnology and BioMedicine, Graduate School at Shenzhen, Tsinghua University, Shenzhen, China.
[4]College of Arts, College of Landscape Architecture, Fujian Agriculture and Forestry University, Fuzhou, China.
[5]Institute of Tropical Plant Sciences, National Cheng Kung University, Tainan, Taiwan.
[6]Department of Life Sciences, National Cheng Kung University, Tainan, Taiwan.
[7]Orchid Research and Development Center, National Cheng Kung University, Tainan, Taiwan.

*Correspondence: tsaiwc@mail.ncku.edu.tw
†These authors contributed equally

https://doi.org/10.21775/9781910190654.04

## Abstract

Next-generation sequencing (NGS) technologies are revolutionizing biology by permitting transcriptome sequencing, whole-genome sequencing and resequencing, and genome-wide single nucleotide polymorphism profiling. Orchid research has benefited from this breakthrough, and a few orchid genomes are now available; new biological questions can be approached and new breeding strategies can be designed. The first part of this review describes the unique features of orchid biology. The second part provides an overview of the current NGS platforms, many of which are already used in plant laboratories. The third part summarizes the state of orchid transcriptome and genome sequencing and illustrates current achievements. The genetic sequences currently obtained will not only provide a broad scope for the study of orchid biology, but also serves as a starting point for uncovering the mystery of orchid evolution.

## Introduction

The Chinese have been cultivating fragrant *Cymbidium* species since 500 BC. The earliest book on record about orchids is Shen Nung Pen Tsao Ching, published during the Han dynasty. This book refers to well-known orchids used as popular medicines, including Dendrobium, Gsatrodia, and Bletilla. It is generally agreed that the term orchid was first used by the Greek philosopher Theophrastus in his inquiry into plants (Arditti, 1992). Orchid cultivation and growth became popular in the late eighteenth century in Europe. Voyages around the world were sponsored by the wealthy to collect orchids, herbarium species, and other exotic plants. Merchants, government officials, sea captains, plant collectors, explorers, privateers, and other travellers began sending plants to their home countries soon after they discovered them. Some of these orchids were sent to botanical gardens; others reached private growers. Subsequently, the landed gentry, the wealthy, and commercial firms started to accumulate orchid collections. In 1794, 15 epiphytic orchids were cultivated at Kew Gardens in London. To satisfy the needs of growers, large numbers of collectors were sent to faraway places. These collectors destroyed millions of plants, discovered many new species, and suffered and died from diseases and deprivation, but sent many orchids to England. By about 1820, it became possible to heat greenhouses with hot water flowing through pipes. These advances permitted growers to simulate what they considered to be appropriate conditions for orchid culture – heat and humidity. Improved methods such as lower temperature, better ventilation and potting contributed to higher survival of the orchids and became even more popular.

The family Orchidaceae is the largest family of flowering plants and the number of species may exceed 25,000 (Atwood, 1986). Like all other living organisms, present-day orchids have evolved from ancestral forms as a result of selection pressure and adaptation. They show a wide diversity of epiphytic and terrestrial growth forms and have successfully colonized almost every habitat on earth. Factors promoting orchid species richness include specific interaction between the orchid flower and pollinator (Cozzolino and Widmer, 2005), sequential and rapid interplay between drift and natural selection (Tremblay *et al.*, 2005), obligate interaction with mycorrhiza (Otero and Flanagan, 2006), and epiphytism which is true for most of all orchids and probably two-thirds of the epiphytic flora of the world. The radiation of the orchid family has probably taken place in a comparatively short period as compared with that of most flowering plant families, which had already started to diversify in the Mid-Cretaceous period (Crane *et al.*, 1995). The time of origin of orchids is in dispute, although Dressler suggests that they originated 80 to 40 million years ago (Mya; late Cretaceous to late Eocene) (Dressler, 1981). Recently, the origin of the Orchidaceae was dated with a fossil orchid and its pollinator. The authors showed that the most recent common ancestor of extant orchids lived in the late Cretaceous (76 to 84 Mya) (Ramírez *et al.*, 2007). They also suggested the largest orchid subfamilies, which together represent >95% of living orchid species, began to diversify early in the Tertiary (65 Mya) (Ramírez *et al.*, 2007).

According to molecular phylogenetic studies, Orchidaceae comprises five subfamilies: Apostasioideae, Cypripedioideae, Vanilloideae, Orchidoideae and Epidendroideae. The Apostasioideae is considered the sister group to other orchids. Vanilloideae diverged just before Cypripedioideae. Both subfamilies have relatively low numbers of genera and species. Most of the taxonomic diversity in orchids is in two recently expanded sister subfamilies: Orchidoideae and especially Epidendroideae (Górniaka *et al.*, 2010). Orchids are known for their diversity of specialized reproductive and ecological strategies (Tsai *et al.*, 2014). For successful reproduction, the production of labellum and gynostemium (a fused structure of androecium and gynoecium) to facilitate pollination is well documented and the co-evolution of orchid flowers and pollinators is well known (Schiestl *et al.*, 2003). In addition, the especially successful evolutionary progress of orchids may be explained by mature pollen grains packaged as pollinia, pollination-regulated ovary/ovule development, synchronized timing of micro- and mega-gametogenesis for effective fertilization, and the release of thousands or millions of immature embryos (seeds without endosperm) in a mature capsule (Yu and Goh, 2001). However, despite their unique developmental reproductive biology, as well as specialized pollination and ecological strategies, orchids remain under-represented in molecular studies relative to other species-rich plant families (Peakall, 2007). The reasons may be associated with the large genome size, long life cycle, and inefficient transformation system of orchids (Hsiao *et al.*, 2011b).

During the last 30 years DNA sequencing has completely changed our vision of biology and particularly plant biology. It has been possible to characterize a large number of genes by their nucleotide sequences, thus providing a shortcut to the corresponding protein sequences and their functions. Information on gene polymorphisms has facilitated genetic mapping, gene cloning and the understanding of evolutionary relationships and has allowed for the initiation of biodiversity studies. The most popular sequencing method has been the Sanger method (Sanger *et al.*, 1977a). When combined with the use of robotics, bioinformatics, computer databases and instrumentation, the method has allowed for sequencing larger DNA fragments and, finally, complete genomes. As a result, a series of landmark genomes was obtained, such as *Caenorabditis elegans, Drosophila melanogaster, Arabidopsis thaliana, Homo sapiens* and *Oryza sativa* (International Human Genome Sequencing Consortium, 2004; International Rice Genome Sequencing Project, 2005; The Arabidopsis Genome Initiative, 2000). The deciphering of these genomes led to the era of functional genomics and completely modified biological investigation. However, this technology remained tedious and expensive. These limiting factors stimulated the development and commercialization of next-generation sequencing (NGS) technologies, as opposed to the automated Sanger method, which is considered a first-generation technology. When coupled with the appropriate computational algorithms, the development of NGS technologies has opened new avenues on a genome-wide scale to radically alter our understanding of biology that could not be answered with classical sequencing.

## Sequencing platforms

Sanger sequencing, developed by Sanger and his colleagues in 1977 based on the chain-termination method, is predominantly employed for DNA sequencing in the following 30 years (Sanger *et al.*, 1977b). It was first commercialized by Applied Biosystems to launch the automatic sequencing machine, AB370, in 1987. By adopting capillary electrophoresis, the sequencing became faster and more accurate (Liu *et al.*, 2012; Illumina 2016). Although Sanger sequencing was applied to complete the genome projects of human and several model organisms including the first sequenced plant, *Arabidopsis thaliana*, it took plenty of time, cost and resources (The Arabidopsis Genome Initiative, 2000; International Human Genome Sequencing Consortium, 2004; van Dijk *et al.*, 2014). Therefore, National Human Genome Research Institute (NHGRI) initiated a funding programme which aimed to reduce the cost of human genome sequencing to 1000 US dollars within 10 years. It prompted NGS development to create sequencers with fast, cheap, easy-to-operate and accurate features (Liu *et al.*, 2012; van Dijk *et al.*, 2014).

The 454 Genome Sequencer employing the pyrosequencing method is the first commercial NGS platform released by 454 Life Sciences in 2005 (Margulies *et al.*, 2005), followed by the Genome Analyser in 2006 from Solexa, which was purchased by Illumina 1 year later (Liu *et al.*, 2012). The SOLiD (Sequencing by Oligo Ligation Detection) platform is the third NGS technology developed by Applied Biosystems (now Thermo Fisher) in 2007 (Valouev *et al.*, 2008). These Solexa (Illumina) and SOLiD sequencers generated short reads with only 35-bp lengths comparing to the 110-bp reads by the 454 system. However, their numbers of reads (30 and 100 million reads, respectively) are much larger than that of the 454 sequencer (200 thousand reads). In 2010, PGM (Personal Genome Machine), the first NGS platform developed by semiconductor technology which generates 100-bp reads, was released by Ion Torrent (now Thermo Fisher). Without using optical-sensing device, the detection for sequencing by Ion Torren PGM is based on measuring the change of pH during nucleotide incorporation (Rothberg *et al.*, 2011; Liu *et al.*, 2012). These unique features make PGM a smaller size with higher speed and lower cost (Liu *et al.*, 2012; van Dijk *et al.*, 2014).

The sequencing approaches of these short-read NGS platforms can be classified to the sequencing by ligation (SBL) employed by SOLiD and the sequencing by synthesis (SBS) adopted by Illumina, 454 and Ion Torren. The SBS can be further distinguished to cyclic reversible termination (CRT) for Illumina and single-nucleotide addition (SNA) for 454 and Ion Torren (Goodwin *et al.*, 2016). Basically, the NGS systems employing the SBS approaches are similar to the first-generation sequencing in that the fluorescently labelled dNTPs (deoxyribonucleotide triphosphates) are incorporated into a DNA template by DNA polymerase. The incorporated nucleotides are then detected by fluorophore excitation during sequencing (Voelkerding *et al.*, 2009; Illumina, 2016). For the SBL approach, instead of DNA polymerase, it relies on DNA ligase to determine the underlying sequence of the template DNA. The fluorescently labelled interrogation probes hybridize to the template DNA followed by ligation and imaging. The identity of the bases in the interrogation probes is indicated by the emission spectrum of the fluorophore (Shendure *et al.*, 2005; Goodwin *et al.*, 2016). The major improvements of NGS leading to high throughput and sequencing rate include preparation of NGS libraries in a cell free system instead of bacterial cloning, parallel proceeding of a large number of sequencing reactions and direct detection of sequencing output without performing electrophoresis (van Dijk et al., 2014; Chaitankar et al., 2016). Compared to the conventional sequencing, second-generation sequencing technologies generate vast amounts of shorter reads, ranging from 35 to 700 bp, making it possible to produce massive sequencing data in a much shorter time (Goodwin et al., 2016). The current NGS systems can complete human genome sequencing within one day, which consumed 15 years before (Illumina, 2016).

Although these short-read sequencing platforms have created a new sequencing era in the past decade, the short read lengths give rise to a new obstacle in genome assembly, such as improper discard of repeated sequences and assembling in wrong locations or orientations. It is required to develop novel computational algorithms for data analysis (Baker, 2012; van Dijk et al., 2014). However, it is increasingly obvious that genomes are highly complex because of having many long repetitive elements, copy numbers and structural variations. The short-read NGS technologies are insufficient to resolve these complex elements (Goodwin et al., 2016). 'Very long, very high-quality reads will do wonders for assembly, and fix many of these issues', says Adam Felsenfeld, the director of the Large-Scale Sequencing Program at the NHGRI (Baker, 2012). Although we are still not there yet, the so-called third-generation sequencing approaches can now provide an alternative choice (Goodwin et al., 2016). The read lengths of the current long-read sequencing can reach to several kilobases, allowing to span complex or repetitive regions with a single continuous read. In addition, it is also helpful to identify the precise connectivity of exons and discern gene isoforms in transcriptomic research because it can span entire mRNA transcripts. No PCR amplification step before sequencing is one of the main characteristics of the third-generation sequencing technologies. The other is that the sequencing reaction can be detected in real time whether in PacBio by fluorescence or in Nanopore by electric current (Liu et al., 2012). Although the more expensive cost and relatively lower throughput than second-generation sequencing currently limit their widespread application, ultra-long-read sequencing technologies with high outputs and low prices could be expected in the near future. In this review article, the current development and application of second- and third-generation sequencing platforms along with their benefits and drawbacks are introduced and discussed (Table 3.1).

## Roche 454

The 454 Life Sciences, purchased by Roche in 2007, released the first NGS platform, the 454 Genome Sequencer, in 2005 (Margulies et al., 2005). The sequencing workflow is initiated by preparation of a sequencing library from the source nucleic acids. First, the long DNA or RNA molecules are fragmented into a suitable size (around 50 to 500 bp). The fragments are next fused with specific adapters followed by a size selection step to enrich molecules with a desired size and to remove the free adapters. The templates are adhered to micro-beads through adaptors and amplified by emulsion PCR (http://454.com/products/technology.asp; Egan et al., 2012; van Dijk et al., 2014). This sample preparation process is similar to the subsequently developed NGS platforms, Ion Torrent and SOLiD. For the 454 pyrosequencing, the templates is denatured and incubated with sequencing reagents including DNA polymerase, dNTP and several enzymes. Pyrophosphates are released and converted to ATP after the appropriate dNTPs are incorporated into the new strand by DNA polymerase. ATP in turn reacts with luciferase to release oxyluciferin fluorescence detected by a CCD (charge-coupled device) camera (Egan et al., 2012; Goodwin et al., 2016).

The Roche 454 currently offers two pyrosequencing platforms including the bench top GS Junior+ (the upgraded version of GS Junior) and FLX+ system (www.454.com). The advantages of the 454 platforms include the relatively fast run times and the long read lengths with maximum of 1 kb (Table 3.1). The generated long reads are helpful for mapping to a reference genome, *de novo* genome assembly or metagenomics applications. However, it possesses the drawbacks of the relatively low throughput, high reagent cost and high error rates in homopolymer repeats. In addition, an announcement that Roche will shut down 454 and stop the supporting services for the sequencing platform should be noticed (van Dijk et al., 2014).

**Table 3.1** Comparison of performance of the current second- and third-generation sequencing platforms

| Company | Platform | Sequencing chemistry | Maximum output | Maximum reads per run | Maximum read length | Run time |
|---|---|---|---|---|---|---|
| Roche (454) | GS Junior+ | SBS (SNA) | 70 Mb | ~0.1 million | 1000 bp | 18 hours |
| | GS FLX+ | SBS (SNA) | 700 Mb | ~1 million | 1000 bp | 23 hours |
| Illumina | MiniSeq | SBS (CRT) | 8 Gb | 25 million | 2 × 150 bp | 4–24 hours |
| | MiSeq | SBS (CRT) | 15 Gb | 25 million | 2 × 300 bp | 4–55 hours |
| | NextSeq | SBS (CRT) | 120 Gb | 400 million | 2 × 150 bp | 12–30 hours |
| | HiSeq | SBS (CRT) | 1.5 Tb | 5 billion | 2 × 150 bp | 7 hours – 6 days |
| | HiSeq X | SBS (CRT) | 1.8 Tb | 6 billion | 2 × 150 bp | <3 days |
| Thermo Fisher | SOLiD 5500 | SBL | 320 Gb | ~1.4 billion | 50 or 75 bp | 10 days |
| | Ion PGM | SBS (SNA) | 2 Gb | 5.5 million | 200 or 400 bp | 4–7.3 hours |
| | Ion Proton | SBS (SNA) | 10 Gb | 80 million | 200 bp | 2–4 hours |
| | Ion S5 | SBS (SNA) | 15 Gb | 80 million | 200 or 400 bp | 2.5–4 hours |
| SeqLL (Helicos) | Heliscope | SMS | 35 Gb | ~1 billion | 55 bp | 8 days |
| Pacific Biosciences | PacBio RS II | SMRT | 1 Gb* | ~55,000 | 60 kb | 0.5–6 hours |
| | Sequel | SMRT | 7 Gb* | ~370,000 | 60 kb | 0.5–6 hours |
| Oxford Nanopore | MinION | SMRT | 42 Gb | 4.4 million | 230–300 kb | <2 days |
| | PromethION | SMRT | 12 Tb | 1250 million | 230–300 kb | <2 days |

The data are obtained from the homepages of each company and two recent review articles (Chaitankar *et al.*, 2016; Goodwin *et al.*, 2016). *Output per SMRT cell (Number of SMRT cell is 1–16). bp, base pairs; CRT, cyclic reversible termination; Gb, gigabase pairs; kb, kilobase pairs; Mb, megabase pairs; SBL, sequencing by ligation; SBS: sequencing by synthesis; SMRT, single-molecule real-time; SMS, single-molecule sequencing; SNA: single-nucleotide addition; Tb, terabase pairs.

## Illumina

The Illumina sequencing platform, the most widely adopted in the industry, employs the CRT approach of SBS methodology (Liu *et al.*, 2012; Goodwin *et al.*, 2016). The sequencing steps include library preparation, cluster generation and SBS. The polynucleotide samples are randomly fragmented followed by adapter ligation at both 5′ and 3′ ends to prepare sequencing library. After adapter ligation, PCR amplification and gel purification are performed. The library is then loaded into a flow cell for cluster generation. The nucleotide fragments are immobilized on the flow cell surface through hybridization of oligos and adapters. By bridge amplification, each fragment is amplified into a clonal cluster. The flow cell is in turn incubated with the sequencing reagents and the four fluorescent dNTPs bound with reversible terminators. The incorporated bases are identified according to emission wavelength and intensity (Illumina, 2016).

Solexa, acquired by Illumina in 2007, launched its first sequencer (the Genome Analyser) which can generate roughly 1 gigabase (Gb) of data per sequencing run in 2006. Following the Genome Analyser, the Illumina sequencers with different application and sequencing power have been developed. The current platforms include MiniSeq, MiSeq, NextSeq, HiSeq Series and HiSeq X Series (www.illumina.com). The outputs range from 1.8 to 7.8 Gb for targeted sequencing studies by the benchtop MiniSeq system to 1.6 to 1.8 terabase (Tb) for population-scale studies by the HiSeqX Series (Table 3.1). The HiSeqX Ten, released in 2014, is currently the sequencer with the highest throughput. It was upgraded to sequence 1.8 Tb per run, leading to the possibility of sequencing over 45 human genomes in a single day. In addition, the cost was down to approximately US$1000 for one human genome. It takes genome sequencing entering a period to see the differences among thousands of people and discover the critical genes causing cancer or other diseases (Illumina, 2016). However, the HiSeq X Ten is a set of 10 HiSeq X system with a price of US$10 million. In addition, the low cost claimed by Illumina is an average that can be reached by running full capacity of all HiSeq

X machines for one year. In other words, it is limited to be used only when large institutions carry out population-scale genome sequencing (Goodwin *et al.*, 2016). To overcome the limitations caused by short-read sequencing, Illumina recently released a Synthetic Long-Read Sequencing technology that can generate reads of around 10 kb. By using this technology, synthetically long fragments can be constructed from shorter sequencing reads generated by the HiSeq platform for accurate genome assembly and genome finishing (www.illumina.com/products/by-type/sequencing-kits/library-prep-kits/truseq-synthetic-long-read.html; Li *et al.*, 2015a).

### Thermo Fisher Scientific

The SOLiD sequencing technology based on SBL approach was developed by George Church and his colleagues. It was published in 2005 for the application in the resequencing of the *Escherichia coli* genome and was later improved and released by Applied Biosystems (now Thermo Fisher) in 2007 (Shendure *et al.*, 2005; Voelkerding *et al.*, 2009). The sequencing procedures include library and template preparation, bead deposition and sequencing. Two types of libraries including fragment or mate-paired library can be constructed according to different research applications. A mixture of short fragments flanked by adaptors are generated and attached to beads followed by emulsion PCR amplification. The beads with the amplified templates are then immobilized onto a glass slide or FlowChip by covalent attachment. SBL is begun by annealing a primer to the complementary adapter. In contrast to SBS mediated by polymerase (3′ hydroxyl group), the primer offer a 5′ phosphate group for ligation with one of the fluorescently labelled interrogation probes which are octamers consisting of two specific bases followed by six degenerate bases. There are 16 combinations for the first two specific bases in the interrogation probes, such as AA, AT and so on. In the first SBL step, these probes compete for hybridization with the templates followed by ligation with the primer and detection of the fluorescence signal. The fluorophore along with three bases is in turn cleaved from the probe, leaving a 5-bp fragment with a 5′ phosphate group for the next ligation of the interrogation probes. Multiple cycles of these processes are performed to complete one round of sequencing. The extension product is then denatured and a second round of sequencing is started with a new primer complementary to the n-1 position of the adapter. Sequencing by five rounds of these primer resets (until n-4) completes each read and each base of the template is sequenced twice (www.thermofisher.com/; Voelkerding *et al.*, 2009). The accuracy rate of the SOLiD system can reach 99.94% which is higher than most other NGS systems because each base is read by twice (Liu *et al.*, 2012; Goodwin *et al.*, 2016). However, the very short read lengths (75 bp), the much longer runtime (at least 6 days) and the less well-developed kits of sample preparation limit its wide application, such as for genome assembly and structural variant detection (van Dijk *et al.*, 2014; Goodwin *et al.*, 2016).

Another well-known NGS platform purchased by Thermo Fisher is Ion Torrent. Although the Ion Torrent system adopts SBS methods similar to most NGS technologies, it is a unique sequencing platform employing an integrated complementary metal oxide semiconductor (CMOS) and an ion-sensitive field-effect transistor (ISFET) as the detection system (Goodwin *et al.*, 2016). The detection is based on measurement of the pH change resulting from proton release during nucleotide incorporation but not fluorescence (Rothberg *et al.*, 2011). For sequencing, the chip is flooded with one nucleotide each time and the incorporated nucleotide is in turn detected. If the incorrect nucleotide is added, no voltage will be detected. In case two nucleotides are added, there will be double voltage (Liu *et al.*, 2012). For different research requirements, several types of chips and instruments are offered by Ion Torrent. Their throughputs range from ~50 Mb to 15 Gb and the runtimes are from 2 to 7 hours (Table 3.1; Goodwin *et al.*, 2016). The Ion PGM is the first commercial sequencer released by Ion Torrent, targeting to clinical applications and small labs. It possesses features of higher speed, lower cost and smaller size because it is not required to perform fluorescence labelling and camera scanning (Liu *et al.*, 2012). With the latest released 318 chips, the Ion PGM improves the output to over 1 Gb. As to the higher-throughput Ion Proton system, the adopted Proton-I chip is manufactured by the 110 nm CMOS technology to increase the number of wells to ~165 million. It can produce 60 to 80 million reads per run with an output of 10 Gb (Egan *et al.*, 2012; Buermans and den Dunnen,

2014). Aiming to develop NGS platform for clinical sequencing, Ion Torrent released its dedicated diagnostic instruments, the Ion PGM Dx and the Ion S5 series. The S5 series coupled with the Ion Chef library preparation and chip loading device could be one of the platforms with the simplest operation. In this combination system, argon required in other Ion Torrent instruments is unnecessary and the plug-and-play protocols have been established. However, the higher-throughput S5 devices along with the Ion Proton have limitations for elucidating long-range genomic or transcriptomic structure, because they cannot be applied for paired-end sequencing (Goodwin *et al.*, 2016).

### Helicos Biosciences

The Heliscope, released by Helicos Biosciences, is the first sequencer for single-molecule sequencing (SMS) derived from the technology developed by Braslavsky *et al.* (2003). It was considered as the interface between second- and third-generation sequencing. In this technology, DNA polymerase is used to acquire sequence information during synthesis of the complementary strand of a single DNA template. The SMS approach is attractive because it can directly sequence nucleic acids in an unbiased manner and prepare samples in a simple way. Without steps for cloning or PCR amplification, the GC-content and size biases appeared in other NGS could be avoided (Pushkarev *et al.*, 2009; Thompson and Steinmann, 2010). The sample preparation includes DNA fragmentation, addition of poly(A) tail at the 3′ end and fluorescence labelling of the final adenosine. The poly(A) tail of the DNA templates are hybridized to Poly(dT) oligonucleotides randomly immobilized on a flow-cell surface by covalent bonding. These random sequencing positions are recorded by the fluorescence of the captured DNA templates. Before sequencing gets started, the fluorescent labels are cleaved and the flow cells are incubated with DNA polymerase and one of the four Cy5-labelled dNTPs. The 'virtual terminator' included in each nucleotide can prevent a further incorporation. The excitation of Cy5 from the incorporated dNTP is in turn detected at 647 nm. The process is repeated to determine the next incorporation of nucleotides (Harris *et al.*, 2008; Goodwin *et al.*, 2016). In addition to single DNA and cDNA molecules, the Heliscope is also the first system that can directly sequence RNA without reverse transcription (Ozsolak *et al.*, 2009). It can prevent the biases from cDNA synthesis by using other RNA sequencing technologies. Moreover, both SMS of short and long RNAs can be done together without performing different sample manipulation steps (Ozsolak, 2016). However, Helicos BioSciences filed for Chapter 11 bankruptcy in 2012 and the properties were acquired by SeqLL in 2014. The SeqLL currently offers customized services for quantitative RNA and specialty DNA sequencing by using the True Single Molecule Sequencing technology (tSMS) of HeliScope Genetic Analysis System (http://seqll.com).

### Pacific Biosciences

The PacBio RS platform, released by Pacific Biosciences in 2010, is the first third-generation sequencing platform employing the Single-Molecule Real-Time (SMRT) sequencing technology. It enables parallel and real-time detection of thousands of single-molecule sequencing reactions (Eid *et al.*, 2009; Liu *et al.*, 2012). The SMRT technology was developed based on the zero-mode waveguide (ZMW) technology published at Science in 2003 (Levene *et al.*, 2003). In the conventional approaches, pico- to nanomolar concentrations of fluorophores are suitable for optical observation of dynamics of individual molecules. However, ligand concentration at micromolar is usually required for biological reactions that make it necessary to reduce sample volume by three orders of magnitude for optical observation of single molecules. ZMWs are tiny nanoholes with a diameter of 70 nm and depth of 100 nm in a metal film. It was successfully applied to observe activity of a single DNA polymerase molecule at micromolar concentrations with microsecond temporal resolution (Levene *et al.*, 2003; McCarthy, 2010). In the current PacBio RS II and Sequel systems, each SMRT cell consists of 150,000 and 1 million of ZMWs, respectively, with a single DNA polymerase at the bottom of each nanohole. The Sequel system thus can produce seven times as many reads as the PacBio RS II (www.pacb.com/products-and-services/pacbio-systems/). During sequencing by synthesis, the DNA polymerase incorporates one of the four nucleotides labelled by different fluorescent dye into the complementary strand of the template DNA. The signal is immediately captured and recorded as a movie format by camera inside the sequencer for real-time observation. The dNTP-bound fluorophore is cleaved by DNA polymerase before the next incorporation of dNTP (McCarthy, 2010; Liu *et al.*, 2012).

Comparing to the second-generation sequencers, the PacBio platforms have several advantages including fast sample preparation (4 to 6h), short run times (0.5 to 6h) and long read lengths. Without PCR step in the sample preparation, the bias and error caused by PCR is reduced. In both of PacBio RS II and Sequel systems, half of the reads are over 20kb with an average of 10kb making PacBio ideal for genome assembly and improvement of the existing draft genomes (Liu *et al.*, 2012; www.pacb.com/products-and-services/pacbio-systems/). In addition, by using unique circular DNA templates, the ones shorter than 3kb can be sequenced multiple times to generate a consensus read of insert, the so-called circular consensus sequence (Goodwin *et al.*, 2016). However, the PacBio platforms have drawbacks of relatively low throughput and high cost, currently limiting the range of applications (van Dijk *et al.*, 2014).

## Oxford Nanopore

Oxford Nanopore is another third-generation sequencing technology because it also sequences single molecules in a real-time manner (van Dijk *et al.*, 2014). Instead of monitor of incorporations or hybridizations of nucleotides employed by other sequencing technologies, the Nanopore platform can directly detect the nucleotide composition of single-stranded DNAs (Goodwin *et al.*, 2016). To carry out sequencing, the bases are identified by the change in electrical conductivity when a DNA molecule is transited through a tiny biopore with diameter in nanoscale (Clarke *et al.*, 2009; Liu *et al.*, 2012).The detection of single molecules based on the nanopore method has emerged from a PNAS paper published in 1996 (Kasianowicz *et al.*, 1996). It was reported that single-stranded DNA and RNA molecules can be driven by an electric field through an ion channel formed by *S. aureus* α-hemolysin across a lipid bilayer. When each polynucleotide molecule translocates through the channel, it can be detected by a transient decrease or block of ionic conductance due to occupy of the pore's volume. Therefore, the possibility of direct and rapid sequencing of single molecules of DNA or RNA by further improving this nanopore method was proposed and investigated (Kasianowicz *et al.*, 1996; Deamer and Akeson, 2000). It was proved that a single adenine nucleotide at a specific location can be identified by the characteristic reductions of ionic current in the α-hemolysin nanopore (Ashkenasy *et al.*, 2005). However, the technique still cannot discriminate each base because the polynucleotide translocation rate is too high. By using an exonuclease enzyme to cleave individual nucleotides from DNA and covalent attachment of an adapter molecule to the protein nanopore, continuous detection of unlabelled individual nucleotide has been achieved (Clarke *et al.*, 2009).

The first commercial nanopore sequencer for sequencing single DNA molecules is MinION released by Oxford Nanopore Technologies. It is an inexpensive portable device connecting to a PC or laptop by USB and capable of producing reads of up to 10kb (van Dijk *et al.*, 2014; Brown and Clarke, 2016). The initial ASIC (application-specific integrated circuit) chip designed for MinION Mk1 flow cell has 512 individual channels enabling to sequence at ~70bp per second. A new 3000-channel ASIC was developed for the new released MinION Mk1B (with an expected increase to 500bp per second) and PromethION, an ultra-high-throughput platform possessing 48 individual flow cells with running at 500bp per second (Goodwin *et al.*, 2016; https://nanoporetech.com/products/minion). Very recently the protocol for direct RNA sequencing by the MinIon device has been developed. It is currently the only platform available for directly sequencing the original RNA strands without cDNA synthesis and PCR reaction. Although the direct RNA sequencing method was firstly reported by Helicos in 2009, it depends on the synthetic copies of the native RNA strands through the SBS reaction. The RNA modifications cannot be detected by this approach (Garalde *et al.*, 2016).

Without performing PCR and fluorescent labelling steps before sequencing, the Nanopore system can reduce costs and increase sequencing speeds (Clarke *et al.*, 2009; Laver *et al.*, 2015). In addition, except for exonuclease, it is not required to use polymerase and ligase, making Oxford Nanopore less temperature sensitive than other platforms (Liu *et al.*, 2012). Because sequence quality is high in the long reads sequenced by the Nanopore system, it benefits to *de novo* sequencing, long-range haplotype mapping and the high-resolution analysis of chromosomal structure variation (Clarke *et al.*, 2009; Laver *et al.*, 2015).

## An overview of current of orchid genome project

### EST and BAC

Genomics studies for the orchids are just in their infancy. A survey of the literature revealed that genome size data for Orchidaceae are comparatively rare, representing just 327 species (Leitch *et al.*, 2009). Nevertheless, they reveal that Orchidaceae are currently the most variable angiosperm family with genome sizes ranging 168-fold (1C = 0.33–55.4 pg). Large scale sequence analysis of orchid genomes was first revealed by bacterial artificial chromosome (BAC) end sequences analysis in *Phalaenopsis* orchid (Hsu *et al.*, 2011). This work offers the first insights into the composition of the Phalaenopsis genome in terms of GC content, transposable elements present, protein-encoding regions, simple sequence repeats, and potential microsynteny between *Phalaenopsis* and other plant species (Hsu *et al.*, 2011). In addition to the nuclear genome, the entire chloroplast genome of *Phalaenopsis* orchid is also sequenced. The chloroplast genome of *P. aphrodite* subsp. *formosana* is about 150 kb, which encode 110 different known genes, including 74 protein-coding genes, four rRNA genes, 30 tRNA genes and two conserved reading frames of unknown function (Chang *et al.*, 2006). Furthermore, the transcripts of 74 protein-coding genes from the chloroplast genome of *P. aphrodite* subsp. *formosana* were used to study extensively the pattern of RNA editing in chloroplasts. A total of 44 editing sites are identified in the 24 transcripts of *P. aphrodite* chloroplast genes, and all are of the C-to-U conversion type (Zeng *et al.*, 2007). On the basis of the above information, the chloroplast genome of several orchids were sequenced, including *Oncidium* Gower Ramsey, *P. equestris*, *Erycina pusilla*, seven species in *Cymbidium*, *Dendrobium officinale* and *Cypripedium macranthos* (Wu *et al.*, 2010; Jheng *et al.*, 2012; Pan *et al.*, 2012; Yang *et al.*, 2013; Luo *et al.*, 2014). Further plastome sequencing of orchids will be necessary to clarify the diversity of chloroplast genomes and to improve our understanding of the relationships within the Orchidaceae.

Large-scale EST sequencing provides a gateway into the genome of organisms owing to the massive information buried in the genome-scale expression data. Before NGS technology has been developed, the most popular sequencing method has been the Sanger method applied to the EST sequencing project. A subtractive EST library was constructed from the pseudobulb of *O.* Gower Ramsey, and 1080 subtractive ESTs were obtained. Most ESTs were annotated as being involved in carbohydrate metabolism, in mannose, pectin and starch biosynthesis, transportation, and stress-related and regulatory function (Tan *et al.*, 2005). To study gene expression in *Phalaenopsis* reproductive organs, a cDNA library was constructed from mature flower buds of *P. equestris*; 5593 ESTs were sequenced and assembled into 3688 unigenes (Tsai *et al.*, 2006). In addition, a cDNA library has been constructed from scented *P. bellina* flower buds with the column removed; 2359 ESTs were sequenced and assembled into 1187 unigenes (Hsiao *et al.*, 2006). The set of floral scent-producing enzymes in the biosynthetic pathway from glyceraldehyde-3-phosphate to geraniol and linalool is recognized through these ESTs and distinguished by comparing their expression patterns in *P. bellina* and a scentless species, *P. equestris* (Hsiao *et al.*, 2006). A similar strategy was adopted for *Vanda* Mimi Palmer principally to mine any potential fragrance-related EST-SSRs as markers in the identification of fragrant vandaceous orchids endemic to Malaysia (Teh *et al.*, 2011).

### Orchid transcriptomes generated by NGS technologies

The sudden rise of relatively low-cost and rapid NGS technologies is dramatically advancing our ability to comprehensively interrogate the nucleic-acid-based information in a cell at unparalleled resolution and depth (Delseny *et al.*, 2010). The technologies were rapidly adopted for orchid transcriptome analysis (Table 3.2). 206,960 ESTs were released from the pool containing *P. equestris*, *P. aphrodite*, and *P. bellina* and a total of 50,908 contig sequences were from six different tissues of *O.* Gower Ramsey (Chang *et al.*, 2011) by 454 technology respectively to expansively cover the *Phalaenopsis* and *Oncidium* orchid transcriptome and facilitate identifying sets of genes involved

**Table 3.2** Characteristics of findings in the literature for the application of next-generation sequencing to orchid transcriptomes

| Subfamily | Species | Sequencing platform | Tissue | Study aim | Reference |
|---|---|---|---|---|---|
| Apostasioideae | *Apostasia shenzhenica* | Illumina/Solexa | Mature flower buds | Study of floral development and evolutionary trends of orchid flowers | Tsai *et al.*, 2013 |
| Apostasioideae | *Neuwiedia malipoensis* | Illumina/Solexa | Mature flower buds | | |
| Vanilloideae | *Vanilla shenzhenica* | Illumina/Solexa | Mature flower buds | | |
| Vanilloideae | *Galeola faberi* | Illumina/Solexa | Mature flower buds | | |
| Cypripedioideae | *Paphiopedilum armeniacum* | Illumina/Solexa | Mature flower buds | | |
| Cypripedioideae | *Cypripedium singchii* | Illumina/Solexa | Mature flower buds | | |
| Orchidoideae | *Habenaria delavayi* | Illumina/Solexa | Mature flower buds | | |
| Orchidoideae | *Hemipilia forrestii* | Illumina/Solexa | Mature flower buds | | |
| Epidendroideae | *Phalaenopsis equestris* | Illumina/Solexa | Mature flower buds | | |
| Epidendroideae | *Cymbidium sinense* | Illumina/Solexa | Mature flower buds | | |
| Vanilloideae | *Vanilla planifolia* | Illumina/Solexa; Roche/454 | Pod tissues, seeds | Study of biosynthetic routes to flavour components | Rao *et al.*, 2014 |
| Cypripedioideae | *Paphiopedilum concolor* | Illumina Hiseq 2000 | Roots | Identify the genes that control root growth and development | Li *et al.*, 2015b |
| Orchidoideae | *Ophrys* species | Roche/454; Illumina/Solexa | Flowers, labellums, leaves, flower organ from open flowers and buds | Identify genes responding for pollinator attraction | Sedeek *et al.*, 2013 |
| Orchidoideae | *Orchis italica* | Illumina/Solexa (MiSeq) | Inflorescences | The roles of small RNAs on the flower development | Aceto *et al.*, 2014 |
| Orchidoideae | *Orchis italica* | Illumina Hiseq 2500 | Florets of inflorescence before anthesis | Analysing transcripts potentially involved in flower development | Paolo *et al.*, 2014 |
| Orchidoideae | *Serapias vomeracea* | Roche/454 | Protocorms | Investigate the molecular bases of the orchid response to mycorrhizal invasion | Perotto *et al.*, 2014 |
| Orchidoideae | *Anoectochilus roxburghii* (Wall.) Lindl. | Illumina HiSeq 4000 | Dry seeds, seeds from asymbiotic or symbiotic germination | Study of seed germination process | Liu *et al.*, 2015 |
| Orchidoideae | *Gastrodia elata* Blume | Illumina Hiseq 2000 | Vegetative tissues, corms, juvenile tubers | Address the gene regulation mechanism in gastrodin biosynthesis | Tsai *et al.*, 2016 |
| Epidendroideae | *Phalaenopsis aphrodite* | Sanger: EST | Protocorms | Gene discovery and genomic annotation | Fu *et al.*, 2011; Hsiao *et al.*, 2011 |
| Epidendroideae | *Phalaenopsis equestris* | Sanger: EST | Mature flower buds | | |
| Epidendroideae | *Phalaenopsis bellina* | Sanger: EST | Mature flower buds without column | | |

**Table 3.2** Continued

| Subfamily | Species | Sequencing platform | Tissue | Study aim | Reference |
|---|---|---|---|---|---|
| Epidendroideae | *Phalaenopsis aphrodite; Phalaenopsis equestris; Phalaenopsis bellina* | Roche/454 | Mixed tissues | Gene discovery and genomic annotation | Fu *et al.*, 2011; Hsiao *et al.*, 2011 |
| Epidendroideae | *Phalaenopsis equestris* | Illumina/Solexa | Leaves | | |
| Epidendroideae | *Phalaenopsis aphrodite* | Roche/454; Illumina/Solexa | Leaves, stems, roots, young inflorescences, stalks, flower buds, flowers, germinating seeds | Investigate expressed genes involved in many biological processes of orchids | Su *et al.*, 2011 |
| Epidendroideae | *Phalaenopsis aphrodite* | Illumina/Solexa | Leaves, stalks, flower buds | Study the roles of small RNAs on the regulation of flowering | An *et al.*, 2011; An and Chan, 2012 |
| Epidendroideae | *Phalaenopsis aphrodite* | Illumina/Solexa | Leaves, roots, flowers, germinating seeds, young inflorescences | Identify species- and tissue-specific miRNAs | Chao *et al.*, 2014 |
| Epidendroideae | *Phalaenopsis* Brother Spring Dancer 'KHM190' | Illumina Hiseq 2000 | Petals, sepals or labellums from flower buds of wild-type and peloric petal mutant plants | Study regulation of floral- organ development | Huang *et al.*, 2015 |
| Epidendroideae | *Phalaenopsis* sp. | Illumina Hiseq 2000 | Explants | Examine *Phalaenopsis* leaf explant browning | Xu *et al.*, 2015 |
| Epidendroideae | *Oncidium* 'Gower Ramsey' | Roche/454 | Leaves, pseudobulbs, young inflorescences, inflorescences, flower buds, mature flowers | Identify genes associated with flowering time | Chang *et al.*, 2011 |
| Epidendroideae | *Oncidium* 'Gower Ramsey' | Illumina/Solexa | Roots with or without fungus | Study the roles of small RNAs on the interaction between root and the fungus | Ye *et al.*, 2014 |
| Epidendroideae | *Erycina pusilla* | Illumina/Solexa | Roots, leaves, peduncles, flowers, capsules | Investigate photoperiod-dependent flowering genes | Chou *et al.*, 2013 |
| Epidendroideae | *Erycina pusilla* | Illumina/Solexa | Roots, leaves, peduncles, flowers, capsules | Study the roles of small RNAs on the regulation of flowering | Lin *et al.*, 2013 |
| Epidendroideae | *Cymbidium ensifolium* 'Tiegusu' | Illumina HiSeq 2000 | Flower buds, mature flower | Identify genes associated with floral development | Li *et al.*, 2013 |
| Epidendroideae | *Cymbidium sinense* 'Qi Jian Bai Mo' | Illumina HiSeq 2000 | Plants in vegetative phase/floral differentiation phase/ reproductive phase | Identify genes associated with floral development | Zhang *et al.*, 2013 |
| Epidendroideae | *Cymbidium hybridum* 'Golden Boy' | Illumina HiSeq 2000 | Roots with or without fungus | Study of orchid-mycorrhizal fungi interactions | Zhao *et al.*, 2014 |

**Table 3.2** Continued

| Subfamily | Species | Sequencing platform | Tissue | Study aim | Reference |
|---|---|---|---|---|---|
| Epidendroideae | *Cymbidium ensifolium* 'Tiegusu' | Illumina/Solexa | Flower bud | Identify miRNAs related to floral development | Li *et al.*, 2015c |
| Epidendroideae | *Cymbidium ensifolium* 'tianesu' | Roche/454 | Sepals, petals, labellums, gynostemia from flower buds and mature flowers | Reveal genes associated with floral organ differentiation | Yang and Zhu, 2015 |
| Epidendroideae | *Cymbidium sinense* 'Dharma' | Roche/454 | Roots, leaves, pseudobulbs, flowers | Analyse molecular mechanism underlying leaf-colour variations | Zhu *et al.*, 2015 |
| Epidendroideae | *Cymbidium sinense; Cymbidium atropurpureum; Cymbidium mannii* | Illumina HiSeq 2000 | Leaves | Explore the evolution and molecular regulation of CAM plants | Zhang *et al.*, 2016c |
| Epidendroideae | *Dendrobium officinale* | Roche/454 | Stems | Study of alkaloid biosynthesis | Guo *et al.*, 2013 |
| Epidendroideae | *Dendrobium officinale* | Illumina HiSeq 2000 | Juvenile and adult plants | Identify genes associated with polysaccharide synthesis | Zhang *et al.*, 2016b |
| Epidendroideae | *Dendrobium officinale* | Illumina HiSeq 2500 | Flower, roots, leaves, stems | Study of the regulatory networks of the production and accumulation of the medicinal constituents | Meng *et al.*, 2016 |

in a broad range of biological processes (Hsiao *et al.*, 2011a; Chang *et al.*, 2011). A total of 121,917 unique transcripts were obtained from the *Ophrys* species by using 454 pyrosequencing and Illumina (Solexa) technologies to identify genes responding for pollinator attraction (Sedeek *et al.*, 2013). To study the genes involved in alkaloid biosynthetic pathway and polysaccharide biosynthesis in *Dendrobium officinale*, an important traditional Chinese herb, 454 pyrosequencing and Illumina technology was respectively applied to generate plentiful ESTs (Guo *et al.*, 2013; Zhang *et al.*, 2016b). To provide a general resource for studying on the pod development of *Vanilla plantifolia*, one of the most valued flavour species for its flavour qualities and is therefore widely cultivated and used for the production of food additives, the combined 454/Illumina RNA-seq platforms produced high quality *de novo* transcriptome assembly for this non-model crop species (Rao *et al.*, 2014). In addition, to improve the horticultural value of *Phalaenopsis* and *Cymbidium*, transcriptome derived from browning leaf of *Phalaenopsis* explant (sequencing by Illumina HiSeq 2000), and variable colour of *Cymbidium* leaf (sequencing by 454 pyrosequencing) were investigated (Xu *et al.*, 2015; Zhu *et al.*, 2015). Orchids are unique among plants in that mycorrhizal symbioses with soil fungi are required throughout the life history stages, from seed germination to adulthood. To understand the molecular mechanism of orchid seed germination and the symbiotic orchid–fungus relationship, 454 and Illumina were adopted to explore transcriptomes derived from *Serapias vomeracea* (Perotto *et al.*, 2014), *C. hybridium* (Zhao *et al.*, 2014), *Anoectochilus roxburghii* (Liu *et al.*, 2015), and *Gastrodia elata* (Tsai *et al.*, 2016). In Orchidaceae, about 40% species adopt crassulacean acid metabolism (CAM) to fix carbon dioxide suggesting the Orchidaceae is the largest CAM clade (Silvera *et al.*, 2009). To illuminate the origin and evolution of CAM pathway, transcriptomes derived from leaves of CAM orchids *P. equestris*, *D. terminale* and *C. mannii* were sampled at different time interval and sequencing by Illumine HiSeq

2000 (Deng *et al.*, 2016; Zhang *et al.*, 2016c). To study the development of spectacular orchid flower morphology, developing floral transcriptomes originating from *Cymbidium* (Li *et al.*, 2013; Yang *et al.*, 2015; Zhang *et al.*, 2013), *Orchis* (De Paolo *et al.*, 2014), and *Phalaenopsis* (Huang *et al.*, 2015) were applied to identify genes associated with floral development. Recently, root transcriptome from *Paphiopedilum concolor* was also produced to explore genes involved in orchid root development (Li *et al.*, 2015b). The accumulated transcribed sequences could be directly used to develop microarray platform, and be the resource for phylogenetic analysis. For example, an oligomicroarray containing 14,732 unigenes based on the information of expressed sequence tags derived from *Phalaenopsis* orchids was developed and applied to compare transcriptome among different types of floral organs including sepal, petal and labellum (Hsiao *et al.*, 2013). 315 single-copy orthologous genes extracted from the transcriptomes of species covering five subfamilies of Orchidaceae were applied to reconstruct a more robust phylogeny of orchids, and the results indicated that this method is more efficient and reliable than methods based on a few gene markers for phylogenic analyses, especially for the holomycotrophic species or those whose DNA sequences have been difficult to amplify (Deng *et al.*, 2015).

Next-generation sequencing technologies are not only applied to characterize orchid transcriptomes but also used to systematically analyse small RNAs in orchids (Table 3.2). The roles of small RNAs were studied on the regulation of flowering in *P. aphrodite* and *E. pusilla* (An *et al.*, 2011; An and Chan, 2012; Lin *et al.*, 2013), flower development in *Orchis italica* (Aceto *et al.*, 2014; De Paolo *et al.*, 2014) and *Cymbidium ensifolium* (Li *et al.*, 2015c), and interaction between the fungus *Piriformospora indica* and the root of an *Oncidium* hybrid orchid (Ye *et al.*, 2014). Later, comprehensive collection of small RNAs derived from *P. aphrodite* (Chao *et al.*, 2014), and *D. officinale* (Meng *et al.*, 2016) were performed. These efforts provide valuable information about the composition, expression and function of small RNAs and will aid functional genomics studies of orchids.

Recently, OrchidBase has collected the transcriptome sequences from 11 Phalaenopsis cDNA libraries and flower tissue of 10 species distributed in five subfamilies of Orchidaceae (Fu *et al.*, 2011; Tsai *et al.*, 2013; Niu *et al.*, 2016). The EST sequences collected in OrchidBase were obtained through both deep sequencing with ABI 3730, Roche 454 and Illumina/Solexa. OrchidBase is freely available at http://orchidbase.itps.ncku.edu.tw/ and provides researchers with a high-quality genetic resource for data mining and efficient experimental studies of orchid biology and biotechnology. Another orchid transcriptomic database, Orchidstra (http://orchidstra.abrc.sinica.edu.tw), was constructed from the 233,924 unique contigs of the transcriptome sequences of *P. aphrodite* by use of a Roche 454 and Illumina/Solexa platform, and the genes of tissue-specific expression were categorized by profiling analysis with RNA-seq (Su *et al.*, 2011). Oncidium cDNA libraries for six different organs, including leaves, pseudobulbs, young inflorescences, inflorescences, flower buds and mature flowers, were generated from 50,908 contig sequences by use of the Roche 454 platform and were constructed into the OncidiumOrchidGenomeBase (http://predictor.nchu.edu.tw/oogb/) (Chang *et al.*, 2011). All this EST information will be very useful for gene annotation in genomic sequencing, specificity of orchids, and organization of the orchid genome. The plentiful collection of ESTs and BESs in *Phalaenopsis* makes them reasonable candidates for orchid whole-genome sequencing. The two native *Phalaenopsis* species in Taiwan, *P. equestris* and *P. aphrodite* subsp. *formosana*, are usually used as parents for breeding and have a relatively small genome size of 3.37 pg/2C and 2.80 pg/2C, respectively. The basic studies and genomics information collected have laid the groundwork for *P. equestris* to serve as a model orchid plant for whole-genome sequencing.

## Orchid genome project

With the quick development and lower cost of NGS, whole genome sequencing of non-model species, like orchids, can be realized. The first milestone is sequencing the tropical epiphytic orchid *Phalaenopsis equestris*, a frequently used parent species for orchid breeding (Cai *et al.*, 2015). The *P. equestris* genome is sequenced via a whole-genome shotgun strategy (Illumina technology) and the genome size is estimated to be 1.16 Gb contains with 29,431 predicted protein-coding genes. Analysing the *P. equestris* genome showed that repetitive DNAs, mostly transposable elements (TEs), account for the majority of the genome, at 62%. The authors find evidence for an orchid-specific paleopolyploidy event that preceded the radiation of most orchid clades. This species is also the first whole-genome-sequenced CAM plant and a gene family (α carbonic anhydrase) involved in CAM pathway is found having an obvious expansion which suggests that gene duplication might have contributed to the evolution of CAM photosynthesis in *P. equestris*. In addition, genes located at the heterozygous regions might relate to self-incompatibility. Genes in type II MADS-box clades, including the E-class, C/D-class, B-class AP3 and AGL6 clades, are found contained more genes than other species. These expanded clades are involved in orchid floral organs which can support the unique evolutionary routes of these floral organ identity genes associated with the unique labellum and gynostemium innovation in orchids. Furthermore, the *Phalaenopsis* genome sequence was applied to identify MYB genes controlling floral pigmentation patterning (Hsu *et al.*, 2015), and TCP genes involved in ovule development (Lin *et al.*, 2016).

Having both ornamental value and a broad range of therapeutic effects, *Dendrobium officinale* is the other Orchidaceae plant which was sequenced by combining the second-generation Illumina Hiseq 2000 and third-generation PacBio sequencing technologies (Yan *et al.*, 2015). The assembled genome of *D. officinale* has a predicted 35,567 protein-coding genes. The number of predicted genes in *D. officinale* is higher than that in *Phalaenopis*. For example, the number of B-class MADS-box genes presented in *D. officinale* is much higher than that in *Phalaenopsis*. In *Phalaenopsis*, there are four members in B-class AP3-like subfamily, and one member in B-class PI-like subfamily. In contrast, there are 19 AP3-like genes and five PI-like genes presented in the *Dendrobium* genome. It is possible that the plants used for the whole genome sequencing are not native species, but hybrids. Later, another *Dendrobium* species, *Dendrobium catenatum* (鐵皮石斛), was whole genome sequenced by Illumina HiSeq 2000 platform (Zhang *et al.*, 2016a). The predicted 28,910 protein-coding genes are comparable with those of *Phalaenopsis*, and a whole genome duplication event could be share with *Phalaenopsis*. The expansion of many resistance-related genes in *Dendrobium* suggests a powerful immune system responsible for adaptation to a wide range of ecological niches. In addition, extensive duplication of genes involved in glucomannan synthase activities is likely related to the synthesis of medicinal polysaccharides. Expansion of MADS-box gene clades ANR1, StMADS11, and MIKC*, involved in the regulation of development and growth, suggests that these expansions are associated with the astonishing diversity of plant architecture in the genus *Dendrobium* (Zhang *et al.*, 2016a).

These complete genome sequences of Orchidaceae species will facilitate future research on the diversity and evolution of orchid plants. The genome sequences will also be an important resource for genetic engineering, such as molecular marker-assisted breeding and the production of transgenic plants, which are necessary to increase the efficiency of orchid breeding and aid orchid horticulture research.

## Future perspective

About 2500 years ago Confucius wrote 'Lan (orchid) that grows in the deep valleys never withholds its fragrances even without being appreciated'. Then, 300 years ago, Charles Darwin, in a letter to Joseph Hooker, wrote 'I never was more interested in any subject in my life, than in this of Orchids'. Because of the unique reproduction strategy in orchids, their origin has been a recurring question in botany and evolutionary biology since the nineteenth century. The study of orchid biology by using NGS technologies, although still young, has already offered new and exciting perspectives on this intriguing plant family. Recent advances in sequencing technologies and functional genomics methodologies have allowed studies on orchid biology to become a standard scientific research topic accessible to many investigators, which has in turn resulted in many exciting new discoveries. With the whole genome sequences of *P. equestris*, *D. catenatum*, and *D. officinale* available, the genetic blueprint of orchids provides a basic understanding of the genetic basis of orchids. Furthermore, the genome sequences of the primitive orchid *Apostasia* and one of the most popular aromatic orchids, *Vanilla*, will be available soon. The efforts by many scientists to use a plethora of genome information and genomics tools will lead to a promising understanding of the biological, physiological, molecular and genetic mechanisms of orchids in years to come. In addition, we will have access to a greater portion of their genetic diversity, thus allowing orchid breeders to associate this diversity with phenotypic traits and to continue to engineer new varieties better adapted to a changing environment. Clearly a new era of orchid biology is now open because of the sequencing revolution.

## Acknowledgements

This work was supported by Grants 103-2313-B-006-001-MY3, 104-2321-B-006-025-, and 105-2321-B-006-026- from Ministry of Science and Technology, Taiwan, and was supported by the 948 Program of State Forestry Administration, China (No. 2011-4-53), Development Special Fund of Biological Industry of Shenzhen Municipality (No. JC201005310692A), Development Funds for Emerging Industries of Strategic Importance of Shenzhen (No. JCYJ20140402093332029, No. NYSW20140331010039), Fundamental Research Project of Shenzhen Municipality (No. JCYJ20150403150235943), and Forestry Science and Technology Innovation Fund Project of Guangdong Province (No. 2016KJCX025).

## References

Aceto, S., Sica, M., De Paolo, S., D'Argenio, V., Cantiello, P., Salvatore, F., and Gaudio, L. (2014). The analysis of the inflorescence miRNome of the orchid *Orchis italica* reveals a *DEF*-like MADS-box gene as a new miRNA target. PLOS ONE *9*, e97839. http://dx.doi.org/10.1371/journal.pone.0097839

An, F.M., and Chan, M.T. (2012). Transcriptome-wide characterization of miRNA-directed and non-miRNA-directed endonucleolytic cleavage using Degradome analysis under low ambient temperature in *Phalaenopsis aphrodite* subsp. *formosana*. Plant Cell Physiol. *53*, 1737–1750. http://dx.doi.org/10.1093/pcp/pcs118

An, F.M., Hsiao, S.R., and Chan, M.T. (2011). Sequencing-based approaches reveal low ambient temperature-responsive and tissue-specific microRNAs in *Phalaenopsis* orchid. PLOS ONE *6*, e18937. http://dx.doi.org/0.1371/journal.pone.0018937

(The reference is the same as The Arabidopsis Genome Initiative. (2000). Analysis of the genome sequence of the flowering plant *Arabidopsis thaliana*. Nature *408*, 796–815. http://dx.doi.org/10.1038/35048692 which listed in Page 4)Arditti, J. (1992). Fundamentals of Orchid Biology. John Wiley, New York, USA.

Ashkenasy, N., Sánchez-Quesada, J., Bayley, H., and Ghadiri, M.R. (2005). Recognizing a single base in an individual DNA strand: a step toward DNA sequencing in nanopores. Angew. Chem. Int. Ed. Engl. *44*, 1401–1404. http://dx.doi.org/10.1002/anie.200462114

Atwood, J.T. (1986). The size of Orchidaceae and the systematic distribution of epiphytic orchids. Selbyana *9*, 171–186.

Baker, M. (2012). De novo genome assembly: what every biologist should know. Nature Met. *9*, 333–337. http://dx.doi.org/10.1038/nmeth.1935

Brown, C.G., and Clarke, J. (2016). Nanopore development at Oxford Nanopore. Nat. Biotechnol. *34*, 810–811. http://dx.doi.org/10.1038/nbt.3622

Buermans, H.P., and den Dunnen, J.T. (2014). Next generation sequencing technology: Advances and applications. Biochim. Biophys. Acta *1842*, 1932–1941.

Cai, J., Liu, X., Vanneste, K., Proost, S., Tsai, W.C., Liu, K.W., Chen, L.J., He, Y., Xu, Q., Bian, C., *et al.* (2015). The genome sequence of the orchid *Phalaenopsis equestris*. Nature Genet. *47*, 65–72. http://dx.doi.org/10.1038/ng.3149

Chaitankar, V., Karakülah, G., Ratnapriya, R., Giuste, F.O., Brooks, M.J., and Swaroop, A. (2016). Next generation sequencing technology and genomewide data analysis: Perspectives for retinal research. Prog. Retin. Eye. Res. *55*, 1–31.

Chang, C.C., Lin, H.C., Lin, I.P., Chow, T.Y., Chen, H.H., Chen, W.H., Cheng, C.H., Lin, C.Y., Liu, S.M., Chang, C.C., *et al.* (2006). The chloroplast genome of *Phalaenopsis aphrodite* (Orchidaceae): comparative analysis of evolutionary rate with that of grasses and its phylogenetic implications. Mol. Biol. Evol. *23*, 279–291. http://dx.doi.org/10.1093/molbev/msj029

Chang, Y.Y., Chu, Y.W., Chen, C.W., Leu, W.M., Hsu, H.F., and Yang, C.H. (2011). Characterization of *Oncidium* 'Gower Ramsey' transcriptomes using 454 GS-FLX pyrosequencing and their application to the identification of genes associated with flowering time. Plant. Cell. Physiol. *52*, 1532–1545. http://dx.doi.org/10.1093/pcp/pcr101

Chao, Y.T., Su, C.L., Jean, W.H., Chen, W.C., Chang, Y.C., and Shih, M.C. (2014). Identification and characterization of the microRNA transcriptome of a moth orchid *Phalaenopsis aphrodite*. Plant. Mol. Biol. *84*, 529–548. http://dx.doi.org/10.1007/s11103-013-0150-0

Chou, M.L., Shih, M.C., Chan, M.T., Liao, S.Y., Hsu, C.T., Haung, Y.T., Chen, J.J., Liao, D.C., Wu, F.H., and Lin, C.S. (2013). Global transcriptome analysis and identification of a *CONSTANS*-like gene family in the orchid *Erycina pusilla*. Planta *237*, 1425–1441. http://dx.doi.org/10.1007/s00425-013-1850-z

Clarke, J., Wu, H.C., Jayasinghe, L., Patel, A., Reid, S., and Bayley, H. (2009). Continuous base identification for single-molecule nanopore DNA sequencing. Nat. Nanotechnol. *4*, 265–270. http://dx.doi.org/10.1038/nnano.2009.12

Cozzolino, S., and Widmer, A. (2005). Orchid diversity: an evolutionary consequence of deception? Trends Ecol. Evol. *20*, 487–494.

Crane, P.R., Friis, E.M., and Pedersen, K.R. (1995). The origin and early diversification of angiosperms. Nature *374*, 27–33. http://dx.doi.org/10.1038/374027a0

De Paolo, S., Salvemini, M., Gaudio, L., and Aceto, S. (2014). *De novo* transcriptome assembly from inflorescence of *Orchis italica*: analysis of coding and non-coding transcripts. PLOS ONE *9*, e102155. http://dx.doi.org/10.1371/journal.pone.0102155

Deamer, D.W., and Akeson, M. (2000). Nanopores and nucleic acids: prospects for ultrarapid sequencing. Trends Biotechnol. *18*, 147–151.

Delseny, M., Han, B., and Hsing, Y.I.e. (2010). High throughput DNA sequencing: The new sequencing revolution. Plant. Sci. *179*, 407–422. http://dx.doi.org/10.1016/j.plantsci.2010.07.019

Deng, H., Zhang, G.Q., Lin, M., Wang, Y., and Liu, Z.J. (2015). Mining from transcriptomes: 315 single-copy orthologous genes concatenated for the phylogenetic analyses of Orchidaceae. Ecol. Evol. *5*, 3800–3807. http://dx.doi.org/10.1002/ece3.1642

Deng, H., Zhang, L.S., Zhang, G.Q., Zheng, B.Q., Liu, Z.J., and Wang, Y. (2016). Evolutionary history of PEPC genes in green plants: Implications for the evolution of CAM in orchids. Mol. Phylogenet. Evol. *94*, 559–564. http://dx.doi.org/10.1016/j.ympev.2015.10.007

Dressler, R.L. (1981). The orchids: Natural history and classification. Cambridge, Massachusetts, USA: Harvard University Press.

Egan, A.N., Schlueter, J., and Spooner, D.M. (2012). Applications of next-generation sequencing in plant biology. Am. J. Bot. *99*, 175–185. http://dx.doi.org/10.3732/ajb.1200020

Eid, J., Fehr, A., Gray, J., Luong, K., Lyle, J., Otto, G., Peluso, P., Rank, D., Baybayan, P., Bettman, B., *et al.* (2009). Real-time DNA sequencing from single polymerase molecules. Science *323*, 133–138. http://dx.doi.org/10.1126/science.1162986

Fu, C.H., Chen, Y.W., Hsiao, Y.Y., Pan, Z.J., Liu, Z.J., Huang, Y.M., Tsai, W.C., and Chen, H.H. (2011). OrchidBase: a collection of sequences of the transcriptome derived from orchids. Plant. Cell. Physiol. *52*, 238–243. http://dx.doi.org/10.1093/pcp/pcq201

Garalde, D.R., Snell, E.A., Jachimowicz, D., Heron, A.J., Bruce, M., Lloyd, J., Warland, A., Pantic, N., Admassu, T., Ciccone, J.,*et al.* (2016). Highly parallel direct RNA sequencing on an array of nanopores. http://dx.doi.org/10.1101/068809.

Goodwin, S., McPherson, J.D., and McCombie, W.R. (2016). Coming of age: ten years of next-generation sequencing technologies. Nat. Rev. Genet. *17*, 333–351. http://dx.doi.org/10.1038/nrg.2016.49

Górniak, M., Paun, O., and Chase, M.W. (2010). Phylogenetic relationships within Orchidaceae based on a low-copy nuclear coding gene, *Xdh*: Congruence with organellar and nuclear ribosomal DNA results. Mol. Phylogenet. Evol. *56*, 784–795. http://dx.doi.org/10.1016/j.ympev.2010.03.003

Guo, X., Li, Y., Li, C., Luo, H., Wang, L., Qian, J., Luo, X., Xiang, L., Song, J., Sun, C., *et al.* (2013). Analysis of the *Dendrobium officinale* transcriptome reveals putative alkaloid biosynthetic genes and genetic markers. Gene *527*, 131–138. http://dx.doi.org/10.1016/j.gene.2013.05.073

Harris, T.D., Buzby, P.R., Babcock, H., Beer, E., Bowers, J., Braslavsky, I., Causey, M., Colonell, J., Dimeo, J., Efcavitch, J.W., *et al.* (2008). Single-molecule DNA sequencing of a viral genome. Science *320*, 106–109. http://dx.doi.org/10.1126/science.1150427

Hsiao, Y.Y., Chen, Y.W., Huang, S.C., Pan, Z.J., Fu, C.H., Chen, W.H., Tsai, W.C., and Chen, H.H. (2011a). Gene discovery using next-generation pyrosequencing to develop ESTs for *Phalaenopsis* orchids. BMC Genomics *12*, 360. http://dx.doi.org/10.1186/1471-2164-12-360

Hsiao, Y.Y., Pan, Z.J., Hsu, C.C., Yang, Y.P., Hsu, Y.C., Chuang, Y.C., Shih, H.H., Chen, W.H., Tsai, W.C., and Chen, H.H. (2011b). Research on orchid biology and biotechnology. Plant. Cell. Physiol. *52*, 1467–1486. http://dx.doi.org/10.1093/pcp/pcr100

Hsiao, Y.Y., Huang, T.H., Fu, C.H., Huang, S.C., Chen, Y.J., Huang, Y.M., Chen, W.H., Tsai, W.C., and Chen, H.H. (2013). Transcriptomic analysis of floral organs from *Phalaenopsis* orchid by using oligonucleotide microarray. Gene *518*, 91–100. http://dx.doi.org/10.1016/j.gene.2012.11.069

Hsiao, Y.Y., Tsai, W.C., Kuoh, C.S., Huang, T.H., Wang, H.C., Leu, Y.L., Wu, T.S., Chen, W.H., and Chen, H.H. (2006). Comparison of transcripts in *Phalaenopsis bellina* and *Phalaenopsis equestris* (Orchidaceae) flowers to deduce the monoterpene biosynthesis pathway. BMC Plant Biol. *6*, 18. http://dx.doi.org/10.1186/1471-2229-6-14

Illumina. (2016). An introduction to next-generation sequencing technology. (www.illumina.com/content/dam/illumina-marketing/documents/products/illumina_sequencing_introduction.pdf).

International Human Genome Sequencing Consortium. (2004). Finishing the euchromatic sequence of the human genome. Nature *431*, 931–945. http://dx.doi.org/10.1016/j.prrv.2009.04.003

International Rice Genome Sequencing Project. (2005). The map-based sequence of the rice genome. Nature *436*, 793–800. http://dx.doi.org/10.1038/nature03895

Jheng, C.F., Chen, T.C., Lin, J.Y., Chen, T.C., Wu, W.L., and Chang, C.C. (2012). The comparative chloroplast genomic analysis of photosynthetic orchids and developing DNA markers to distinguish *Phalaenopsis* orchids. Plant Sci. *190*, 62–73. http://dx.doi.org/10.1016/j.plantsci.2012.04.001

Kasianowicz, J.J., Brandin, E., Branton, D., and Deamer, D.W. (1996). Characterization of individual polynucleotide molecules using a membrane channel. Proc. Natl. Acad. Sci. U.S.A. *93*, 13770–13773.

Laver, T., Harrison, J., O'Neill, P.A., Moore, K., Farbos, A., Paszkiewicz, K., and Studholme, D.J. (2015). Assessing the performance of the Oxford Nanopore Technologies MinION. Biomol. Detect. Quantif. 3, 1–8. http://dx.doi.org/10.1016/j.bdq.2015.02.001

Leitch, I.J., Kahandawala, I., Suda, J., Hanson, L., Ingrouille, M.J., Chase, M.W., and Fay, M.F. (2009). Genome size diversity in orchids: consequences and evolution. Ann. Bot. *104*, 469–481. http://dx.doi.org/10.1093/aob/mcp003

Levene, M.J., Korlach, J., Turner, S.W., Foquet, M., Craighead, H.G., and Webb, W.W. (2003). Zero-mode waveguides for single-molecule analysis at high concentrations. Science *299*, 682–686. http://dx.doi.org/10.1126/science.1079700

Li, R., Hsieh, C.L., Young, A., Zhang, Z., Ren, X., and Zhao, Z. (2015a). Illumina Synthetic Long Read Sequencing Allows Recovery of Missing Sequences even in the 'Finished' *C. elegans* Genome. Sci. Rep. 5, 10814. http://dx.doi.org/10.1038/srep10814

Li, D.M., Zhao, C.Y., Liu, X.R., Liu, X.F., Lin, Y.J., Liu, J.W., Chen, H.M., and Lŭ, F.B. (2015b). *De novo* assembly and characterization of the root transcriptome and development of simple sequence repeat markers in *Paphiopedilum concolor*. Genet. Mol. Res. *14*, 6189–6201. http://dx.doi.org/10.4238/2015.June.9.5

Li, X., Jin, F., Jin, L., Jackson, A., Ma, X., Shu, X., Wu, D., and Jin, G. (2015c). Characterization and comparative profiling of the small RNA transcriptomes in two phases of flowering in *Cymbidium ensifolium*. BMC Genomics *16*, 622. http://dx.doi.org/10.1186/s12864-015-1764-1

Li, X., Luo, J., Yan, T., Xiang, L., Jin, F., Qin, D., Sun, C., and Xie, M. (2013). Deep sequencing-based analysis of the *Cymbidium ensifolium* floral transcriptome. PLOS ONE *8*, e85480. http://dx.doi.org/10.1371/journal.pone.0085480

Lin, C.S., Chen, J.J., Huang, Y.T., Hsu, C.T., Lu, H.C., Chou, M.L., Chen, L.C., Ou, C.I., Liao, D.C., Yeh, Y.Y., *et al.* (2013). Catalog of *Erycina pusilla* miRNA and categorization of reproductive phase-related miRNAs and their target gene families. Plant. Mol. Biol. *82*, 193–204. http://dx.doi.org/10.1007/s11103-013-0055-y

Lin, Y.F., Chen, Y.Y., Hsiao, Y.Y., Shen, C.Y., Hsu, J.L., Yeh, C.M., Mitsuda, N., Ohme-Takagi, M., Liu, Z.J., and Tsai, W.C. (2016). Genome-wide identification and characterization of TCP genes involved in ovule development of *Phalaenopsis equestris*. J. Exp. Bot. *67*, 5051–5066. http://dx.doi.org/10.1093/jxb/erw273

Liu, L., Li, Y., Li, S., Hu, N., He, Y., Pong, R., Lin, D., Lu, L., and Law, M. (2012). Comparison of next-generation sequencing systems. J. Biomed. Biotechnol. *2012*, 251364. http://dx.doi.org/10.1155/2012/251364

Liu, S.S., Chen, J., Li, S.C., Zeng, X., Meng, Z.X., and Guo, S.X. (2015). Comparative transcriptome analysis of genes involved in GA-GID1-DELLA regulatory module in symbiotic and asymbiotic seed germination of *Anoectochilus roxburghii* (Wall.) Lindl. (Orchidaceae). Int. J. Mol. Sci. *16*, 30190–30203. http://dx.doi.org/10.3390/ijms161226224

Luo, J., Hou, B.W., Niu, Z.T., Liu, W., Xue, Q.Y., and Ding, X.Y. (2014). Comparative chloroplast genomes of photosynthetic orchids: insights into evolution of the orchidaceae and development of molecular markers for phylogenetic applications. PLOS ONE *9*, e99016. http://dx.doi.org/10.1371/journal.pone.0099016

McCarthy, A. (2010). Third-generation DNA sequencing: pacific biosciences' single molecule real time technology. Chem. Biol. *17*, 675–676. http://dx.doi.org/10.1016/j.chembiol.2010.07.004

Margulies, M., Egholm, M., Altman, W.E., Attiya, S., Bader, J.S., Bemben, L.A., Berka, J., Braverman, M.S., Chen, Y.J., Chen, Z., *et al.* (2005). Genome sequencing in microfabricated high-density picolitre reactors. Nature *437*, 376–380.

Meng, Y., Yu, D., Xue, J., Lu, J., Feng, S., Shen, C., and Wang, H. (2016). A transcriptome-wide, organ-specific regulatory map of *Dendrobium officinale*, an important traditional Chinese orchid herb. Sci. Rep. *6*, 18864. http://dx.doi.org/10.1038/srep18864

Niu, S.C., Xu, Q., Zhang, G.Q., Zhang, Y.Q., Tsai, W.C., Hsu, J.L., Liang, C.K., Luo, Y.B., and Liu, Z.J. (2016). *De novo* transcriptome assembly databases for the butterfly orchid *Phalaenopsis equestris*. Sci. Data 3, 160083. http://dx.doi.org/10.1038/sdata.2016.83

Otero, J.T., and Flanagan, N.S. (2006). Orchid diversity – beyond deception. Trends Ecol. Evol. *21*, 64–65.

Ozsolak, F., Platt, A.R., Jones, D.R., Reifenberger, J.G., Sass, L.E., McInerney, P., Thompson, J.F., Bowers, J., Jarosz, M., and Milos, P.M. (2009). Direct RNA sequencing. Nature *461*, 814–818. http://dx.doi.org/10.1038/nature08390

Ozsolak, F. (2016). Attomole-level Genomics with Single-molecule Direct DNA, cDNA and RNA Sequencing Technologies. Curr. Issues. Mol. Biol. *18*, 43–48.

Pan, I.C., Liao, D.C., Wu, F.H., Daniell, H., Singh, N.D., Chang, C., Shih, M.C., Chan, M.T., and Lin, C.S. (2012). Complete chloroplast genome sequence of an orchid model plant candidate: *Erycina pusilla* apply in tropical *Oncidium* breeding. PLOS ONE 7, e34738. http://dx.doi.org/10.1371/journal.pone.0034738

Peakall, R. (2007). Speciation in the Orchidaceae: confronting the challenges. Mol. Ecol. *16*, 2834–2837.

Perotto, S., Rodda, M., Benetti, A., Sillo, F., Ercole, E., Rodda, M., Girlanda, M., Murat, C., and Balestrini, R. (2014). Gene expression in mycorrhizal orchid protocorms suggests a friendly plant-fungus relationship. Planta *239*, 1337–1349. http://dx.doi.org/10.1007/s00425-014-2062-x

Pushkarev, D., Neff, N.F., and Quake, S.R. (2009). Single-molecule sequencing of an individual human genome. Nat. Biotechnol. 27, 847–850. http://dx.doi.org/10.1038/nbt.1561

Ramírez, S.R., Gravendeel, B., Singer, R.B., Marshall, C.R., and Pierce, N.E. (2007). Dating the origin of the Orchidaceae from a fossil orchid with its pollinator. Nature *448*, 1042–1045.

Rao, X., Krom, N., Tang, Y., Widiez, T., Havkin-Frenkel, D., Belanger, F.C., Dixon, R.A., and Chen, F. (2014). A deep transcriptomic analysis of pod development in the vanilla orchid (*Vanilla planifolia*). BMC Genomics *15*, 964. http://dx.doi.org/10.1186/1471-2164-15-964

Rothberg, J.M., Hinz, W., Rearick, T.M., Schultz, J., Mileski, W., Davey, M., Leamon, J.H., Johnson, K., Milgrew, M.J., Edwards, M., *et al.* (2011). An integrated semiconductor device enabling non-optical genome sequencing. Nature *475*, 348–352. http://dx.doi.org/10.1038/nature10242

Sanger, F., Air, G.M., Barrell, B.G., Brown, N.L., Coulson, A.R., Fiddes, J.C., Hutchison III, C.A., Slocombe, P.M., and Smith, M. (1977a). Nucleotide sequence of bacteriophage phi X174 DNA. Nature *265*, 687–695. http://dx.doi.org/10.1038/265687a0

Sanger, F., Nicklen, S., and Coulson, A.R. (1977b). DNA sequencing with chain-terminating inhibitors. Proc. Natl. Acad. Sci. U.S.A. *74*, 5463–5467.

Schiestl, F.P., Peakall, R., Mant, J.G., Ibarra, F., Schulz, C., Franke, S., and Francke, W. (2003). The chemistry of sexual deception in an orchid-wasp pollination system. Science *302*, 437–438. http://dx.doi.org/10.1126/science.1087835

Sedeek, K.E., Qi, W., Schauer, M.A., Gupta, A.K., Poveda, L., Xu, S., Liu, Z.J., Grossniklaus, U., Schiestl, F.P., and Schlüter, P.M. (2013). Transcriptome and proteome data reveal candidate genes for pollinator attraction in sexually deceptive orchids. PLOS ONE *8*, e64621. http://dx.doi.org/10.1371/journal.pone.0064621

Shendure, J., Porreca, G.J., Reppas, N.B., Lin, X., McCutcheon, J.P., Rosenbaum, A.M., Wang, M.D.,

Zhang, K., Mitra, R.D., and Church, G.M. (2005). Accurate multiplex polony sequencing of an evolved bacterial genome. Science *309*, 1728–1732.

Silvera, K., Santiago, L.S., Cushman, J.C., and Winter, K. (2009). Crassulacean acid metabolism and epiphytism linked to adaptive radiations in the Orchidaceae. Plant Physiol. *149*, 1838–1847. http://dx.doi.org/10.1104/pp.108.132555

Su, C.L., Chao, Y.T., Alex Chang, Y.C., Chen, W.C., Chen, C.Y., Lee, A.Y., Hwa, K.T., and Shih, M.C. (2011). De novo assembly of expressed transcripts and global analysis of the *Phalaenopsis aphrodite* transcriptome. Plant. Cell. Physiol. *52*, 1501–1514. http://dx.doi.org/10.1093/pcp/pcr097

Tan, J., Wang, H.L., and Yeh, K.W. (2005). Analysis of organ-specific, expressed genes in *Oncidium* orchid by subtractive expressed sequence tags library. Biotechnol. Lett *27*, 1517–1528. http://dx.doi.org/10.1007/s10529-005-1468-8

Teh, S.L., Chan, W.S., Abdullah, J.O., and Namasivayam, P. (2011). Development of expressed sequence tag resources for *Vanda* Mimi Palmer and data mining for EST-SSR. Mol. Biol. Rep. *38*, 3903–3909. http://dx.doi.org/10.1007/s11033-010-0506-3

The *Arabidopsis* Genome Initiative. (2000). Analysis of the genome sequence of the flowering plant *Arabidopsis thaliana*. Nature *408*, 796–815. http://dx.doi.org/10.1038/35048692

Thompson, J.F., and Steinmann, K.E. (2010). Single molecule sequencing with a HeliScope genetic analysis system. Curr. Protoc. Mol. Biol. *Chapter 7*, Unit7.10. http://dx.doi.org/10.1002/0471142727.mb0710s92

Tremblay, R.L., Ackerman, J.D., Zimmerman, J.K., and Calvo, R.N. (2005). Variation in sexual reproduction in orchids and its evolutionary consequence: a spasmodic journey to diversification. Biol. J. Linn. Soc. *84*, 1–54. http://dx.doi.org/10.1111/j.1095-8312.2004.00400.x

Tsai, C.C., Wu, K.M., Chiang, T.Y., Huang, C.Y., Chou, C.H., Li, S.J., and Chiang, Y.C. (2016). Comparative transcriptome analysis of *Gastrodia elata* (Orchidaceae) in response to fungus symbiosis to identify gastrodin biosynthesis-related genes. BMC Genomics *17*, 212. http://dx.doi.org/10.1186/s12864-016-2508-6

Tsai, W.C., Fu, C.H., Hsiao, Y.Y., Huang, Y.M., Chen, L.J., Wang, M., Liu, Z.J., and Chen, H.H. (2013). OrchidBase 2.0: comprehensive collection of Orchidaceae floral transcriptomes. Plant. Cell. Physiol. *54*, e7. http://dx.doi.org/10.1093/pcp/pcs187

Tsai, W.C., Hsiao, Y.Y., Lee, S.H., Tung, C.W., Wang, D.P., Wang, H.C., Chen, W.H., and Chen, H.H. (2006). Expression analysis of the ESTs derived from the flower buds of *Phalaenopsis equestris*. Plant Sci. *170*, 426–432. http://dx.doi.org/10.1016/j.plantsci.2005.08.029

Tsai, W.C., Pan, Z.J., Su, Y.Y., and Liu, Z.J. (2014). New insight into the regulation of floral morphogenesis. Int. Rev. Cell. Mol. Biol. *311*, 157–182. http://dx.doi.org/10.1016/B978-0-12-800179-0.00003-9

Valouev, A., Ichikawa, J., Tonthat, T., Stuart, J., Ranade, S., Peckham, H., Zeng, K., Malek, J.A., Costa, G., McKernan, K., *et al.* (2008). A high-resolution, nucleosome position map of *C. elegans* reveals a lack of universal sequence-dictated positioning. Genome Res. *18*, 1051–1063. http://dx.doi.org/10.1101/gr.076463.108

van Dijk, E.L., Auger, H., Jaszczyszyn, Y., and Thermes, C. (2014). Ten years of next-generation sequencing technology. Trends Genet. *30*, 418–426. http://dx.doi.org/10.1016/j.tig.2014.07.001

Voelkerding, K.V., Dames, S.A., and Durtschi, J.D. (2009). Next-generation sequencing: from basic research to diagnostics. Clin. Chem. *55*, 641–658. http://dx.doi.org/10.1373/clinchem.2008.112789

Wu, F.H., Chan, M.T., Liao, D.C., Hsu, C.T., Lee, Y.W., Daniell, H., Duvall, M.R., and Lin, C.S. (2010). Complete chloroplast genome of *Oncidium* Gower Ramsey and evaluation of molecular markers for identification and breeding in Oncidiinae. BMC Plant Biol. *10*, 68. http://dx.doi.org/10.1186/1471-2229-10-68

Xu, C., Zeng, B., Huang, J., Huang, W., and Liu, Y. (2015). Genome-wide transcriptome and expression profile analysis of *Phalaenopsis* during explant browning. PLOS ONE *10*, e0123356. http://dx.doi.org/10.1371/journal.pone.0123356

Yan, L., Wang, X., Liu, H., Tian, Y., Lian, J., Yang, R., Hao, S., Wang, X., Yang, S., Li, Q., *et al.* (2015). The genome of *Dendrobium officinale* illuminates the biology of the important traditional chinese orchid herb. Mol. Plant *8*, 922–934. http://dx.doi.org/10.1016/j.molp.2014.12.011

Yang, F., and Zhu, G. (2015). Digital gene expression analysis based on *de novo* transcriptome assembly reveals new genes associated with floral organ differentiation of the orchid plant *Cymbidium ensifolium*. PLOS ONE *10*, e0142434. http://dx.doi.org/10.1371/journal.pone.0142434

Yang, J.B., Tang, M., Li, H.T., Zhang, Z.R., and Li, D.Z. (2013). Complete chloroplast genome of the genus *Cymbidium*: lights into the species identification, phylogenetic implications and population genetic analyses. BMC Evol. Biol. *13*, 84. http://dx.doi.org/10.1186/1471-2148-13-84

Ye, W., Shen, C.H., Lin, Y., Chen, P.J., Xu, X., Oelmüller, R., Yeh, K.W., and Lai, Z. (2014). Growth promotion-related miRNAs in *Oncidium* orchid roots colonized by the endophytic fungus *Piriformospora indica*. PLOS ONE *9*, e84920. http://dx.doi.org/10.1371/journal.pone.0084920

Yu, H., and Goh, C.J. (2001). Molecular genetics of reproductive biology in orchids. Plant Physiol. *127*, 1390–1393.

Zeng, W.H., Liao, S.C., and Chang, C.C. (2007). Identification of RNA editing sites in chloroplast transcripts of *Phalaenopsis aphrodite* and comparative analysis with those of other seed plants. Plant. Cell. Physiol. *48*, 362–368.

Zhang, G.Q., Xu, Q., Bian, C., Tsai, W.C., Yeh, C.M., Liu, K.W., Yoshida, K., Zhang, L.S., Chang, S.B., Chen, F., *et al.* (2016a). The *Dendrobium catenatum* Lindl. genome sequence provides insights into polysaccharide synthase, floral development and adaptive evolution. Sci. Rep. *6*, 19029. http://dx.doi.org/10.1038/srep19029

Zhang, J., He, C., Wu, K., Teixeira da Silva, J.A., Zeng, S., Zhang, X., Yu, Z., Xia, H., and Duan, J. (2016b). Transcriptome analysis of *Dendrobium officinale* and its application to the identification of genes associated with polysaccharide synthesis. Front. Plant Sci. *7*, 5. http://dx.doi.org/10.3389/fpls.2016.00005

Zhang, J., Wu, K., Zeng, S., Teixeira da Silva, J.A., Zhao, X., Tian, C.E., Xia, H., and Duan, J. (2013). Transcriptome analysis of *Cymbidium sinense* and its application to the identification of genes associated with floral development. BMC Genomics *14*, 279. http://dx.doi.org/10.1186/1471-2164-14-279

Zhang, L., Chen, F., Zhang, G.Q., Zhang, Y.Q., Niu, S., Xiong, J.S., Lin, Z., Cheng, Z.M., and Liu, Z.J. (2016c). Origin and mechanism of crassulacean acid metabolism in orchids as implied by comparative transcriptomics and genomics of the carbon fixation pathway. Plant J. *86*, 175–185. http://dx.doi.org/10.1111/tpj.13159

Zhao, X., Zhang, J., Chen, C., Yang, J., Zhu, H., Liu, M., and Lv, F. (2014). Deep sequencing-based comparative transcriptional profiles of *Cymbidium hybridum* roots in response to mycorrhizal and non-mycorrhizal beneficial fungi. BMC Genomics *15*, 747. http://dx.doi.org/10.1186/1471-2164-15-747

Zhu, G., Yang, F., Shi, S., Li, D., Wang, Z., Liu, H., Huang, D., and Wang, C. (2015). Transcriptome characterization of *Cymbidium sinense* 'Dharma' using 454 pyrosequencing and its application in the identification of genes associated with leaf color variation. PLOS ONE *10*, e0128592. http://dx.doi.org/10.1371/journal.pone.0128592

# Bioinformatics Resources for Plant Genomics: Opportunities and Bottlenecks in the -omics Era

4

Luca Ambrosino†, Chiara Colantuono†, Francesco Monticolo and Maria Luisa Chiusano*

Department of Agricultural Sciences, University of Naples Federico II, Portici, Italy.

*Correspondence: chiusano@unina.it
†These authors contributed equally

https://doi.org/10.21775/9781910190654.05

## Abstract
The sudden exponential increase of biological data concerning genome structure and functionalities, also fostered by the advent of next-generation sequencing technologies, while expanding the opportunity to highlight still uncovered molecular aspects, challenges bioinformatics in several respects. Data management, processing, updating, dissemination and integration are the major areas of concern.

The rapid increase in various omics technologies causes two major issues, which may even appear contrasting: the dissemination of poorly curated datasets, still in the form of raw collections or preliminary draft results, and the fast updating of information that, as a consequence, affects the establishment of stable reliable resources. These issues are mainly caused by the lower rate of bioinformatics in extracting added value information from the large number of data, when compared to the faster technologies involved in data production.

This review describes main bioinformatics resources for plants genomics to underline the heterogeneity of the available collections, coherent with the multifaceted complexity of plant sciences. It aims to provide an in-depth report highlighting bottlenecks that may significantly affect a fluent progress in the field and attempts to suggest possible solutions to the various issues.

## The impact of plant genomics
The deciphering of the molecular mechanisms which determine plant diversity and adaptation to different environments and the impact that this knowledge may provide for sustainable productivity in the food industry, energy production or biotechnological applications (Blanchfield, 2004; Ma *et al.*, 2003; Wilson and Roberts, 2014; Yuan *et al.*, 2008) can be strongly supported by structural and functional genomics. These are among the main reasons why the scientific community is increasingly demanding fully sequenced plant genomes with the aim to exploit the advantages and the opportunities that genomics may offer in plant sciences. The release of the genome of *Arabidopsis thaliana* (The Arabidopsis Genome Initiative, 2000), represents a milestone in the field, making this relevant species in plant biology a model reference for plant genomics (Becker *et al.*, 2011; Gan *et al.*, 2011; Kim *et al.*, 2007). Nevertheless, the expanding of genomics has further highlighted the multifaceted complexity of plants, where complex genomes, often larger than those of mammals (Mayer *et al.*, 2012), present peculiar organizations and extensive duplications and reshuffling (Bowers *et al.*, 2003; Cui *et al.*, 2006; Flagel and Wendel, 2009; Hughes, 2005; Jiao *et al.*, 2012; Lynch and Conery, 2000; Maere *et al.*, 2005; Magadum *et al.*, 2013), revealing complex evolutionary histories

often involving poliploidization events followed by diploidization and gene reduction (Blanc *et al.*, 2003; Jaillon *et al.*, 2007; Jiao *et al.*, 2011; Moniz de Sa and Drouin, 1996; Wolfe, 2001).

The recent establishment of novel, low-cost and easily accessible technologies is further encouraging the increase in genome sequencing efforts. Indeed, the recent introduction of next-generation sequencing (NGS) technologies, which replaced the conventional Sanger strategy (Sanger *et al.*, 1977), deeply reshaped the omics research areas (Bateman and Quackenbush, 2009; Esposito *et al.*, 2016; Wang *et al.*, 2015), contributing to several species-specific efforts usually curated by dedicated consortia (Fig. 4.1). The result is usually in the form of assembled draft genome sequences with quality levels established by the consortia requirements and by funding opportunities. It is common practice to accompany new genome assemblies with consortium curated gene annotations, which benefit from specific competences from the interested scientific community that is contributing to the consortium. Moreover, dedicated web resources are usually made available to convey the Consortia efforts into a unified framework. These efforts are consistently contributing to the research of some of the most relevant crops (Brenchley *et al.*, 2012; Choulet *et al.*, 2010; Jia *et al.*, 2013; Ling *et al.*, 2013; Nystedt *et al.*, 2013; Wang *et al.*, 2011). Besides, the reference database for the model plant *Arabidopsis thaliana* represented by The Arabidopsis Information Resource (TAIR) (Lamesch *et al.*, 2012), numerous consortia gave rise to species-specific platforms including results from the genome sequencing and/or gene annotations efforts. As an example, among the major efforts for relevant crops, the MSU Rice Genome Annotation Project (Kawahara *et al.*, 2013), funded by the National Science Foundation, provides sequence and annotation data for the rice genome, which was released as a first version in 2002 (Goff *et al.*, 2002). The grapevine genome was released by an Italian-French consortium and made available through two main websites maintained by its members: the Genoscope Institute website (Jaillon *et al.*, 2007) (www.genoscope.cns.fr/spip/) and the CRIBI website (http://genomes.cribi.unipd.it/grape/). CRIBI recently released an updated gene annotation (Vitulo *et al.*, 2014) on the same genome version. Worthy to note that this novel annotation version is not reported in the Genoscope website. On the other hand, the genome analysis of a heterozygous grapevine variety was also published (Velasco *et al.*, 2007), though the genome assembly was never publicly released. The *Sorghum bicolor* genome sequencing and annotation (Paterson *et al.*, 2009) was released by the Joint Genome

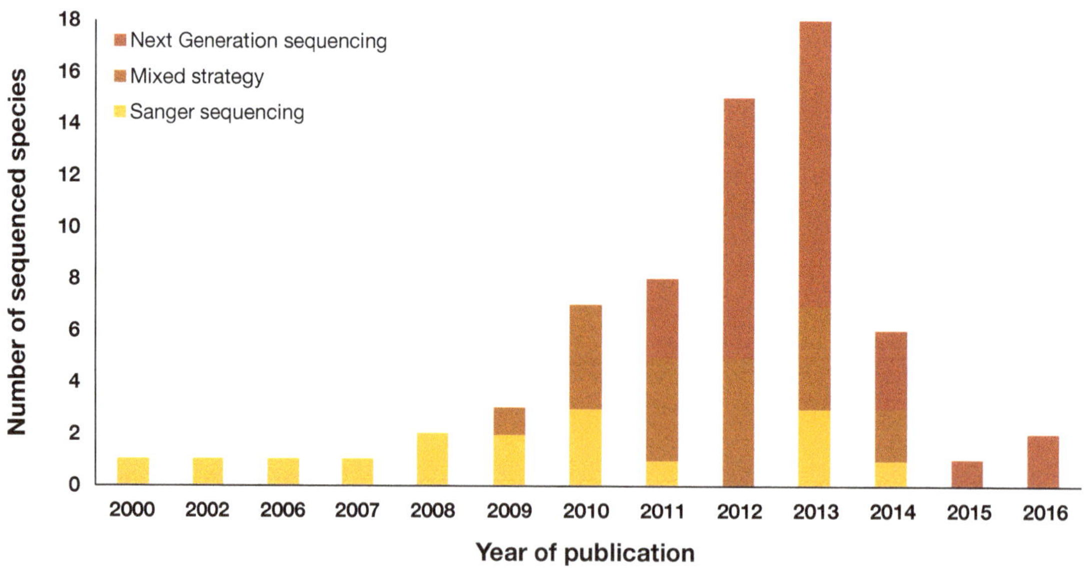


**Figure 4.1** Number of plant genomes sequenced from 2000 (publication year of *Arabidopsis thaliana*) until today. The sequencing strategy is also highlighted.

Institute (JGI), which maintains a protected website. Thanks to the Maize Genome Sequencing Project, funded by the National Science Foundation, the complete genome sequence of *Zea mays* cv. B73 (Schnable *et al.*, 2009) was made available in MaizeGDB (Andorf *et al.*, 2016), and included in the collection available in Gramene (Tello-Ruiz *et al.*, 2016). The Potato Genome Sequencing Consortium (PGSC) released the first draft of the potato genome (Xu *et al.*, 2011), that was made available on the SpudDB website (Hirsch *et al.*, 2014) and in the Solanaceae Genomics Network collection. The International Tomato Genome Sequencing Project and the International Tomato Annotation Group (ITAG) have defined the sequence and the annotation of the tomato genome (The Tomato Genome Consortium, 2012), respectively, both released and maintained by the leading website of the Solanaceae Genomics Networks (SGN) (Fernandez-Pozo *et al.*, 2015), though offered also through several parallel dedicated platforms (Chiusano *et al.*, 2008; Hirsch *et al.*, 2014). The same consortium also defined the genome of *Solanum pimpinellifolium*, a wild species of the domesticated tomato, available in the form of genome scaffolds in the SGN platform (https://solgenomics.net/organism/Solanum_pimpinellifolium/genome). An international group of scientists from Korea, Israel and the USA sequenced and annotated the hot pepper genome (Kim *et al.*, 2014), which is available in the Pepper Genome Database (http://peppersequence.genomics.cn/page/species/index.jsp) and also in the SGN platform.

## From genome structure to function

The rise of several independent projects for the sequencing of diverse plant genomes offers hints to understand their organization and functionality, revealing unknown molecular information and supporting scientific knowledge and the technological transfer of useful information (Esposito *et al.*, 2016). The understanding of genome functionalities, previously mainly supported by EST sequencing (Blair *et al.*, 2011; Izzah *et al.*, 2014; Wang *et al.*, 2005) and microarray technologies (Bülow *et al.*, 2007; Mukherjee *et al.*, 2005; Page and Coulibaly, 2008), is being unexpectedly favoured by the evolution of parallel -omics efforts providing enriched information to support the definition of genome structure organization and the investigation on its functionality (Bateman and Quackenbush, 2009). To this aim, advances in transcriptomics, epigenomics, proteomics and metabolomics, are consistently contributing novel data sources, useful to unravel hidden molecular aspects.

In particular, the amount of information provided by novel RNA sequencing technologies (RNA-seq), beyond contributing a deeper and expanded overview of gene expression levels, even for poorly expressed genes, supports their function profiling in different tissues and developmental stages, as well as in stress and pathological conditions. This improves the gene annotation, the identification of variants and the definition of expression patterns, useful to highlight specificities and peculiarities of the control and regulation of gene expression, providing an enriched snapshot of the transcriptome plasticity. Moreover, novel applications from NGS technologies in plants support the characterization of small and microRNAs, of genome methylated regions or of chromatin organization, also depicting protein binding sites (Becker *et al.*, 2003; Bokszczanin *et al.*, 2015; Horner *et al.*, 2010; MacLean *et al.*, 2009; Mardis, 2008a,b, 2009; Morozova and Marra, 2008a,b; Morrissy *et al.*, 2009; Schuster, 2008). These approaches (Adams *et al.*, 1991; Brenner *et al.*, 2000; Kodzius *et al.*, 2006; Velculescu *et al.*, 1995) generally accompanied the flourishing of genome sequencing projects of reference plant species, such as *Arabidopsis thaliana* (The Arabidopsis Genome Initiative, 2000). Similar efforts on other model or non-model species of agricultural interest soon followed. Some examples are rice (Goff *et al.*, 2002), grapevine (Jaillon *et al.*, 2007), maize (Schnable *et al.*, 2009), potato (Xu *et al.*, 2011), tomato (The Tomato Genome Consortium, 2012) genome sequencing which expanded the number of plant genomes currently available to the scientific community with enriched information favoured by alternative -omics approaches from transcriptomics, proteomics, epigenomics and metagenomics projects (Esposito *et al.*, 2016).

## Bioinformatics data resources for plant genomics

The -omics efforts must be necessarily accompanied by bioinformatics (Schatz *et al.*, 2012) to translate data into knowledge. However, bioinformatics, followed by human curated result interpretations, represent the slowest phases of -omics research, when compared to the fast sequencing rate. The management, the analysis, the integration and the comparison of the large data number under production are, indeed, a major challenge in bioinformatics and in plant genomics as well (De Luca *et al.*, 2012). Computational methods and suitable bioinformatics are being evolving to accompany the evolution of the technologies and to face the need of added-value information, continuously improving and adapting to the increase of the amount of biological data. However, further novel solutions are still required (Esposito *et al.*, 2016).

Bioinformatics has been always fundamental also for the organization and the integration of molecular data collections, to offer access and suitable data views to all the interested scientific community, even to non-experts in -omics data management. Indeed, the release of integrated molecular information through dedicated web-based services or platforms has been always pushing the evolution of -omics research representing, since the beginning, an essential source of information in science. This is why general reference bioinformatics resources were attempted since the initial production of molecular information (Dayhoff *et al.*, 1965) and policies to unify and share the data were established to benefit the whole scientific community (Brunak *et al.*, 2002). Since then, scientists have been well aware of the responsibility and the opportunities offered by the release of their published data in general reference databases (Cochrane *et al.*, 2016; Brunak *et al.*, 2002). Indeed, conveying comprehensive collections in a common computational platform establishes references for all scientists for a one-stop shop, independently from the main scientific interests. Specialized, secondary resources, on the other hand, aim to offer curated (UniProt Consortium, 2015; Kanehisa and Goto, 2000) and dedicated collections (Goodstein *et al.*, 2012; Dong *et al.*, 2004). Table 4.1 summarizes some of

**Table 4.1** List of the major bioinformatics resources available for plant genomics. Description and website are specified. General databases include also collections from non-plant species

| Database | Description | Website |
|---|---|---|
| **General** | | |
| INSDC | Unified DDBJ, EMBL-EBI and NCBI repository | www.insdc.org/ |
| UniProt | Database of functional annotated protein sequences | www.uniprot.org/ |
| Protein Data Bank | 3D structure of proteins and nucleic acids | www.rcsb.org/pdb/home/home.do |
| RFAM | RNA family collections | http://rfam.xfam.org/ |
| Gene Ontology | Database of ontologies and gene annotations | http://geneontology.org/page/go-database |
| KEGG | Metabolic pathway database | www.kegg.jp/ |
| EggNOG | Comparative genomics | http://eggnog.embl.de/version_4.0.beta/ |
| InParanoid | Comparative genomics | http://inparanoid.sbc.su.se/cgi-bin/index.cgi |
| **Plant specific** | | |
| Ensembl Plants | Plant genomics database | http://plants.ensembl.org/index.html |
| Phytozome | Plant genomics database | www.phytozome.net/ |
| PlantGDB | Plant genomics database | www.plantgdb.org/ |
| Plant Metabolic Network | Plant metabolic pathway database | www.plantcyc.org/ |
| Plant Reactome | Plant metabolic pathway database | http://plantreactome.gramene.org/ |
| GreenPhyl | Plant comparative genomics | www.greenphyl.org/cgi-bin/index.cgi |
| Plaza | Plant comparative genomics | http://bioinformatics.psb.ugent.be/plaza/ |

the major resources, general or specialized, offering comprehensive collections for plant genomics.

The International Nucleotide Sequence Database Collaboration (INSDC) holds raw sequence data, accommodating also experimental design details and NGS reads (Cochrane *et al.*, 2016). Ensembl Plants is part of the Ensembl genome project, which includes genome sequences, protein annotations, transcriptional data, genetic variation and comparative results from different taxa (Kersey *et al.*, 2016). It contains reference genome assemblies from 33 plant species (Bolser *et al.*, 2016). Phytozome is a comparative platform for plants. It provides access to sequences and functional annotations of complete plant genomes (currently 65), and a view of the evolutionary history of every plant gene (Goodstein *et al.*, 2012). PlantGDB is a reference genomic database for plants (Dong *et al.*, 2004), including 50 plant species. The Universal Protein Resource (UniProt) is a general reference platform of protein sequences and their annotation. UniProt is divided in two sections: the Reviewed (Swiss-Prot) manually annotated database, in which proteins and their information are manually curated based also on literature data, and the Unreviewed (TrEMBL) computationally analysed database, containing automatically annotated proteins from general nucleotide databases (UniProt Consortium, 2015). The Protein Data Bank (PDB) is a database of tridimensional structure data. This database stores X-ray crystal structures, nuclear magnetic resonance (NMR) structures, cryo-electron microscopy and theoretical modelling (Berman *et al.*, 2000). Rfam is a collection of non-coding RNA families depicted by manually curated sequence alignments, annotation and consensus of secondary structures (Nawrocki *et al.*, 2015). Gene Ontology Consortium (GOC) is a project dedicated to the definition of consistent descriptions of gene products, incorporating many collections from plant, animal and microbial genomes (Ashburner *et al.*, 2000). KEGG (Kyoto Encyclopaedia of Genes and Genomes) is a database for systematic analysis of biological systems (pathways included) (Kanehisa and Goto, 2000). Plant Metabolic Network (PMN) is a database of metabolic pathways for plants (Dreher, 2014). PMN currently groups one multispecies reference database called PlantCyc and 22 species/taxon-specific databases. Plant Reactome is a database of pathways and reactions from plants (Fabregat *et al.*, 2016). It supplies molecular details of DNA replication, metabolism, signal transduction, gene expression, diseases, transmembrane transport of small molecules and vesicle-mediated transport. EggNOG is a database of orthologous groups of proteins from different taxonomic levels with functional annotations from 20 plant species. Furthermore, it provides scaffolds for quickly mapping novel sequences to orthologous groups based on HMM profiles (Huerta-Cepas *et al.*, 2016). InParanoid is a collection of pairwise orthologue groups including 20 plant species (O'Brien *et al.*, 2005). GreenPhyl is a web-based resource for comparative and functional genomics in plants (Rouard *et al.*, 2011), containing a catalogue of gene families based on gene predictions, covering a broad taxonomy of green plants. Plaza is a public resource containing 37 plant species genomes (Proost *et al.*, 2015), with the aim to facilitate inspection on structural and functional annotations, gene families, protein domains, and detailed information about genome organization and phylogenetic trees. All these collections resemble the amount of resources and facilities today available for plant genomics. All of them are endowed with peculiarities and specificities that should be appropriately addressed by interested users for an appropriate exploitation.

## NGS-based resources

Next-generation sequencing, though fast and cheaper, is computationally expensive. Indeed, it requires suitable and dedicated software and platforms able to manage, organize, analyse and disseminate huge quantities of short nucleotide reads, reaching commonly at least one gigabase per analytical run (Bateman and Quackenbush, 2009; Horner *et al.*, 2010; Tang and Zhao, 2015; Yang *et al.*, 2009). Data management and -omics data integration has always represented a challenge for bioinformatics (Benson *et al.*, 2000; Bita *et al.*, 2011; Chiusano *et al.*, 2008; Dong *et al.*, 2004; Edgar *et al.*, 2002; Flicek *et al.*, 2013; Kodama *et al.*, 2012; Leinonen *et al.*, 2011; O'Leary *et al.*, 2015), and the spreading of NGS data made the challenge even harder, increasing the need for suitable storage, methods for the processing and mining, and platforms for immediate access to results from novel massive data generated by these technologies (Magi *et al.*, 2010; Wang *et al.*, 2015).

The Sequence Read Archive (SRA, available at www.ncbi.nlm.nih.gov/sra) (Kodama *et al.*, 2012), for example, is a public resource established by the NCBI with the aim to gather raw collections from NGS efforts, including data from Roche 454 GS System (Droege and Hill, 2008), Illumina Genome Analyzer (Bennett, 2004), Applied Biosystems SOLiD System (Porreca *et al.*, 2006) and Helicos Heliscope (Harris *et al.*, 2008). Another resource at the NCBI, GEO (Gene Expression Omnibus) (Edgar *et al.*, 2002), which was initially established to favour dissemination of microarray results, today also provides results from NGS-based collections, such as gene expression or methylation profiles. The 'plant' keyword was used in order to identify NGS data from plants from both resources (Table 4.2), resulting in 131,207 matches in SRA and 1829 matches in GEO. The table shows that the results in GEO are far distant from representing the whole NGS collections publicly available in SRA. This highlights the gap in general resources to access results from NGS approaches. Indeed, the inestimable source of information that transcriptomics, epigenomics and metagenomics (Esposito *et al.*, 2016) may provide to elucidate genome organization and functionalities still needs appropriate platforms to be made available to the scientific community, though results from dedicated efforts are somewhat made available. Table 4.3 lists general NGS resources and dedicated platforms that include plant RNA-seq collections.

**Table 4.2** Number of query matches (hits) per species searching for 'plant' keyword in NCBI's SRA (www.ncbi.nlm.nih.gov/sra) and GEO (www.ncbi.nlm.nih.gov/gds). In the case of GEO, a filter for high-throughput sequencing was applied. In both cases, hits associated to non-plant species were omitted

| Database | Species | Number of hits |
|---|---|---|
| SRA | *Arabidopsis thaliana* | 9824 |
| | Soil metagenome | 7918 |
| | *Oryza sativa* | 3561 |
| | *Manihot esculenta* | 3047 |
| | *Zea mays* | 2448 |
| | *Triticum aestivum* | 2334 |
| | Plant metagenome | 2280 |
| | *Hordeum vulgare* | 2138 |
| | *Solanum lycopersicum* | 2025 |
| | *Brassica napus* | 1931 |
| | *Glycine max* | 1826 |
| | Root metagenome | 1697 |
| | *Solanum tuberosum* | 1650 |
| | *Populus trichocarpa* | 1619 |
| | *Erythranthe guttata* | 1549 |
| | *Boechera stricta* | 1529 |
| | *Miscanthus sinensis* | 1480 |
| | All other taxa | 58,234 |
| GEO | *Arabidopsis thaliana* | 508 |
| | *Zea mays* | 98 |
| | *Oryza sativa* | 88 |
| | *Glycine max* | 68 |
| | *Solanum lycopersicum* | 41 |
| | *Chlamydomonas reinhardtii* | 35 |
| | *Vitis vinifera* | 31 |
| | *Gossypium hirsutum* | 25 |
| | *Brassica rapa* | 22 |
| | *Arabidopsis lyrata* | 21 |
| | *Triticum aestivum* | 18 |
| | *Brassica napus* | 17 |
| | *Medicago truncatula* | 17 |
| | *Manihot esculenta* | 15 |
| | *Capsicum annuum* | 14 |
| | *Physcomitrella patens* | 14 |

## Comparative genomics resources

A great opportunity for genomics is represented by comparative efforts. They drive investigations on differences and similarities among species, therefore contributing to the deciphering of the evolutionary forces that shaped genomic features, beyond supporting the transfer of information from model organisms to newly sequenced genomes. The detection of orthologue genes among different species is a key approach for comparative analyses (Altenhoff and Dessimoz, 2009, 2012; Altenhoff *et al.*, 2011; Ambrosino and Chiusano, 2013; Dessimoz *et al.*, 2012; Kristensen *et al.*, 2011; Trachana *et al.*, 2014). Collections of orthologues are organized in widely used comparative platforms that today include numerous plant species (Table 4.4).

It is evident from the overview provided here the high number of similar resources, including different collections and based on different approaches to define the respective results.

**Table 4.3** Summary of general and species-specific RNA-seq resources. Description of contents and links are also included

| Database | Description | Website |
|---|---|---|
| **General** | | |
| SRA (Kodama *et al*., 2012) | Sequence Read Archive; includes raw data reads | www.ncbi.nlm.nih.gov/sra |
| ENA (Leinonen *et al*., 2011) | European Nucleotide Archive; includes raw data, assembly and functional annotation | www.ebi.ac.uk/ena |
| GEO (Edgar *et al*., 2002) | Gene Expression Omnibus; includes microarray and NGS data results | www.ncbi.nlm.nih.gov/gds |
| Expression Atlas (Petryszak *et al*., 2016) | Provides information about gene expression patterns; includes both microarray and RNA-seq data | www.ebi.ac.uk/gxa/home |
| Next-Gen Sequence Databases (Nakano *et al*., 2006) | NGS databases of 19 plant species; selection of small-RNA, RNA-seq, MethylC-seq and Chip-seq for plant species are made available | https://mpss.danforthcenter.org/index.php |
| MedPlant – RNAseq Database | Sequence Read Archive; includes raw data reads | www.medplantrnaseq.org/ |
| **Species specific** | | |
| AGED (O'Rourke *et al*., 2015) | Alfalfa expression atlas database; RNA-seq of two subspecies, *Medicago sativa* ssp. *sativa* and *Medicago sativa* ssp. *falcate*. Query per gene expression and per differentially expressed genes are allowed | http://plantgrn.noble.org/AGED/index.jsp |
| Genome Database for Rosaceae (Jung *et al*., 2013) | Database dedicated to Rosaceae family; provides links to SRA | www.rosaceae.org/ |
| morexGenes | Barley gene expression levels database; includes results from RNA-seq from different tissues and developmental stages includes also microarray data | https://ics.hutton.ac.uk/morexGenes/ |
| MOROKOSHI (Makita *et al*., 2015) | Sorghum transcriptome database; includes results from different tissues and developmental stages based on 26 RNA-seq samples. Links to the raw data used are also provided | http://sorghum.riken.jp/morokoshi/Home.html |
| NexGenEx-Tom (Bostan and Chiusano, 2015) | Tomato gene expression atlas; includes RNA-seq results from different tissues and developmental stages. The platform offers expression matrix, profiles and correlations and reads mapping onto the tomato genome | http://cab.unina.it/NexGenEx-Tom |
| PvGEA (O'Rourke *et al*., 2014) | Bean database; includes gene expression profiles in different tissues and developmental stages based on 24 RNA-seq samples. Links to the raw data used are also provided | http://plantgrn.noble.org/PvGEA/ |
| Rice Gene Expression (Kawahara *et al*., 2013) | Rice database of gene expression profiles based on RNA-seq downloaded from SRA. Links to the raw data used are also provided | http://rice.plantbiology.msu.edu/expression.shtml |
| SGN (Fernandez-Pozo *et al*., 2015) | Solanaceae Genome Network; includes raw RNA-seq data for tomato and mapping of reads on the tomato genome | https://solgenomics.net/ |
| SoyBase (Severin *et al*., 2010) | Soybean genomic database; includes RNA-seq from 14 tissues. The database offers the opportunity to query for differentially expressed genes between two tissues and to search for tissue-specific gene expression. Links to the raw data used are also provided | http://soybase.org/soyseq/ |
| SpinachBase (Xu *et al*., 2015) | Spinach genomic database; provides links to RNA-seq data | www.spinachbase.org/cgi-bin/spinach/index.cgi |
| SpudDB (Hirsch *et al*., 2014) | Potato genomic database; includes raw RNA-seq data and mapping of reads on the potato genome | http://solanaceae.plantbiology.msu.edu/ |

**Table 4.3** Continued

| Database | Description | Website |
|---|---|---|
| TENOR (Kawahara *et al.*, 2016) | Rice database of expression profiles; information of cis-regulatory elements in promoter regions and co-expressed transcript based on RNA-seq data from 140 environmental stress experiments and plant hormone treatments. Links to the raw data used are also provided | http://tenor.dna.affrc.go.jp/ |
| TomExpress | Tomato database of gene expression profiles per different tissues and developmental stages, based on RNA-seq data downloaded from ENA and SRA | http://gbf.toulouse.inra.fr/tomexpress/www/welcomeTomExpress.php |
| TRAVA (Klepikova *et al.*, 2015) | *Arabidopsis thaliana* database of gene expression profiles from different tissues and developmental stages based on 79 RNA-seq samples | http://travadb.org/ |
| Vespucci (Moretto *et al.*, 2015) | Grapevine expression compendium obtained by publicly available transcriptome experiments from RNA-seq and microarray data | http://vespucci.colombos.fmach.it/ |
| WheatExp (Pearce *et al.*, 2015) | Wheat database of gene expression profiles per different tissues and developmental stages, based on RNA-seq data downloaded from ENA, SRA and GEO. Links to the raw data used are also provided | http://wheat.pw.usda.gov/WheatExp/ |

## Bottlenecks and challenges

Plant sciences are typically characterized by high heterogeneity, multiple different species, an incredible amount of crops and variants, and distinct and widespread scientific communities. Therefore, the easy accessibility to sequencing technologies drove far beyond the sequencing of reference species (Goff *et al.*, 2002; Jaillon *et al.*, 2007; Schnable *et al.*, 2009; The Arabidopsis Genome Initiative, 2000; The Tomato Genome Consortium, 2012; Xu *et al.*, 2011) and paved the way to the production of multiple genomes from variants, wild species and community-specific collections (Aflitos *et al.*, 2014; Aversano *et al.*, 2015; Ercolano *et al.*, 2014; Lam *et al.*, 2010; Weigel and Mott, 2009), giving rise to multifaceted genomics data sources.

Similar efforts are even more widespread when considering transcriptomes or other -omics approaches: physiological, stress or pathological conditions for all the possible variants enrich molecular databases of heterogeneous collections. However, these collections often need to be mapped on the genome sequences, and therefore they may suffer the drawback of not being appropriately exploitable by a reference genome sequence representing the genome of a distinct genotype. Moreover, the collections can derive from limited experimental design for contributing as suitable data sources for gene expression atlases and/or for gene co-expression analyses (Bostan and Chiusano, 2015; Di Salle *et al.*, 2016; Schmid *et al.*, 2005). These scientific trends, therefore, give rise to overwhelming data that need selection and reconciliation to contribute as consistent source of information in integrative analyses that could support structure and functional genomics.

The advent of revolutionary experimental technologies and novel computational approaches are evidently not accompanied by a comparable progress in genome characterizations of plant species. Indeed, although the attitude of the whole scientific community is being consistently affected by the interest for solving primary structures of genomes of different species, genotypes or cultivars (Aflitos *et al.*, 2014; Lam *et al.*, 2010; Weigel and Mott, 2009), generally driven by international consortiums and pushed by fast and low-cost technologies, only 10% of the genomes have been today confidently deciphered. This highlights that, despite 70 years having passed from the discovery of the DNA structure, the genomics era is still in its early stage, and extensive bioinformatics is still required in order to exploit molecular models of complex biological organisms (Esposito *et al.*, 2016).

Indeed, genome sequencing efforts handed to the scientific community an increasing number of newly sequenced plant genomes. However, several of them are still in the form of drafts with a still

**Table 4.4** Summary of the major comparative genomics platform available for plants. The number of plant species out of total species and the methods for the detection of orthologues are included

| Orthologues database | Plant species/total | Methods |
|---|---|---|
| EggNog (Huerta-Cepas *et al.*, 2016) | 20/2031 | Orthologous groups inferred by the SIMAP (Similarity MAtrix of Proteins) approach and processed by the EggNOG orthology prediction pipeline |
| | | Phylogenetic reconstruction for all groups was performed using the ETE toolkit |
| Ensembl Plants (Bolser *et al.*, 2016) | 44/44 | Comparative genomics based on protein sequences providing gene trees and orthology information |
| | | Whole genome alignments between selected genomes (based on LastZ and translated BLAT) |
| | | Syntenies calculated from genome or peptide alignments |
| | | Gene families constructed from classification of proteins |
| GreenPhyl (Rouard *et al.*, 2011) | 37/37 | Clustering performed on the protein-coding gene using TribesMCL |
| | | Phylogenetic analyses and ortholog inference based on MAFFT, PhyML and RAPGreen v54 |
| InParanoid (O'Brien *et al.*, 2005) | 20/273 | Homology detected by BLAST program |
| | | Phylogenetic tree generated by UPGMA clustering of pairwise species distances |
| OrthoMCL database (Chen *et al.*, 2006) | 11/150 | All-versus-all BLASTp of the protein sequences |
| | | Putative inparalog, ortholog and co-ortholog pairs inferred using the OrthoMCL Pairs program |
| | | MCL program to cluster the protein sequences pairs into groups |
| Plaza (Proost *et al.*, 2015) | 64/64 | Orthologous gene families (ORTHO) inferred using OrthoMCL |
| | | Tree-based orthologs (TROG) inferred using tree reconciliation of the phylogenetic tree of a gene family |
| | | Best-Hits-and-Inparalogs (BHI) inferred from Blast hits against the PLAZA protein database |
| | | Anchor points refer to gene-based colinearity between species |

preliminary gene annotation. On the other hand, others among all may evolve faster, this depending on the quality of the first release and on the available opportunities in terms of support for dedicated efforts. Both aspects, however, affect the establishment of reliable reference resources that could drive associated -omics efforts (e.g. RNA-seq, proteomics, epigenomics) and their appropriate exploitation by non-expert users.

Main reasons for poorly annotated draft genomes are related to bottlenecks from bioinformatics and human curation, when compared to the faster data production. In particular, correct assembling of large amount of complex, redundant sequence data and sufficient in-depth studies on multiple heterogeneous collections require time. This slows down the reasonable deciphering of the major information content that could enable the understanding of the intricate aspects of genome functionality. Moreover, publication of novel genome sequences is often more attractive than the care for updates of already published information. Funding agencies may also consider more appealing the sequencing of a novel genome than the expansion and the integration of information contributing to the in-depth analyses of already sequenced ones, especially when they do not represent a reference international species. As a consequence, preliminary drafts often risk to remain in the preliminary status. On the counterpart, data production is also moving too fast, affecting the updating rate and the definition of stable reference resources that could support the whole interested community (Chiusano, 2015). Table 4.5 reports the genome annotation versions for different species of agronomic interest, as they are available in some of the most popular public resources.

With the exception of *Arabidopsis* and rice, whose genomes were sequenced more than 10 years ago (Goff *et al.*, 2002; The *Arabidopsis* Genome Initiative, 2000), there is an evident non-uniformity among the data included in public resources. This limit are related not only to recently sequenced genomes, such as tomato (The Tomato Genome Consortium, 2012) and potato (Xu *et al.*, 2011), but also to older genomes releases, such as the ones of maize (Schnable *et al.*, 2009) and grapevine (Jaillon *et al.*, 2007) (Table 4.5). This may be due to slower update rates, but also to the fast release of novel versions. As an example, the tuber crop potato, published in 2011 and initially released in the form of scaffolds, has been already endowed of six different genome annotation versions related to five different genome assemblies (Table 4.6), each with a level of heterogeneity further highlighted when considering differences in the number of predicted genes and alternative transcripts from each version (Table 4.6).

In addition, the dissemination of genome sequencing results is strongly affected also by other aspects. As an example, sequenced plant genomes, such as coffee (Denoeud *et al.*, 2014), available only in a dedicated resource (http://coffee-genome.org/), tobacco (Sierro *et al.*, 2014) or other genome releases (Ercolano *et al.*, 2014; The Tomato Genome Consortium, 2012; Velasco *et al.*, 2007) are not present in any of the reference databases here considered, NCBI and Ensembl Plants included. Undoubtedly, these limits and ambiguities do not support scientists, expert or non-expert users as well. The set-up of reference comprehensive public resources, supporting user friendly investigations for the whole scientific community, represents, indeed, the key strategy to make -omics really profitable. The recent story of molecular biology underlines that reference collections and data sharing thanks to bioinformatics resources have been fundamental, representing the backbone that moulded the progress in -omics sciences we are currently living. However, the flourishing of an increasing number of public databases with heterogeneous collections, providing different genome versions associated to varied independent annotations, as well as the lack of dissemination of published data, affect the establishment of well-accepted references and undoubtedly limit scientific applications.

Different gene annotations of the same genome slow down the work of the end user, often forced to carefully investigate among the available resources to assess the reliability of a collection, this also requiring appropriate expertise. Moreover, this may compromise the reproducibility of results, since they are affected by the materials and the methodologies employed (Di Salle *et al.*, 2016). As an example, well-known platforms for comparative genomics are based on different gene annotation versions from the same plant species (Fig. 4.2) and, as a consequence, on different results, compromising comparability among the different methodologies employed (as reported in Table 4.4) and reliability of analyses pending from the different efforts. Moreover, the heterogeneity of data, the lack of coordination to fix methodologies and/or collections, as well as the not clearly declared obsolescence of published resources, result in diversified information that limit non-expert users and do not facilitate uniform and reproducible scientific approaches. As an example, the presence of too many resources available in support of gene expression analyses for the reference species *A. thaliana* misleads the users also with discordant results (as reviewed in Di Salle *et al.*, 2016). Moreover, the establishment of platforms including species-specific selected collections of expression data is spreading for several plant species (Table 4.3), but these independent efforts, not always accompanied by evident coordination and international support, will not produce, presumably, stable, long term impact in plant sciences.

One last dramatic trend to consider, moreover, is the poor crosslinking of plant genome results with general databases worldwide recognized as well established resources. Dedicated annotations from independent consortium are precious, since they benefit from specific knowledge that typically comes from the scientific community supporting the consortium. However, many of the consortia results are not cross-linked with the general efforts undertaken by reference databases. As an example,

**Table 4.5** List of plant species of relevant agronomical interest, sorted by the year of publication of their genomes. The annotation version available for these species in different reference databases is also reported

| Species | Year of publication | Annotation version | | | | | | |
|---|---|---|---|---|---|---|---|---|
| | | NCBI | Ensembl Plants | PlantGDB | Phytozome | Plaza | Eggnog | KEGG |
| *Arabidopsis* | 2000 | TAIR 10 | TAIR 10 | TAIR 10 | TAIR 10 | TAIR 10 | TAIR 10 | TAIR 10 |
| Rice | 2002 | RGAP 7 | RGAP 7 | RGAP 7 | RGAP 7 | RGAP 7 | RGAP 7 | RGAP 7 |
| Grapevine | 2007 | GCF_000003745.3 | V1 Cribi | V2 Genoscope | V2 Genoscope | V2 Genoscope | V1 Cribi | GCF_000003745.3 |
| Maize | 2009 | B73_RefGen_v3 | B73_RefGen_v4 | B73_RefGen_v2 | B73_RefGen_v3 | B73_RefGen_v4 | B73_RefGen_v4 | B73_RefGen_v3 |
| Potato | 2011 | GCF_000226075.1 | GCF_000226075.1 | PGSC v.3 2.1.10 | PGSC v. 3.4 | iTAG v. 1 | GCF_000226075.1 | GCF_000226075.1 |
| Tomato | 2012 | GCF_000188115.3 | iTAG v. 2.4 | NA | iTAG v. 2.3 | iTAG v. 2.3 | iTAG v. 2.3 | GCF_000188115.3 |
| Sweet orange | 2013 | GCF_000317415.1 | NA | NA | JGI v1 | JGI v1 | NA | GCF_000317415.1 |
| Pepper | 2014 | GCF_000710875.1 | NA | NA | NA | NA | NA | NA |
| Amborella | 2013 | GCF_000471905.1 | GCF_000471905.1 | NA | AmTr_v_0.10 | AmTr_v_0.10 | NA | GCF_000471905.1 |
| Soybean | 2006 | GCF_000004515.4 | GCF_000004515.1 | Wm82.a2.v1 | Wm82.a2.v1 | Wm82.a2.v1 | GCF_000004515.1 | GCF_000004515.4 |
| Apple | 2010 | GCF_000148765.1 | NA | NA | M. Domestica v1.0 | M. Domestica v1.0 | NA | GCF_000148765.1 |
| Sorghum | 2009 | GCF_000003195.2 | GCF_000003195.2 | Sbi1.4 | Sbi3.1 | NA | GCF_000003195.2 | GCF_000003195.2 |
| Barrel medic | 2011 | GCF_000219495.2 | Mt4.0 | Mt3.5 | Mt4.0 | Mt4.0 | NA | GCF_000219495.2 |
| Banana | 2012 | GCF_000313855.1 | GCF_000313855.1 | NA | M. acuminata v 1 | NA | GCF_000313855.1 | GCF_000313855.1 |
| Cocoa | 2011 | GCF_000208745.1 | GCF_000208745.1 | NA | C. Matina v1.1 | C. Matina v1.1 | NA | GCF_000208745.1 |

**Table 4.6** List of different genome assemblies and annotation versions available for potato. Number of representative genes and alternative transcripts for each annotation version are also reported

| Genome assembly version | Annotation version | Year of release | Number of representative genes | Number of alternative transcripts |
|---|---|---|---|---|
| v3 superscaffold | iTAG v.1 | 2011 | 35,004 | NA |
| | PGSC 3.4 | 2011 | 39,031 | 56,218 |
| | GCF_000226075.1 | 2011 | 33,608 | 37,885 |
| v3 2.1.10 pseudomolecules | PGSC v. 3 2.1.10 | 2012 | NA | 52,228 |
| v3 2.1.11 pseudomolecules | PGSC v. 3 2.1.11 | 2012 | NA | 52,228 |
| v4.03 pseudomolecules | PGSC v. 4.03 | 2013 | 39,146 | 56,980 |
| v4.04 pseudomolecules | NA | 2016 | NA | NA |

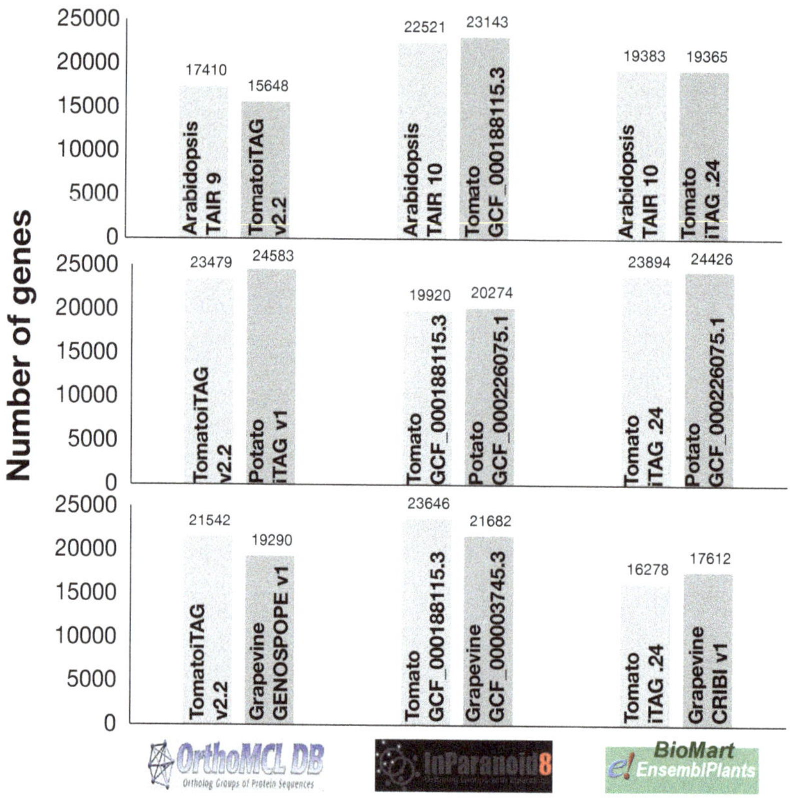


**Figure 4.2** Pairs of orthologue genes stored in three widespread comparative genomics platforms. For each species (*Arabidopsis*, tomato, potato and grapevine), the annotation version and the number of genes that have an orthologue relationship with the compared counterpart is shown.

the RefSeq annotation (O'Leary *et al.*, 2015) comes out from a unified gene annotation pipeline for all public sequenced genomes provided by the NCBI. Refseq annotations are cross-linked to relevant resources such as UniProt (UniProt Consortium, 2015) and KEGG (Kanehisa and Goto, 2000). Very few genome platforms for plant species also include the RefSeq annotations. On the other hand, RefSeq analyses does not appear to integrate the community related gene annotation for a species. This is quite common in plant genomics and determines two main limits: the presence of added-value information in specific websites not shared with general reference database, and limited information access to non-plant users, that usually refer to general databases. This is evidently against the basic concept of sharing and global exchange of information established by the International Nucleotide Sequence Database Collaboration (INSDC) in 2002 (Brunak *et al.*, 2002) and definitely remarked in 2015 (Cochrane *et al.*, 2016). On the other hand, this lack of direct links with well-established reference collections makes plant users isolated from the general trends in bioinformatics, which are fast expanding for reference animal species. Furthermore, because of the multifaceted efforts in plant genomics, driven by different communities focused on specific species and usually growing in specific and specialized context, crosslinks are usually hard and difficult to be exploited. While interconnections and comparative efforts would surely support profitable science.

However, although these limitations, -omics in the NGS era together with bioinformatics innovations largely moulded the experimental design in plant molecular biology, consistently contributing to the scientific knowledge in plants molecular biology and positively affecting many applications of agriculture sciences (Esposito *et al.*, 2016). Diverse plants research fields, such as breeding, environmental sciences and microbiology, are benefiting from the available knowledge and are contributing data, favouring scientific advances, improving sustainability, products quality and strategies for stress or disease treatments (Deusch *et al.*, 2015; Tringe and Coleman-Derr, 2014).

Undoubtedly, NGS approaches are driving more efficient solutions from bioinformatics, continuously stimulating novel computational approaches. The need for innovative interfaces and even more intuitive tools to address the main biological questions emphasized by the novel technologies will not decline for many years to come (Horner *et al.*, 2010).

The most critical issues that bioinformatics should face to meet the major trends in plant biology and overcome some of the bottlenecks here highlighted are mainly related to a proper dissemination and stabilization of resources, favouring integration of information from different levels of the cell functionality and from different species. To this aim, the development of comprehensive collections and the favouring of in-depth knowledge through education and training in omics-based technologies and in bioinformatics, opportunely integrating experimental and computational efforts, and possibly different scientific communities, are essential (Chiusano, 2015; Esposito *et al.*, 2016).

Beyond all, coordinated efforts preserving data curation should be consistently supported, with the main purpose of avoiding that profitable -omics could be restricted to major experts and limited to major species.

The need for 'tailoring' bioinformatics to dedicated efforts that could improve the quality of draft results and make them appropriately accessible, usable and reconciled with related resources, is fundamental to reducing the risk that fast data production could compromise the sharing of reliable and updated information. Project reviewers and publication policies may also play a relevant role in facing the presented issues, since they can drive towards appropriate cross-referencing, data sharing and high quality standards in both software and resources.

The care for appropriate knowledge dissemination should remain the priority in Science. Enhanced bioinformatics is the strategy that can fulfil the scope.

## Acknowledgements

This work was supported by the Genopom PRO and the GenHORT Projects [Ministero dell'Istruzione, dell'Università e della Ricerca (MIUR), Italy] and in the frame of the COST ACTION (FA1106).

## References

Adams, M.D., Kelley, J.M., Gocayne, J.D., Dubnick, M., Polymeropoulos, M.H., Xiao, H., Merril, C.R., Wu, A., Olde, B., Moreno, R.F., *et al.* (1991). Complementary DNA sequencing: expressed sequence tags and human genome project. Science *252*, 1651–1656. http://dx.doi.org/10.1126/science.2047873

Aflitos, S., Schijlen, E., de Jong, H., de Ridder, D., Smit, S., Finkers, R., Wang, J., Zhang, G., Li, N., Mao, L., *et al.* (2014). Exploring genetic variation in the tomato (Solanum section Lycopersicon) clade by whole-genome sequencing. Plant J. *80*, 136–148. http://dx.doi.org/10.1111/tpj.12616

Altenhoff, A.M., and Dessimoz, C. (2009). Phylogenetic and functional assessment of orthologs inference projects and methods. PLoS Comput Biol *5*, e1000262. http://dx.doi.org/10.1371/journal.pcbi.1000262

Altenhoff, A.M., and Dessimoz, C. (2012). Inferring orthology and paralogy. Methods Mol. Biol. *855*, 259–279. http://dx.doi.org/10.1007/978-1-61779-582-4_9

Altenhoff, A.M., Schneider, A., Gonnet, G.H., and Dessimoz, C. (2011). OMA 2011: orthology inference

among 1000 complete genomes. Nucleic Acids Res. *39*, D289–294. http://dx.doi.org/10.1093/nar/gkq1238
Ambrosino, L., and Chiusano, M.L. (2013). In quest of orthologs. In BBCC 2013 (Avellino).
Andorf, C.M., Cannon, E.K., Portwood, J.L., 2nd, Gardiner, J.M., Harper, L.C., Schaeffer, M.L., Braun, B.L., Campbell, D.A., Vinnakota, A.G., Sribalusu, V.V., *et al.* (2016). MaizeGDB update: new tools, data and interface for the maize model organism database. Nucleic Acids Res. *44*, D1195–1201. http://dx.doi.org/10.1093/nar/gkv1007
Ashburner, M., Ball, C.A., Blake, J.A., Botstein, D., Butler, H., Cherry, J.M., Davis, A.P., Dolinski, K., Dwight, S.S., Eppig, J.T., *et al.* (2000). Gene ontology: tool for the unification of biology. The Gene Ontology Consortium. Nat. Genet. *25*, 25–29. http://dx.doi.org/10.1038/75556
Aversano, R., Contaldi, F., and Ercolano, M.R. (2015). The *Solanum commersonii* genome sequence provides insights into adaptation to stress conditions and genome evolution of wild potato relatives. Plant Cell *27*, 954–968. http://dx.doi.org/10.1105/tpc.114.135954
Bateman, A., and Quackenbush, J. (2009). Bioinformatics for next generation sequencing. Bioinformatics *25*, 429. http://dx.doi.org/10.1093/bioinformatics/btp037
Becker, C., Hagmann, J., Muller, J., Koenig, D., Stegle, O., Borgwardt, K., and Weigel, D. (2011). Spontaneous epigenetic variation in the *Arabidopsis thaliana* methylome. Nature *480*, 245–249. http://dx.doi.org/10.1038/nature10555
Becker, J.D., Boavida, L.C., Carneiro, J., Haury, M., and Feijó, J.A. (2003). Transcriptional profiling of *Arabidopsis* tissues reveals the unique characteristics of the pollen transcriptome. Plant Physiol. *133*, 713-725. http://dx.doi.org/10.1104/pp.103.028241
Bennett, S. (2004). Solexa Ltd. Pharmacogenomics *5*, 433–438. http://dx.doi.org/10.1517/14622416.5.4.433
Benson, D.A., Karsch-Mizrachi, I., Lipman, D.J., Ostell, J., Rapp, B.A., and Wheeler, D.L. (2000). GenBank. Nucleic Acids Res. *28*, 15–18.
Berman, H.M., Westbrook, J., Feng, Z., Gilliland, G., Bhat, T.N., Weissig, H., Shindyalov, I.N., and Bourne, P.E. (2000). The Protein Data Bank. Nucleic Acids Res. *28*, 235–242. http://dx.doi.org/10.1093/nar/28.1.235
Betryszak, R., Keays, M., Tang, Y.A., Fonseca, N.A., Barrera, E., Burdett, T., Füllgrabe, A., Fuentes, A.M.-P., Jupp, S., Koskinen, S., *et al.* (2016). Expression Atlas update—an integrated database of gene and protein expression in humans, animals and plants. Nucleic Acids Res. *44*, D746–D752.
Bita, C.E., Zenoni, S., Vriezen, W.H., Mariani, C., Pezzotti, M., and Gerats, T. (2011). Temperature stress differentially modulates transcription in meiotic anthers of heat-tolerant and heat-sensitive tomato plants. BMC Genomics *12*, 384. http://dx.doi.org/10.1186/1471-2164-12-384
Blair, M.W., Fernandez, A.C., Ishitani, M., Moreta, D., Seki, M., Ayling, S., and Shinozaki, K. (2011). Construction and EST sequencing of full-length, drought stress cDNA libraries for common beans (*Phaseolus vulgaris* L.). BMC Plant Biol. *11*, 171. http://dx.doi.org/10.1186/1471-2229-11-171
Blanc, G., Hokamp, K., and Wolfe, K.H. (2003). A recent polyploidy superimposed on older large-scale duplications in the *Arabidopsis* genome. Genome Res. *13*, 137–144. http://dx.doi.org/10.1101/gr.751803
Blanchfield, J.R. (2004). Genetically modified food crops and their contribution to human nutrition and food quality. J. Food Sci.*69*, CRH28-CRH30. http://dx.doi.org/10.1111/j.1365-2621.2004.tb17846.x
Bokszczanin, K., Krezdorn, N., Fragkostefanakis, S., Muller, S., Rycak, L., Chen, Y., Hoffmeier, K., Kreutz, J., Paupiere, M., Chaturvedi, P., *et al.* (2015). Identification of novel small ncRNAs in pollen of tomato. BMC Genomics *16*, 714. http://dx.doi.org/10.1186/s12864-015-1901-x
Bolser, D., Staines, D.M., Pritchard, E., and Kersey, P. (2016). Ensembl plants: integrating tools for visualizing, mining, and analyzing plant genomics data. Methods Mol. Biol. *1374*, 115–140. http://dx.doi.org/10.1007/978-1-4939-3167-5_6
Bostan, H., and Chiusano, M.L. (2015). NexGenEx-Tom: a gene expression platform to investigate the functionalities of the tomato genome. BMC Plant Biol. *15*, 48. http://dx.doi.org/10.1186/s12870-014-0412-2
Bowers, J.E., Chapman, B.A., Rong, J., and Paterson, A.H. (2003). Unravelling angiosperm genome evolution by phylogenetic analysis of chromosomal duplication events. Nature *422*, 433-438. http://dx.doi.org/10.1038/nature01521
Brenchley, R., Spannagl, M., Pfeifer, M., Barker, G.L., D'Amore, R., Allen, A.M., McKenzie, N., Kramer, M., Kerhornou, A., Bolser, D., *et al.* (2012). Analysis of the bread wheat genome using whole-genome shotgun sequencing. Nature *491*, 705–710. http://dx.doi.org/10.1038/nature11650
Brenner, S., Johnson, M., Bridgham, J., Golda, G., Lloyd, D.H., Johnson, D., Luo, S., McCurdy, S., Foy, M., Ewan, M., *et al.* (2000). Gene expression analysis by massively parallel signature sequencing (MPSS) on microbead arrays. Nat. Biotechnol. *18*, 630–634. http://dx.doi.org/10.1038/76469
Brunak, S., Danchin, A., Hattori, M., Nakamura, H., Shinozaki, K., Matise, T., and Preuss, D. (2002). Nucleotide Sequence Database Policies. Science *298*, 1333.
Bülow, L., Schindler, M., and Hehl, R. (2007). PathoPlant®: a platform for microarray expression data to analyze co-regulated genes involved in plant defense responses. Nucleic Acids Res. *35*, D841–D845. http://dx.doi.org/10.1093/nar/gkl835
Chen, F., Mackey, A.J., Stoeckert, C.J., Jr., and Roos, D.S. (2006). OrthoMCL-DB: querying a comprehensive multi-species collection of ortholog groups. Nucleic Acids Res. *34*, D363–368.
Chiusano, M.L. (2015). On the multifaceted aspects of bioinformatics in the next generation era: the run that must keep the quality. Next Generat Sequenc & Applic. e106. http://dx.doi.org/10.4172/2469-9853.1000e106
Chiusano, M.L., D'Agostino, N., Traini, A., Licciardello, C., Raimondo, E., Aversano, M., Frusciante, L., and Monti, L. (2008). ISOL@: an Italian SOLAnaceae genomics resource. BMC Bioinformatics *9 Suppl 2*, S7. http://dx.doi.org/10.1186/1471-2105-9-s2-s7
Choulet, F., Wicker, T., Rustenholz, C., Paux, E., Salse, J., Leroy, P., Schlub, S., Le Paslier, M.C., Magdelenat, G.,

Gonthier, C., *et al.* (2010). Megabase level sequencing reveals contrasted organization and evolution patterns of the wheat gene and transposable element spaces. Plant Cell *22*, 1686–1701. http://dx.doi.org/10.1105/tpc.110.074187

Cochrane, G., Karsch-Mizrachi, I., and Takagi, T. (2016). The International Nucleotide Sequence Database Collaboration. Nucleic Acids Res. *44*, D48-50. http://dx.doi.org/10.1093/nar/gkv1323

Cui, L., Wall, P.K., Leebens-Mack, J.H., Lindsay, B.G., Soltis, D.E., Doyle, J.J., Soltis, P.S., Carlson, J.E., Arumuganathan, K., Barakat, A., *et al.* (2006). Widespread genome duplications throughout the history of flowering plants. Genome Res. *16*, 738–749. http://dx.doi.org/10.1101/gr.4825606

Dayhoff, M., Eck, R., Chang, M., and Sochard, M. (1965). Atlas of Protein Sequence and Structure, Vol. 1 (MD: Silver Spring).

De Luca, V., Salim, V., Atsumi, S.M., and Yu, F. (2012). Mining the biodiversity of plants: a revolution in the making. Science *336*, 1658–1661. http://dx.doi.org/10.1126/science.1217410

Denoeud, F., Carretero-Paulet, L., Dereeper, A., Droc, G., Guyot, R., Pietrella, M., Zheng, C., Alberti, A., Anthony, F., Aprea, G., *et al.* (2014). The coffee genome provides insight into the convergent evolution of caffeine biosynthesis. Science *345*, 1181–1184. http://dx.doi.org/10.1126/science.1255274

Dessimoz, C., Gabaldon, T., Roos, D.S., Sonnhammer, E.L., and Herrero, J. (2012). Toward community standards in the quest for orthologs. Bioinformatics *28*, 900–904. http://dx.doi.org/10.1093/bioinformatics/bts050

Deusch, S., Tilocca, B., Camarinha-Silva, A., and Seifert, J. (2015). News in livestock research – use of Omics-technologies to study the microbiota in the gastrointestinal tract of farm animals. Comput Struct Biotechnol J *13*, 55–63. http://dx.doi.org/10.1016/j.csbj.2014.12.005

Di Salle, P., Incerti, G., Colantuono, C., and Chiusano, M.L. (2016). Gene co-expression analyses: an overview from microarray collections in *Arabidopsis thaliana*. Brief Bioinform. http://dx.doi.org/10.1093/bib/bbw002

Dong, Q., Schlueter, S.D., and Brendel, V. (2004). PlantGDB, plant genome database and analysis tools. Nucleic Acids Res. *32*, D354-359. http://dx.doi.org/10.1093/nar/gkh046

Dreher, K. (2014). Putting the Plant Metabolic Network Pathway Databases to Work: Going Offline to Gain New Capabilities. In Plant Metabolism: Methods and Protocols, G. Sriram, ed. (Totowa, NJ: Humana Press), pp. 151-171. http://dx.doi.org/10.1007/978-1-62703-661-0_10

Droege, M., and Hill, B. (2008). The Genome Sequencer FLX™ System – Longer reads, more applications, straightforward bioinformatics and more complete data sets. J. Biotechnol. *136*, 3-10. http://dx.doi.org/http://dx.doi.org/10.1016/j.jbiotec.2008.03.021

Edgar, R., Domrachev, M., and Lash, A.E. (2002). Gene Expression Omnibus: NCBI gene expression and hybridization array data repository. Nucleic Acids Res. *30*, 207-210. http://www.ncbi.nlm.nih.gov/geo

Ercolano, M.R., Sacco, A., Ferriello, F., D'Alessandro, R., Tononi, P., Traini, A., Barone, A., Zago, E., Chiusano, M.L., Buson, G., *et al.* (2014). Patchwork sequencing of tomato San Marzano and Vesuviano varieties highlights genome-wide variations. BMC Genomics *15*, 138. http://dx.doi.org/10.1186/1471-2164-15-138

Esposito, A., Colantuono, C., Ruggieri, V., and Chiusano, M.L. (2016). Bioinformatics for agriculture in the Next-Generation sequencing era. Chemical and Biological Technologies in Agriculture *3*, 1-12. http://dx.doi.org/10.1186/s40538-016-0054-8

Fabregat, A., Sidiropoulos, K., Garapati, P., Gillespie, M., Hausmann, K., Haw, R., Jassal, B., Jupe, S., Korninger, F., McKay, S., *et al.* (2016). The Reactome pathway Knowledgebase. Nucleic Acids Res. *44*, D481–487. http://dx.doi.org/10.1093/nar/gkv1351

Fernandez-Pozo, N., Menda, N., Edwards, J.D., Saha, S., Tecle, I.Y., Strickler, S.R., Bombarely, A., Fisher-York, T., Pujar, A., Foerster, H., *et al.* (2015). The Sol Genomics Network (SGN) – from genotype to phenotype to breeding. Nucleic Acids Res. *43*, D1036–1041. http://dx.doi.org/10.1093/nar/gku1195

Flagel, L.E., and Wendel, J.F. (2009). Gene duplication and evolutionary novelty in plants. New Phytol *183*, 557–564. http://dx.doi.org/10.1111/j.1469-8137.2009.02923.x

Flicek, P., Ahmed, I., Amode, M.R., Barrell, D., Beal, K., Brent, S., Carvalho-Silva, D., Clapham, P., Coates, G., Fairley, S., *et al.* (2013). Ensembl 2013. Nucleic Acids Res. *41*, D48–55. http://dx.doi.org/10.1093/nar/gks1236

Gan, X., Stegle, O., Behr, J., Steffen, J.G., Drewe, P., Hildebrand, K.L., Lyngsoe, R., Schultheiss, S.J., Osborne, E.J., Sreedharan, V.T., *et al.* (2011). Multiple reference genomes and transcriptomes for *Arabidopsis thaliana*. Nature *477*, 419–423. http://dx.doi.org/10.1038/nature10414

Goff, S.A., Ricke, D., Lan, T.-H., Presting, G., Wang, R., Dunn, M., Glazebrook, J., Sessions, A., Oeller, P., Varma, H., *et al.* (2002). A Draft Sequence of the Rice Genome (*Oryza sativa* L. ssp. *japonica*). Science *296*, 92–100. http://dx.doi.org/10.1126/science.1068275

Goodstein, D.M., Shu, S., Howson, R., Neupane, R., Hayes, R.D., Fazo, J., Mitros, T., Dirks, W., Hellsten, U., Putnam, N., *et al.* (2012). Phytozome: a comparative platform for green plant genomics. Nucleic Acids Res. *40*, D1178–1186. http://dx.doi.org/10.1093/nar/gkr944

Harris, T.D., Buzby, P.R., Babcock, H., Beer, E., Bowers, J., Braslavsky, I., Causey, M., Colonell, J., DiMeo, J., Efcavitch, J.W., *et al.* (2008). Single-molecule DNA sequencing of a viral genome. Science *320*, 106. http://dx.doi.org/10.1126/science.1150427

Hirsch, C.D., Hamilton, J.P., Childs, K.L., Cepela, J., Crisovan, E., Vaillancourt, B., Hirsch, C.N., Habermann, M., Neal, B., and Buell, C.R. (2014). Spud DB: a resource for mining sequences, genotypes, and phenotypes to accelerate potato breeding. The Plant Genome 7. http://dx.doi.org/10.3835/plantgenome2013.12.0042

Horner, D.S., Pavesi, G., Castrignano, T., De Meo, P.D., Liuni, S., Sammeth, M., Picardi, E., and Pesole, G. (2010). Bioinformatics approaches for genomics and post genomics applications of next-generation sequencing. Brief Bioinform. *11*, 181–197. http://dx.doi.org/10.1093/bib/bbp046

Huerta-Cepas, J., Szklarczyk, D., Forslund, K., Cook, H., Heller, D., Walter, M.C., Rattei, T., Mende, D.R., Sunagawa, S., Kuhn, M., *et al.* (2016). eggNOG 4.5: a hierarchical orthology framework with improved functional annotations for eukaryotic, prokaryotic and viral sequences. Nucleic Acids Res. *44*, D286–293. http://dx.doi.org/10.1093/nar/gkv1248

Hughes, A.L. (2005). Gene duplication and the origin of novel proteins. Proc. Natl. Acad. Sc.i U.S.A. *102*, 8791–8792. http://dx.doi.org/10.1073/pnas.0503922102

Izzah, N.K., Lee, J., Jayakodi, M., Perumal, S., Jin, M., Park, B.-S., Ahn, K., and Yang, T.-J. (2014). Transcriptome sequencing of two parental lines of cabbage (Brassica oleracea L. var. capitata L.) and construction of an EST-based genetic map. BMC Genomics *15*, 149. http://dx.doi.org/10.1186/1471-2164-15-149

Jaillon, O., Aury, J.M., Noel, B., Policriti, A., Clepet, C., Casagrande, A., Choisne, N., Aubourg, S., Vitulo, N., Jubin, C., *et al.* (2007). The grapevine genome sequence suggests ancestral hexaploidization in major angiosperm phyla. Nature *449*, 463–467. http://dx.doi.org/10.1038/nature06148

Jia, J., Zhao, S., Kong, X., Li, Y., Zhao, G., He, W., Appels, R., Pfeifer, M., Tao, Y., Zhang, X., *et al.* (2013). Aegilops tauschii draft genome sequence reveals a gene repertoire for wheat adaptation. Nature *496*, 91–95. http://dx.doi.org/10.1038/nature12028

Jiao, Y., Leebens-Mack, J., Ayyampalayam, S., Bowers, J.E., McKain, M.R., McNeal, J., Rolf, M., Ruzicka, D.R., Wafula, E., Wickett, N.J., *et al.* (2012). A genome triplication associated with early diversification of the core eudicots. Genome Biol. *13*, R3. http://dx.doi.org/10.1186/gb-2012-13-1-r3

Jiao, Y., Wickett, N.J., Ayyampalayam, S., Chanderbali, A.S., Landherr, L., Ralph, P.E., Tomsho, L.P., Hu, Y., Liang, H., Soltis, P.S., *et al.* (2011). Ancestral polyploidy in seed plants and angiosperms. Nature *473*, 97–100. http://dx.doi.org/10.1038/nature09916

Jung, S., Ficklin, S.P., Lee, T., Cheng, C.-H., Blenda, A., Zheng, P., Yu, J., Bombarely, A., Cho, I., Ru, S., *et al.* (2013). The Genome Database for Rosaceae (GDR): year 10 update. Nucleic Acids Res. *42*, D1237–D1244.

Kanehisa, M., and Goto, S. (2000). KEGG: kyoto encyclopedia of genes and genomes. Nucleic Acids Res. *28*, 27-30.

Kawahara, Y., de la Bastide, M., Hamilton, J.P., Kanamori, H., McCombie, W.R., Ouyang, S., Schwartz, D.C., Tanaka, T., Wu, J., Zhou, S., *et al.* (2013). Improvement of the *Oryza sativa* Nipponbare reference genome using next generation sequence and optical map data. Rice (NY) *6*, 4. http://dx.doi.org/10.1186/1939-8433-6-4

Kawahara, Y., Oono, Y., Wakimoto, H., Ogata, J., Kanamori, H., Sasaki, H., Mori, S., Matsumoto, T., and Itoh, T. (2016). TENOR: Database for comprehensive mRNA-Seq experiments in rice. Plant Cell Physiol. 57, e7.

Kersey, P.J., Allen, J.E., Armean, I., Boddu, S., and Bolt, B.J. (2016). Ensembl Genomes 2016: more genomes, more complexity. *44*, D574-580. http://dx.doi.org/10.1093/nar/gkv1209

Kim, S., Park, M., Yeom, S.-I., Kim, Y.-M., Lee, J.M., Lee, H.-A., Seo, E., Choi, J., Cheong, K., Kim, K.-T., *et al.* (2014). Genome sequence of the hot pepper provides insights into the evolution of pungency in *Capsicum* species. Nat. Genet. *46*, 270–278. http://dx.doi.org/10.1038/ng.2877

Kim, S., Plagnol, V., Hu, T.T., Toomajian, C., Clark, R.M., Ossowski, S., Ecker, J.R., Weigel, D., and Nordborg, M. (2007). Recombination and linkage disequilibrium in *Arabidopsis thaliana*. Nat. Genet. *39*, 1151–1155. http://dx.doi.org/10.1038/ng2115

Klepikova, A.V., Logacheva, M.D., Dmitriev, S.E., and Penin, A.A. (2015). RNA-seq analysis of an apical meristem time series reveals a critical point in *Arabidopsis thaliana* flower initiation. BMC Genomics *16*, 466.

Kodama, Y., Shumway, M., and Leinonen, R. (2012). The Sequence Read Archive: explosive growth of sequencing data. Nucleic Acids Res. *40*, D54–56. http://dx.doi.org/10.1093/nar/gkr854

Kodzius, R., Kojima, M., Nishiyori, H., Nakamura, M., Fukuda, S., Tagami, M., Sasaki, D., Imamura, K., Kai, C., Harbers, M., *et al.* (2006). CAGE: cap analysis of gene expression. Nat Methods *3*, 211–222. http://dx.doi.org/10.1038/nmeth0306-211

Kristensen, D.M., Wolf, Y.I., Mushegian, A.R., and Koonin, E.V. (2011). Computational methods for Gene Orthology inference. Brief Bioinform *12*, 379–391. http://dx.doi.org/10.1093/bib/bbr030

Lam, H.M., Xu, X., Liu, X., Chen, W., Yang, G., Wong, F.L., Li, M.W., He, W., Qin, N., Wang, B., *et al.* (2010). Resequencing of 31 wild and cultivated soybean genomes identifies patterns of genetic diversity and selection. Nat. Genet. *42*, 1053–1059. http://dx.doi.org/10.1038/ng.715

Lamesch, P., Berardini, T.Z., Li, D., Swarbreck, D., Wilks, C., Sasidharan, R., Muller, R., Dreher, K., Alexander, D.L., Garcia-Hernandez, M., *et al.* (2012). The *Arabidopsis* Information Resource (TAIR): improved gene annotation and new tools. Nucleic Acids Res. *40*, D1202–1210. http://dx.doi.org/10.1093/nar/gkr1090

Leinonen, R., Akhtar, R., Birney, E., Bower, L., Cerdeno-Tarraga, A., Cheng, Y., Cleland, I., Faruque, N., Goodgame, N., Gibson, R., *et al.* (2011). The European Nucleotide Archive. Nucleic Acids Res. *39*, D28-31. http://dx.doi.org/10.1093/nar/gkq967

Ling, H.Q., Zhao, S., Liu, D., Wang, J., Sun, H., Zhang, C., Fan, H., Li, D., Dong, L., Tao, Y., *et al.* (2013). Draft genome of the wheat A-genome progenitor *Triticum urartu*. Nature *496*, 87–90. http://dx.doi.org/10.1038/nature11997

Lynch, M., and Conery, J.S. (2000). The evolutionary fate and consequences of duplicate genes. Science *290*, 1151-1155. http://dx.doi.org/10.1126/science.290.5494.1151

Ma, J.K., Drake, P.M., and Christou, P. (2003). The production of recombinant pharmaceutical proteins in plants. Nat. Rev. Genet. *4*, 794–805. http://dx.doi.org/10.1038/nrg1177

MacLean, D., Jones, J.D., and Studholme, D.J. (2009). Application of 'next-generation' sequencing technologies to microbial genetics. Nat. Rev. Microbiol. 7, 287–296. http://dx.doi.org/10.1038/nrmicro2122

Maere, S., De Bodt, S., Raes, J., Casneuf, T., Van Montagu, M., Kuiper, M., and Van de Peer, Y. (2005). Modeling gene and genome duplications in eukaryotes. Proc. Natl. Acad. Sci. U.S.A. *102*, 5454–5459. http://dx.doi.org/10.1073/pnas.0501102102

Magadum, S., Banerjee, U., Murugan, P., Gangapur, D., and Ravikesavan, R. (2013). Gene duplication as a major force in evolution. J. Genet. *92*, 155-161.

Magi, A., Benelli, M., Gozzini, A., Girolami, F., Torricelli, F., and Brandi, M.L. (2010). Bioinformatics for Next Generation Sequencing Data. Genes *1*, 294–307. http://dx.doi.org/10.3390/genes1020294

Makita, Y., Shimada, S., Kawashima, M., Kondou-Kuriyama, T., Toyoda, T., and Matsui, M. (2015). MOROKOSHI: transcriptome database in *Sorghum bicolor*. Plant Cell Physiol. *56*, e6.

Mardis, E.R. (2008a). The impact of next-generation sequencing technology on genetics. Trends Genet. *24*, 133–141. http://dx.doi.org/10.1016/j.tig.2007.12.007

Mardis, E.R. (2008b). Next-generation DNA sequencing methods. Annu. Rev. Genomics Hum. Genet. *9*, 387–402. http://dx.doi.org/10.1146/annurev.genom.9.081307.164359

Mardis, E.R. (2009). New strategies and emerging technologies for massively parallel sequencing: applications in medical research. Genome Med *1*, 40. http://dx.doi.org/10.1186/gm40

Mayer, K.F., Waugh, R., Brown, J.W., Schulman, A., Langridge, P., Platzer, M., Fincher, G.B., Muehlbauer, G.J., Sato, K., Close, T.J., *et al.* (2012). A physical, genetic and functional sequence assembly of the barley genome. Nature *491*, 711–716. http://dx.doi.org/10.1038/nature11543

Moniz de Sa, M., and Drouin, G. (1996). Phylogeny and substitution rates of angiosperm actin genes. Mol. Biol. Evol. *13*, 1198–1212.

Moretto, M., Sonego, P., Dierckxsens, N., Brilli, M., Bianco, L., Ledezma-Tejeida, D., Gama-Castro, S., Galardini, M., Romualdi, C., Laukens, K., *et al.* (2015). COLOMBOS v3.0: leveraging gene expression compendia for cross-species analyses. Nucleic Acids Res. *44*, D620–D623.

Morozova, O., and Marra, M.A. (2008a). Applications of next-generation sequencing technologies in functional genomics. Genomics *92*, 255–264. http://dx.doi.org/10.1016/j.ygeno.2008.07.001

Morozova, O., and Marra, M.A. (2008b). From cytogenetics to next-generation sequencing technologies: advances in the detection of genome rearrangements in tumors. Biochem Cell Biol *86*, 81-91. http://dx.doi.org/10.1139/o08-003

Morrissy, A.S., Morin, R.D., Delaney, A., Zeng, T., McDonald, H., Jones, S., Zhao, Y., Hirst, M., and Marra, M.A. (2009). Next-generation tag sequencing for cancer gene expression profiling. Genome Res. *19*, 1825–1835. http://dx.doi.org/10.1101/gr.094482.109

Mukherjee, G., Abeygunawardena, N., Parkinson, H., Contrino, S., Durinck, S., Farne, A., Holloway, E., Lilja, P., Moreau, Y., Oezcimen, A., *et al.* (2005). Plant-Based Microarray Data at the European Bioinformatics Institute. Introducing AtMIAMExpress, a Submission Tool for *Arabidopsis* Gene Expression Data to ArrayExpress. Plant Physiol *139*, 632-636. http://dx.doi.org/10.1104/pp.105.063156

Nakano, M., Nobuta, K., Vemaraju, K., Tej, S.S., Skogen, J.W., and Meyers, B.C. (2006). Plant MPSS databases: signature-based transcriptional resources for analyses of mRNA and small RNA. Nucleic Acids Res. *34*, D731–D735.

Nawrocki, E.P., Burge, S.W., Bateman, A., Daub, J., Eberhardt, R.Y., Eddy, S.R., Floden, E.W., and Gardner, P.P. (2015). Rfam 12.0: updates to the RNA families database. Nucleic Acids Res. *43*, D130–137. http://dx.doi.org/10.1093/nar/gku1063

Nystedt, B., Street, N.R., Wetterbom, A., Zuccolo, A., Lin, Y.C., Scofield, D.G., Vezzi, F., Delhomme, N., Giacomello, S., Alexeyenko, A., *et al.* (2013). The Norway spruce genome sequence and conifer genome evolution. Nature *497*, 579-584. http://dx.doi.org/10.1038/nature12211

O'Brien, K.P., Remm, M., and Sonnhammer, E.L. (2005). Inparanoid: a comprehensive database of eukaryotic orthologs. Nucleic Acids Res. *33*, D476-480. http://dx.doi.org/10.1093/nar/gki107

O'Leary, N.A., Wright, M.W., Brister, J.R., Ciufo, S., Haddad, D., McVeigh, R., Rajput, B., Robbertse, B., Smith-White, B., Ako-Adjei, D., *et al.* (2015). Reference sequence (RefSeq) database at NCBI: current status, taxonomic expansion, and functional annotation. Nucleic Acids Res. http://dx.doi.org/10.1093/nar/gkv1189 http://www.ncbi.nlm.nih.gov/refseq

O'Rourke, J.A., Fu, F., Bucciarelli, B., Yang, S.S., Samac, D.A., Lamb, J.F.S., Monteros, M.J., Graham, M.A., Gronwald, J.W., Krom, N., *et al.* (2015). The *Medicago sativa* gene index 1.2: a web-accessible gene expression atlas for investigating expression differences between *Medicago sativa* subspecies. BMC Genomics *16*, 1–17.

O'Rourke, J.A., Iniguez, L.P., Fu, F., Bucciarelli, B., Miller, S.S., Jackson, S.A., McClean, P.E., Li, J., Dai, X., Zhao, P.X., *et al.* (2014). An RNA-Seq based gene expression atlas of the common bean. BMC Genomics *15*, 866.

Page, G.P., and Coulibaly, I. (2008). Bioinformatic Tools for Inferring Functional Information from Plant Microarray Data: Tools for the First Steps. Int J Plant Genomics *2008*, 147563. http://dx.doi.org/10.1155/2008/147563

Paterson, A.H., Bowers, J.E., Bruggmann, R., Dubchak, I., Grimwood, J., Gundlach, H., Haberer, G., Hellsten, U., Mitros, T., Poliakov, A., *et al.* (2009). The Sorghum bicolor genome and the diversification of grasses. Nature *457*, 551–556. http://dx.doi.org/10.1038/nature07723

Pearce, S., Vazquez-Gross, H., Herin, S.Y., Hane, D., Wang, Y., Gu, Y.Q., and Dubcovsky, J. (2015). WheatExp: an RNA-seq expression database for polyploid wheat. BMC Plant Biol. *15*, 1–8.

Porreca, G.J., Shendure, J., and Church, G.M. (2006). Polony DNA sequencing. Curr. Protocols Mol. Biol., 7.8.1–7.8.22. http://dx.doi.org/10.1002/0471142727.mb0708s76

Proost, S., Van Bel, M., Vaneechoutte, D., Van de Peer, Y., Inze, D., Mueller-Roeber, B., and Vandepoele, K. (2015). PLAZA 3.0: an access point for plant comparative genomics. Nucleic Acids Res. *43*, D974–981. http://dx.doi.org/10.1093/nar/gku986

Rouard, M., Guignon, V., Aluome, C., Laporte, M.A., Droc, G., Walde, C., Zmasek, C.M., Perin, C., and Conte, M.G. (2011). GreenPhylDB v2.0: comparative and functional genomics in plants. Nucleic Acids Res. *39*, D1095-1102. http://dx.doi.org/10.1093/nar/gkq811

Sanger, F., Nicklen, S., and Coulson, A.R. (1977). DNA sequencing with chain-terminating inhibitors. Proc. Natl. Acad. Sci. U.S.A. *74*, 5463–5467.

Schatz, M.C., Witkowski, J., and McCombie, W.R. (2012). Current challenges in *de novo* plant genome sequencing and assembly. Genome Biol *13*, 243. http://dx.doi.org/10.1186/gb4015

Schmid, M., Davison, T.S., Henz, S.R., Pape, U.J., Demar, M., Vingron, M., Scholkopf, B., Weigel, D., and Lohmann, J.U. (2005). A gene expression map of *Arabidopsis thaliana* development. Nat. Genet. *37*, 501-506. http://dx.doi.org/10.1038/ng1543

Schnable, P.S., Ware, D., Fulton, R.S., Stein, J.C., Wei, F., Pasternak, S., Liang, C., Zhang, J., Fulton, L., Graves, T.A., *et al.* (2009). The B73 maize genome: complexity, diversity, and dynamics. Science *326*, 1112–1115. http://dx.doi.org/10.1126/science.1178534

Schuster, S.C. (2008). Next-generation sequencing transforms today's biology. Nat. Methods *5*, 16–18. http://dx.doi.org/10.1038/nmeth1156

Severin, A.J., Woody, J.L., Bolon, Y.-T., Joseph, B., Diers, B.W., Farmer, A.D., Muehlbauer, G.J., Nelson, R.T., Grant, D., Specht, J.E., *et al.* (2010). RNA-Seq Atlas of Glycine max: A guide to the soybean transcriptome. BMC Plant Biol. *10*, 160.

Sierro, N., Battey, J.N.D., Ouadi, S., Bakaher, N., Bovet, L., Willig, A., Goepfert, S., Peitsch, M.C., and Ivanov, N.V. (2014). The tobacco genome sequence and its comparison with those of tomato and potato. Nat. Commun. 5. http://dx.doi.org/10.1038/ncomms4833

Tang, H., and Zhao, Z. (2015). Bioinformatics drives the applications of next-generation sequencing in translational biomedical research. Methods *79–80*, 1–2. http://dx.doi.org/10.1016/j.ymeth.2015.04.035

Tello-Ruiz, M.K., Stein, J., Wei, S., Preece, J., Olson, A., Naithani, S., Amarasinghe, V., Dharmawardhana, P., Jiao, Y., Mulvaney, J., *et al.* (2016). Gramene 2016: comparative plant genomics and pathway resources. Nucleic Acids Res. *44*, D1133–1140. http://dx.doi.org/10.1093/nar/gkv1179

The *Arabidopsis* Genome Initiative (2000). Analysis of the genome sequence of the flowering plant *Arabidopsis thaliana*. Nature *408*, 796-815. http://dx.doi.org/10.1038/35048692

The Tomato Genome Consortium (2012). The tomato genome sequence provides insights into fleshy fruit evolution. Nature *485*, 635–641. http://dx.doi.org/10.1038/nature11119

Trachana, K., Forslund, K., Larsson, T., Powell, S., Doerks, T., von Mering, C., and Bork, P. (2014). A phylogeny-based benchmarking test for orthology inference reveals the limitations of function-based validation. PLOS ONE *9*, e111122. http://dx.doi.org/10.1371/journal.pone.0111122

Tringe, S., and Coleman-Derr, D. (2014). Building the crops of tomorrow: advantages of symbiont-based approaches to improving abiotic stress tolerance (Ernest Orlando Lawrence Berkeley National Laboratory, Berkeley, CA (US)).

UniProt Consortium (2015). UniProt: a hub for protein information. Nucleic Acids Res. *43*, D204–212. http://dx.doi.org/10.1093/nar/gku989

Velasco, R., Zharkikh, A., Troggio, M., Cartwright, D.A., Cestaro, A., Pruss, D., Pindo, M., FitzGerald, L.M., Vezzulli, S., Reid, J., *et al.* (2007). A High Quality Draft Consensus Sequence of the Genome of a Heterozygous Grapevine Variety. PLOS ONE *2*, e1326. http://dx.doi.org/10.1371/journal.pone.0001326

Velculescu, V.E., Zhang, L., Vogelstein, B., and Kinzler, K.W. (1995). Serial analysis of gene expression. Science *270*, 484–487. http://dx.doi.org/10.1126/science.270.5235.484

Vitulo, N., Forcato, C., Carpinelli, E.C., Telatin, A., Campagna, D., D'Angelo, M., Zimbello, R., Corso, M., Vannozzi, A., Bonghi, C., *et al.* (2014). A deep survey of alternative splicing in grape reveals changes in the splicing machinery related to tissue, stress condition and genotype. BMC Plant Biol *14*, 99. http://dx.doi.org/10.1186/1471-2229-14-99

Wang, G., Liu, Y., Zhu, D., Klau, G.W., and Feng, W. (2015). Bioinformatics Methods and Biological Interpretation for Next-Generation Sequencing Data. Biomed Res Int *2015*, 690873. http://dx.doi.org/10.1155/2015/690873

Wang, J.-P.Z., Lindsay, B.G., Cui, L., Wall, P.K., Marion, J., Zhang, J., and dePamphilis, C.W. (2005). Gene capture prediction and overlap estimation in EST sequencing from one or multiple libraries. BMC Bioinformatics *6*, 300. http://dx.doi.org/10.1186/1471-2105-6-300

Wang, X., Wang, H., Wang, J., Sun, R., Wu, J., Liu, S., Bai, Y., Mun, J.H., Bancroft, I., Cheng, F., *et al.* (2011). The genome of the mesopolyploid crop species Brassica rapa. Nat. Genet. *43*, 1035–1039. http://dx.doi.org/10.1038/ng.919

Weigel, D., and Mott, R. (2009). The 1001 genomes project for *Arabidopsis thaliana*. Genome Biol *10*, 107. http://dx.doi.org/10.1186/gb-2009-10-5-107

Wilson, S.A., and Roberts, S.C. (2014). Metabolic engineering approaches for production of biochemicals in food and medicinal plants. Curr. Opin. Biotechnol. *26*, 174–182. http://dx.doi.org/10.1016/j.copbio.2014.01.006

Wolfe, K.H. (2001). Yesterday's polyploids and the mystery of diploidization. Nat Rev Genet *2*, 333–341. http://dx.doi.org/10.1038/35072009

Xu, C., Jiao, C., Zheng, Y., Sun, H., Liu, W., Cai, X., Wang, X., Liu, S., Xu, Y., Mou, B., *et al.* (2015). *De novo* and comparative transcriptome analysis of cultivated and wild spinach. Sci Rep 5, 17706.

Xu, X., Pan, S., Cheng, S., Zhang, B., Mu, D., Ni, P., Zhang, G., Yang, S., Li, R., Wang, J., *et al.* (2011). Genome sequence and analysis of the tuber crop potato. Nature *475*, 189–195. http://dx.doi.org/10.1038/nature10158

Yang, M.Q., Athey, B.D., Arabnia, H.R., Sung, A.H., Liu, Q., Yang, J.Y., Mao, J., and Deng, Y. (2009). High-throughput next-generation sequencing technologies foster new cutting-edge computing techniques in bioinformatics. BMC Genomics *10 Suppl 1*, I1. http://dx.doi.org/10.1186/1471-2164-10-s1-i1

Yuan, J.S., Tiller, K.H., Al-Ahmad, H., Stewart, N.R., and Stewart, C.N., Jr. (2008). Plants to power: bioenergy to fuel the future. Trends Plant Sci *13*, 421–429. http://dx.doi.org/10.1016/j.tplants.2008.06.001

# Applications of Bioinformatics to Plant Biotechnology

5

Diego F. Gomez-Casati[1]*, María V. Busi[1], Julieta Barchiesi[1], Diego A. Peralta[1], Nicolás Hedin[1] and Vijai Bhadauria[2]

[1]Centro de Estudios Fotosintéticos y Bioquímicos (CEFOBI-CONICET), Universidad Nacional de Rosario, Suipacha, Rosario, Argentina.
[2]Crop Development Centre and Department of Plant Sciences, University of Saskatchewan, Saskatoon, SK, Canada.

*Correspondence: gomezcasati@cefobi-conicet.gov.ar

https://doi.org/10.21775/9781910190654.06

## Abstract

Bioinformatics encompasses many tools and techniques that today are essential for all areas of research in the biological sciences. New databases with a wealth of information about genomes, proteins, metabolites, and metabolic pathways appear almost daily. Particularly for scientists who carry out research in plant biology, the amount of information has multiplied exponentially due to the large number of databases available for many individual plant species. In this sense, bioinformatics together with next-generation sequencing and 'omics' approaches, can provide tools for plant breeding and the genetic engineering of plants. In addition, these technologies enable a better understanding of the processes and mechanisms that can lead to plants with increased tolerance to different abiotic stress conditions and resistance to pathogen attack, as well as the development of crop varieties with improved nutritional quality of seeds and fruits.

## Introduction

In recent years, bioinformatics has emerged as a new, interdisciplinary field of science that can be described as the structuring of biological information to enable the understanding of biological phenomena (Edwards and Batley, 2004; Rhee *et al.*, 2006). Bioinformatics is essential in many areas of biology such as molecular biology, genetics and physiology, and plant biotechnology. Bioinformatics also includes the collection of biological information, storage and management of data, and the development of software for analysis of large data sets.

In order to deal with this deluge of available information, scientists need the ability to collect accurate biological data, store it in databases, and organize it in a systematic way. In addition, computer programs and mathematical tools such as algorithms and statistics are necessary for the analysis of the data.

Recent advances in genomic technologies have led to the accumulation of large amounts of data, resulting in a significant increase in the amount of biological information available for plant research and many other areas of biology, particularly the biomedical sciences. The applications of bioinformatics in plant biotechnology have expanded with the emergence of next-generation sequencing (NGS) techniques and the various omics technologies, providing data integration across the various omics platforms to reveal information about the genome, transcriptome, proteome, metabolome, and metabolic pathways of different plant species (Edwards and Batley, 2004; Rhee *et al.*, 2006).

In this review we cover some aspects of the applications of bioinformatics to problems in plant biotechnology, such as the improvement of fruit quality, the use of bioinformatics tools for plant breeding, and the improvement of crop species in their responses to various types of stress and pathogen attack. We also describe the databases available for the study of gene expression and metabolites, especially those containing information on different plant species.

## Gene expression and other omics resource databases

Transcriptomic analysis comprises the study of changes in gene expression profiles due to changes in the normal condition of a cell or tissue. In addition, transcriptomic studies can be used to assess how these changes in gene expression affect the development, growth, or phenotype of a particular biological system.

Microarrays are currently the most widely used technology for monitoring gene expression in plants. This technique allows the simultaneous determination of transcript abundance for many thousands of genes (Rhee *et al.*, 2006). Microarray data may be directly integrated into functional genomic approaches aimed at both assigning function to identified genes and studying the organization and control of genetic pathways that act concurrently to create a functional organism. The rationale behind this approach is that genes with similar expression patterns may be functionally related and subject to the same or similar genetic control mechanisms. Therefore, a common strategy undertaken in early microarray studies was to analyse data by clustering genes into groups based on their expression profiles scored over multiple experiments (Aharoni *et al.*, 2001; Brown and Botstein, 1999). Recently, RNA sequencing (RNA-seq) has been developed to directly measure mRNA profiles in a biological system using NGS technology (Morin *et al.*, 2008; Weber, 2015). This technique has the advantage in that it can generate relative measures of mRNA and also individual exon abundance while simultaneously profiling the prevalence of both annotated and novel exons and exon-splicing events. In addition, it is possible to identify known single nucleotide polymorphisms (SNPs) as well as novel single-base variants (Morin *et al.*, 2008). Microarray and RNA-seq techniques have enabled not only the identification of genes and the networks involved in different developmental processes, but have also helped to identify the downstream genes involved in such metabolic processes (Agarwal *et al.*, 2014). Fortunately, all gene expression data collected under different metabolic and environmental conditions is deposited in databases that are publicly accessible (Table 5.1). These data provide valuable information that could potentially be used to produce improved plant varieties.

On the other hand, bioinformatics is also being applied to the study and analysis of the metabolome. The omics approach known as metabolomics involves the comprehensive, high-throughput analysis of complex metabolite mixtures that, ideally, allows the identification and quantification of every individual metabolite that are the end products of cellular regulatory processes (Gomez-Casati *et al.*, 2013). Metabolomic studies, in conjunction with genomics or proteomics, can reflect changes in the phenotype and function of a particular tissue or organism. As with gene expression data, there are several databases for plant species (some of which also include useful information for biomedical researchers). Some of these databases also provide users with a powerful tool that allows them to generate an unambiguous means to identify structures. This 'tag' is known as the International Chemical Identifier (InChI) developed by the International Union of Pure and Applied Chemistry (IUPAC) and the National Institute of Standards and Technology (NIST). The InChI identifier is based on the chemical structure, and it makes it easier to perform searches on databases since the user can translate from the InChI to the structure, or back-translate. CheBI (www.ebi.ac.uk/chebi/) is one of the databases that allow users to generate InChI identifiers, and PubChem (http://pubchem.ncbi.nlm.nih.gov/) has a structure drawing tool to generate the InChI identifier. In addition, Fiehn Lab Chemical Translation Service (http://cts.fiehnlab.ucdavis.edu/) is an online tool that has the capability to generate InChI identifiers.

Some of the most well-known of the available metabolomics databases are the Golm Metabolome Database (http://gmd.mpimp-golm.mpg.de/), which contains information about the mass spectra from biologically active metabolites, and the Human Metabolome Database www.hmdb.ca/), which contains information on metabolites found in humans, including biochemical, clinical, and chemical data linking known metabolites to several genes and proteins. The Madison Metabolomics

**Table 5.1** Plant-specific gene expression databases

| Plant species | Database name | URL |
|---|---|---|
| *Arabidopsis* | NASC Arrays | http://affymetrix.arabidopsis.info/Arabidopsis |
| *Arabidopsis* | Genevestigator | www.genevestigator.ethz.ch/ |
| *Arabidopsis* | AREX LITE | www.arexdb.org/ |
| *Arabidopsis* | AraNet | www.inetbio.org/aranet/ |
| Barley | Barleybase | www.barleybase.org |
| Citrus | CitrusPLEX | www.plexdb.org/plex.php?database=Citrus |
| Cotton | CottonPLEX | www.plexdb.org/plex.php?database=Cotton |
| Generic | GEO (Gene Expression Omnibus) | www.ncbi.nlm.nih.gov/projects/geo |
| Generic | ArrayExpress | www.ebi.ac.uk/arrayexpress |
| Generic | Plant Co-expression database | http://planex.plantbioinformatics.org/ |
| Generic | Expression Atlas EMBL-EBI | www.ebi.ac.uk/gxa/home |
| Generic | BAR | http://bar.utoronto.ca/ |
| Generic | PLANET | http://aranet.mpimp-golm.mpg.de/ |
| Generic | PathoPlant | www.pathoplant.de/ |
| Generic | Plant Promoter Database | http://linux1.softberry.com/berry.phtml?topic=plantprom&group=data&subgroup=plantprom |
| Generic | PLEXdb | www.barleybase.org/plexdb/html/index.php |
| Grape | GrapePLEX | www.plexdb.org/plex.php?database=Grape |
| Maize | Maize genomics and genetics database | www.maizegdb.org/ |
| Maize | Zeamage | www.maizearray.org |
| Maize | MaizePLEX | www.plexdb.org/plex.php?database=Corn |
| Medicago | MedicagoPLEX | www.plexdb.org/plex.php?database=Medicago |
| Poplar | PoplarPLEX | www.plexdb.org/plex.php?database=Poplar |
| Rice | Rice Expression Database | http://tenor.dna.affrc.go.jp/ |
| Rice | NSF Rice Oligo Array Project Database | www.ricearray.org |
| Rice | Rice expression Profile Database | http://ricexpro.dna.affrc.go.jp/ |
| Rice | RicePLEX | www.plexdb.org/plex.php?database=Rice |
| Solanaceae | Solanaceae Gene Expression Database | www.tigr.org/tdb/potato |
| Solanaceae | SolGenomics Network | https://solgenomics.net/gem/experimental_design.pl?id=2 |
| Soybean | Soybean Genome and Microarray Database | https://omictools.com/soybean-genome-and-microarray-database-tool |
| Soybean | SoyPLEX | www.plexdb.org/plex.php?database=Soybean |
| Sugarcane | SugarPLEX | www.plexdb.org/plex.php?database=Sugarcane |
| Tobacco | TOBFAC | http://infogizmo.com/tobfac_tobacco_transcription_factors.html |
| Tomato | Tomato Functional Genomics Database | http://ted.bti.cornell.edu |
| Tomato | TomPLEX | www.plexdb.org/plex.php?database=Tomato |
| *Vitis vinifera* (grape) | VvGDB | www.plantgdb.org/VvGDB/ |
| Wheat | WheatPLEX | www.plexdb.org/plex.php?database=Wheat |

Database (http://mmcd.nmrfam.wisc.edu/) contains information about metabolites determined using NMR and MS. Another integrative cross-species and cross-technique database is Metabolights (www.ebi.ac.uk/metabolights/), which contains information about metabolite structures, spectra, and their biological roles. Metabolomics at Rothamsted (MeT-RO) is a project that contains several resources that can be applied to plant and microbial metabolomics (www.metabolomics.bbsrc.ac.uk/MeT-RO.htm). The Metlin Metabolite Database contains information on about 240,000 metabolites and nearly 72,000 high-resolution MS/MS spectra and tandem MS experiments (http://metlin.scripps.edu/). PRIMe, a platform for RIKEN metabolomics, is a database that integrates genomic and metabolomics data. PRIMe contains information on metabolites obtained from GC-MS, LC-MS, CE-MS and NMR spectroscopy (http://prime.psc.riken.jp/). ReSpect for Phytochemicals (http://spectra.psc.riken.jp/menta.cgi/index) is a collection of more than 9000 literature and in-house MSn spectra records for research on plant metabolomics. MassBank (www.massbank.jp) is a shared public repository of mass spectral data that, as of this writing (September 2016), contained 41,092 spectra records. These data are useful for chemical identification of compounds detected by mass spectrometry. The MS-MS Fragment Viewer (http://webs2.kazusa.or.jp/msmsfragmentviewer/) is a metabolomics database that contains FT-, IT-MS/MS and FT-MS spectral data with predicted structures of ion fragments observed in LC-FT/ICR-MS analysis. Metabolome Express (www.metabolome-express.org) is a public place where GC/MS metabolomics datasets can be processed, interpreted, and shared. It houses both uncurated repositories as well as quality-controlled databases.

Databases specific for individual plant species have also been developed; an example is the Metabolome Tomato Database (MoTo DB) (www.ab.wur.nl/moto/), which is based on metabolites from *Solanum lycopersicum* obtained by LC-MS. Plant Metabolomics (http://plantmetabolomics.vrac.iastate.edu/ver2/) is a metabolomic and functional genomics tool for determining the roles of *Arabidopsis thaliana* genes. Terpmed (www.terpmed.eu/databases.html) is another plant related database that contains information about plant terpenoids, particularly sesquiterpene lactones and phenolic diterpenes important for use as therapeutic drugs.

There are also several resources that integrate metabolic pathways with metabolite data such as MetaCyc (http://metacyc.org/) that contains information on ~2450 pathways from more than 2780 organisms, and BioCyc (http://biocyc.org) a collection of about 7600 pathway and genome databases. PlantCyc (http://plantcyc.org) is a comprehensive multispecies database that collects biochemical information from many plant pathways. Other related databases are RiceCyc (http://pathway.gramene.org/gramene/ricecyc.shtml), the Sol Genomics Network (http://solgenomics.net/tools/solcyc/index.pl), AraCyc (www.arabidopsis.org/biocyc/), HumanCyc (http://humancyc.org/), Mapman (http://mapman.gabipd.org/web/guest/mapman), ChemSpider (www.chemspider.com/), Reactome (www.reactome.org), the KEGG Pathway database (www.genome.jp/kegg/pathway.html), KappaView (http://kpv.kazusa.or.jp/kpv4/), and KNApSAcK (http://kanaya.naist.jp/KNApSAcK/). These databases are summarized in Table 5.2.

## Applications of bioinformatics to enhance fruit development and ripening

The increasing number of NGS technologies and new bioinformatics algorithms developed since the publication of the *Arabidopsis thaliana* genome sequence in 2000 (The Arabidopsis Genome Initiative, 2000), have enabled a deeper understanding of plant biology in both model species and crops (Gapper *et al.*, 2014).

Current NGS technologies enable whole-genome shotgun DNA sequencing, and have become established as the preferred genome sequencing technology, due to longer read lengths, drastically reduced costs, and improved assembly algorithms. These advantages are reflected in the fact that the genomes sequenced in the last 10 years are mostly from crops or non-model plant species, allowing an increase in the number of applied studies in crops and fruits of interest (Feuillet *et al.*, 2011; Gapper *et al.*, 2014; Hamilton and Buell, 2012).

**Table 5.2** Metabolomics resources databases

| Name | URL | Information/species |
|---|---|---|
| ChEBI | www.ebi.ac.uk/chebi/ | Dictionary of small chemical molecules |
| PubChem | http://pubchem.ncbi.nlm.nih.gov/ | General dictionary of chemicals |
| Fiehn Lab Chemical | http://cts.fiehnlab.ucdavis.edu/ | Chemical structures, names, synonyms and database identifiers |
| **Translation service** | | |
| Golm metabolome database | http://gmd.mpimp-golm.mpg.de/ | GS-MS |
| Human metabolome database | www.hmdb.ca/ | Chemical and biological data of human metabolites |
| Madison metabolomics | http://mmcd.nmrfam.wisc.edu/ | NMR and MS |
| **Database** | | |
| Metabolights | www.ebi.ac.uk/metabolights/ | Metabolite structures, spectra, function/ cross-species |
| Metabolomics at Rothamsted | www.metabolomics.bbsrc.ac.uk/ MeT-RO.htm | Plant and microbial metabolites |
| **MeT-RO** | | |
| Metlin metabolite database | http://metlin.scripps.edu/ | High-resolution MS/MS spectra and tandem MS experiments |
| PRIMe | http://prime.psc.riken.jp/ | Genomic and metabolomics data, NMR spectroscopy, GC-MS, LC-MS and CE-MS |
| ReSpect for Phytochemicals | http://spectra.psc.riken.jp/menta.cgi/index | MSn spectra data for research on plant metabolomics |
| MassBank | www.massbank.jp | Database of mass spectral data |
| MS-MS Fragment Viewer | http://webs2.kazusa.or.jp/ msmsfragmentviewer/ | FT-MS, IT- and FT-MS/MS spectral data |
| MetabolomeExpress | http://metabolome-express.org | GC/MS datasets |
| Metabolome Tomato Database | www.ab.wur.nl/moto/ | Metabolites identified by LC-MS |
| **MoTo DB** | | |
| Plantmetabolomics | http://plantmetabolomics.vrac.iastate.edu/ver2/ | *Arabidopsis* and other plant species |
| Terpmed | www.terpmed.eu/databases.html | Plant terpenoids, natural products, secondary metabolites, therapeutic drugs |
| MetaCyc | http://metacyc.org/ | Integration of metabolite data with metabolic pathways/2000 organisms |
| BioCyc | http://biocyc.org | Collection of pathway/genome database |
| PlantCyc | http://plantcyc.org | Multispecies plant database |
| RiceCyc | http://pathway.gramene.org/ gramene/ricecyc.shtml | Metabolic pathways, enzymes, metabolites |
| Sol Genomics Network | http://solgenomics.net/tools/solcyc/ index.pl | Pathway genome databases/Solanaceae species |
| AraCyc | www.arabidopsis.org/biocyc/ | Metabolic pathways, compounds, *Arabidopsis* |
| HumanCyc | http://humancyc.org/ | Metabolic pathways, genome/human |
| Mapman | http://mapman.gabipd.org/web/ guest/mapman | Datasets (e.g. gene expression data, metabolic pathways) |
| ChemSpider | www.chemspider.com/ | Chemical structure database |
| Reactome | www.reactome.org | Free curated and peer reviewed pathway database |

**Table 5.2** Continued

| Name | URL | Information/species |
|---|---|---|
| KEGG Pathway database | www.genome.jp/kegg/pathway.html | Pathways, metabolism, genetic information, cellular processes, human diseases |
| KappaView | http://kpv.kazusa.or.jp/kpv4/ | Metabolic pathway database aimed at metabolic regulation |
| KNApSAcK | http://kanaya.naist.jp/KNApSAcK/ | Species–metabolite relationship database |

The genome sequences of 27 fruit and nut crops have been published to date; these include grape (Jaillon *et al.*, 2007), papaya (Ming *et al.*, 2008), cucumber (Huang *et al.*, 2009), apple (Velasco *et al.*, 2010), strawberry (Shulaev *et al.*, 2011), tomato (Tomato Genome Consortium, 2012), melon (Garcia-Mas *et al.*, 2012), banana (D'Hont *et al.*, 2012), Chinese plum (Zhang *et al.*, 2012), pear (Wu *et al.*, 2013), watermelon (Guo *et al.*, 2013), peach (Verde *et al.*, 2013), kiwifruit (Huang *et al.*, 2013), orange (Xu *et al.*, 2013a), mandarin (Wang *et al.*, 2015), jujube (Liu *et al.*, 2014), mulberry (He *et al.*, 2013), pineapple (Zhang *et al.*, 2014), blueberry (Gupta *et al.*, 2015), raspberry (VanBuren *et al.*, 2016), mango (Azim *et al.*, 2014), macadamia nut (Nock *et al.*, 2014), cocoa (Motamayor *et al.*, 2013), coffee (Denoeud *et al.*, 2014), hazelnut (Hampson *et al.*, 1996), aubergine (Hirakawa *et al.*, 2014) and pepper (Qin *et al.*, 2014).

Nevertheless, the additional development of physical and functional annotation and improved genome assembly is required. For instance, the assembly of the tomato genome is the most complete of all fruit species sequenced, but functional annotation of its genes is still incomplete (Gapper *et al.*, 2014).

The new genomics approaches enable the identification of genes responsible for mutant phenotypes in a way that is much faster than ever before. One example is the cloning of a GOLDEN-LIKE2 Myb superfamily transcription factor responsible for the uniform ripening mutant in tomato. This mutation has been bred into many commercial cultivars resulting in more uniform ripening (Nguyen *et al.*, 2014; Powell *et al.*, 2012).

The availability of a collection of mutants with ripening inhibition phenotypes has been essential to clarifying the transcriptional control of fruit ripening (Gapper *et al.*, 2014). Tomato has one of the best characterized mutant collections available (http://tgrc.ucdavis.edu/) (Gapper *et al.*, 2014; Seymour *et al.*, 2013). Similar collections are the European Prunus Database (www.bordeaux.inra.fr/eupeachdb/) and the Citrus Genome Database (www.citrusgenomedb.org/).

Gene expression during fruit ripening is also regulated by epigenetic modifications, which can consist of genome structure, packaging, and non-sequence-based changes such as DNA methylation, acetylation, and histone modifications (Gallusci *et al.*, 2016; Gapper *et al.*, 2014). The entire set of epigenetic modifications, known as the epigenome, can be elucidated using NGS approaches. Recently, using bisulfite sequencing of the tomato genome, Liu and colleagues demonstrated for the first time that active DNA demethylation is an absolute requirement for fruit ripening to occur, and show a direct cause and effect relationship between hypermethylation at specific promoters and repression of gene expression (Liu *et al.*, 2015). However, even in tomato, the extent and role of the epigenetic regulation of fruit ripening is still relatively poorly understood (Gallusci *et al.*, 2016).

Several transcriptomic studies have identified new regulators of fruit development and ripening. Zhu and colleagues, using RNA sequencing and functional analysis, demonstrated the regulatory role of long non-coding RNAs in tomato fruit ripening (Zhu *et al.*, 2015). In another example, Asif and collaborators examined the global changes that occurred during fruit ripening in banana, sequencing cDNA libraries from unripe and ripe banana fruit pulp using the 454-GS platform (Asif *et al.*, 2014).

Frequently used proteomics techniques are now being adapted to high throughput screening methods that allow for the increased use of annotated genome sequences and robust bioinformatics tools.

Mass spectrometry-based methods have gained importance compared with antibody-based techniques owing to their higher specificity, precision and reproducibility, and the capability to quickly analyse numerous peptide transitions in a single test (Findeisen and Neumaier, 2009; Gapper *et al.*, 2014).

Using high-throughput iTRAQ (isobaric tag for relative and absolute quantization of tryptic peptides) and high-resolution mass spectrometry, Li and colleagues performed comparative analyses of the proteome and transcriptome during pear fruit development and identified 35 differentially expressed proteins related to fruit quality, including proteins related to sugar formation, aroma synthesis, and to the formation of lignin (Li *et al.*, 2015). Studies of this type can provide resources for improvement of fruit quality in pear.

Fruits are an extremely rich source of primary and secondary metabolites. Metabolomics studies on the fruit ripening process have been undertaken in several species including peach (Lombardo *et al.*, 2011), melon (Moing *et al.*, 2011), tomato (Oms-Oliu *et al.*, 2011), apple (Rudell *et al.*, 2009), pear (Pedreschi *et al.*, 2009), avocado (Pedreschi *et al.*, 2014) and pepper (Osorio *et al.*, 2012). These studies have provided new insights into conserved and unique metabolic pathways associated with fruit maturation (Gapper *et al.*, 2014).

One interesting recent study was a comparative investigation to identify common and/or species-specific modes of regulation in sugar accumulation. For that purpose, Dai and collaborators designed a process-based mathematical framework to compare soluble sugar accumulation in three fruits: grape, tomato, and peach (Dai *et al.*, 2016). These authors demonstrated that at maturity, grape presented the highest soluble sugar concentrations, followed by peach and tomato. They also showed that the higher soluble sugar concentration in grape compared to tomato is a result of higher sugar importation, although the higher soluble sugar concentration in grape compared to peach is due to lower water dilution. On the other hand, they observed that carbon utilization for synthesis of non-soluble sugar compounds was conserved among the three fruit species. These results increase our understanding of the origins of differences in soluble sugar concentration between fleshy fruits.

## Biotechnology and bioinformatics for plant breeding

At present, there exists a continuous challenge in plant breeding to improve crops that produce fodder, fuel, food, or other products through the application of crop genetics (Ray and Satya, 2014). These challenges are also related to the fact that some crop genomes are not yet fully sequenced and annotated, either because these crops have been under-researched, or due to the structural complexity of some very large genomes (Sikhakhane *et al.*, 2016).

The accessibility of whole genome sequences allows for the identification of sequence-based single nucleotide polymorphisms (SNPs) as DNA markers instead of DNA fragment-based polymorphism identification, which has substantially increased the number of informative markers that can be developed (Stapley *et al.*, 2010). In this way, although there are various strategies for plant breeding, the use of genomics-assisted breeding is an effective and economic strategy that is becoming widely used (Sikhakhane *et al.*, 2016). Moreover, the genetics of particular traits in species with large and complex genomes may now be studied in related plants with smaller genomes that share conserved regions through comparative genomics. This will potentially identify genes or quantitative trait loci (QTL) and putative SNP markers to be annotated for genome-wide association mapping. The application of NGS techniques in several crops is revolutionizing and accelerating the pace of plant breeding (Sikhakhane *et al.*, 2016). Fig. 5.1 shows a scheme for plant breeding that incorporates NGS.

Since the introduction of the Sanger sequencing technique nearly 40 years ago (Sanger *et al.*, 1977), several significant advances have been made that address the limitations of this early technology. These advances complement the development of more sophisticated DNA sequencing methodologies that allow *de novo* genome sequencing by rapidly producing huge amounts of sequence information at very low cost. Table 5.3 shows a summary of the advances made in DNA sequencing technology, from dideoxy chain termination sequencing to modern NGS technologies including 454 sequencing (Roche), Illumina

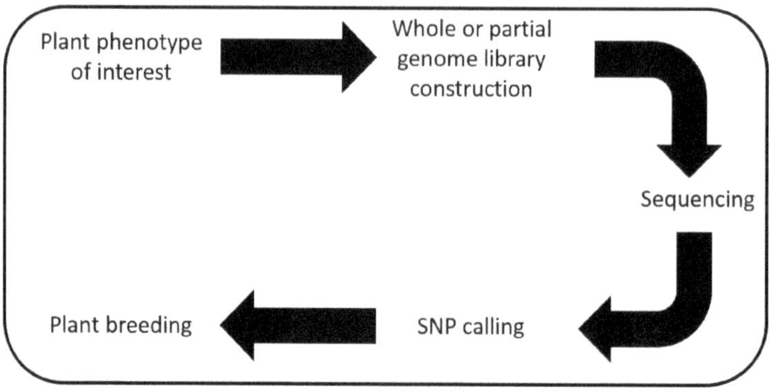


**Figure 5.1** Plant breeding assisted by NGS.

**Table 5.3** NGS evolution timeline

| Year introduced | Method/Platform | Read length | Performance (data yield) |
|---|---|---|---|
| 1977 | Sanger sequencing | Up to 1000 bp | Up to 84 kb per 3 hours |
| 2004 | Roche/454 | Up to 1000 bp | 700 Mb per day |
| 2006 | Illumina | Up to 300 bp | 130 Gb per 30 h |
| 2006 | ABI SOLiD | Up to 50 bp | 20 Gb per 72 hours |
| 2009 | Helicos | 35 bp (average) | 35 Gb per 8 days |
| 2010 | Ion Torrent/Pacific Biosciences | 35 to 400 bp | 2 Gb per 5 hours (average) |

sequencing-by-synthesis, SOLiD sequencing by oligonucleotide ligation and detection, single molecule sequencing (Helicos Biosciences and Pacific Biosciences), and Ion Torrent sequencing. There are available several modifications for each of these technologies and protocols are continuously being developed to overcome some of the current limitations.

Despite the beneficial prospects of the various NGS platforms, many limitations must be addressed to maximize their potential. These limitations are related to bioinformatics data analysis and the storage and integration of the large number of data generated by these technologies. This constant challenge is also related to the advent of new techniques and protocols in order to make data treatment and analysis easier. Therefore, high-throughput bioinformatics equipment, computational investments, and human resources combined with several NGS technologies will permit the data generated from different sources to be used universally.

Genotyping-by-sequencing (GBS), a widely used SNP discovery method, is feasible for species with large genomes and species with high levels of nucleotide diversity due to advances in genomic technologies that have decreased the cost of high throughput DNA sequencing. GBS is based on NGS technology and takes advantage of reducing genomic complexity to enable high-throughput genotyping of large numbers of samples at a large number of SNP marker loci (Glaubitz *et al.*, 2014). Repetitive regions of the genome may be avoided and lower copy regions targeted with high efficiency by using restriction enzymes (REs). This approach widely simplifies bioinformatic sequence alignment issues in species that possess a high level of genetic diversity (Elshire *et al.*, 2011). This robust, cost-effective, and straightforward method is widely used for genotyping and linkage mapping in many diverse plant species. Although the methods used to analyse raw GBS sequence data are constantly being improved, there are two main commonly used methods at present: genome-wide association studies (GWAS) (Cantor *et al.*, 2010; Visscher *et al.*, 2012) and TASSEL-GBS (Glaubitz *et al.*, 2014). These methods are designed for the efficient processing of raw GBS sequence data for SNP genotyping. GWAS and TASSEL-GBS procedures largely satisfy the following design criteria: (1) the ability to run on modest computing resources, including desktop or laptop machines without a lot of RAM, (2) scalability from small to large studies (large breeding programs), and (3) applicability in an accelerated breeding context. Wide adoption of these approaches across different methods will speed up the production of useful databases and datasets for effective plant breeding (Sikhakhane *et al.*, 2016).

## Bioinformatics for studying stress resistance in plants

The molecular regulatory networks involved in stress resistance and adaptation in plants can be decoded using a combination of omics studies (Unamba *et al.*, 2015). NGS technologies and potent computational pipelines have reduced the cost of whole genome and transcriptomic sequencing, allowing their application to model and non-model plants. This generates a large number of functional genomics resources, which allows us to increase our understanding of the molecular mechanisms underlying plant responses to stress conditions (Xu *et al.*, 2013b).

Plants constantly adjust their transcriptome profile in response to different stresses (Debnath *et al.*, 2011); thus, techniques such as microarrays or RNA-seq can provide a vast amount of information about differential gene expression. One example is the transcriptomic analysis of salt and extreme drought tolerance in *Ipomoea imperati*, a sweet potato relative, that was performed using 454 pyrosequencing by Solis and collaborators. This study identified several genes related to salt tolerance in this species. The information generated in this study could be useful in marker-assisted breeding to improve salt tolerance in sweet potato (Solis *et al.*, 2016).

In another report, Bhardwaj and colleagues performed the first comprehensive transcriptome study of *B. juncea* (the oilseed crop Indian mustard) under conditions of high temperature and drought stress (Bhardwaj *et al.*, 2015). They constructed three transcriptome libraries, sequenced them on the Illumina GA IIx platform, and assembled the high quality reads obtained using the SOAP *de novo* assembler. Bhardwaj *et al.* identified a subset of about 19,000 transcripts that are differentially regulated by either high temperature and/or drought stress. Most of the up-regulated transcription factor genes belonged to heat shock factors and dehydration-responsive element-binding gene families. This information will enhance the understanding of the molecular mechanisms behind the plant responses to abiotic stresses, and will enable efficient strategies to improve tolerance to high temperature and drought stress.

Considerable progress has been made towards engineering transgenic plants that express abiotic stress tolerance. Analyses of the large quantity of abiotic stress-related transcriptomic data generated in several plant species has revealed the importance of osmoprotectants in stress resistance. These osmoprotectants help protect plants against the damaging effects of osmotic and ionic stress. For example, several drought- and salt-resistant transgenic potato plants developed using this strategy have been reported. The tubers were successfully engineered to express an osmotin-like protein (Evers *et al.*, 1999), glyceraldehyde-3-phosphate dehydrogenase (Jeong *et al.*, 2001), and nucleoside diphosphate kinase 2 (Tang *et al.*, 2008).

The many efforts to improve abiotic stress tolerance in plants have resulted in important achievements (Khan *et al.*, 2015). However, due to the complexity of stress tolerance, the success of these strategies has been limited because most of the resulting transgenic plants have been tested only under controlled laboratory conditions. In such conditions, transgenic plants are evaluated at the early growth stage and are exposed to the stress condition for a short period of time. Short exposure experiments at early growth stages performed in laboratories may not predict the response of the plant under true field conditions. In addition, transgenic plants may be exposed to multiple stresses in the field, which may suppress the protective effects of the introduced transgene.

Future investigations into improving plant stress tolerance should put more emphasis on combining different approaches, such as multigenic strategies to simultaneously incorporate more than one gene in transgenic plants. For example, the genes involved in osmoprotectant synthesis could be co-expressed with other stress resistance-related genes such as ion transporters and transcription factors (Khan *et al.*, 2015).

## Bioinformatics approaches to study resistance to plant pathogens

Plants have evolved a diverse range of strategies to counteract pathogen attacks that involve the modification of gene expression, activation of several metabolic pathways, and posttranslational modification of proteins (Gomez-Casati *et al.*, 2016).

The new DNA sequencing technologies, combined with sophisticated bioinformatics, are having a major influence on research in the field of plant pathology. These technologies are gaining importance in studies of the genomics, metagenomics, proteomics, metabolomics, and transcriptomics of both the host plant and the pathogen, and are also extremely useful in epidemiology and diagnostics, especially with respect to viruses (Studholme *et al.*, 2011).

Whole genome sequences are available for numerous plants and a diverse group of microbial phytopathogens that includes bacteria, viruses, fungi, and oomycetes. Many of these microbial genome sequences were obtained using the traditional Sanger sequencing method. The first plant pathogen genome sequenced using second-generation sequencing technologies was *P. syringae* pathovar *oryzae*, a bacterium that causes halo blight on rice (Reinhardt *et al.*, 2009; Studholme *et al.*, 2011). Reinhardt and colleagues sequenced the *P. syringae* pv. *oryzae* genome by combining data generated from the Roche 454 and Illumina sequencing platforms and using a new bioinformatics pipeline based on their VCAKEv1.5 assembly algorithm (Jeck *et al.*, 2007). Since that time, several other phytopathogen genomes have been sequenced (Harrison *et al.*, 2016; Studholme *et al.*, 2011; Wasukira *et al.*, 2014). These draft genome sequences are expected to contain errors and be incomplete. However, they can provide valuable information about the molecular basis for infection of plant hosts, including he sequences of potential novel virulence factors, like T3SS effectors (Studholme *et al.*, 2009).

For disease diagnostics, NGS allows the direct identification of bacteria or viruses in an unbiased way without requiring previous knowledge of the pathogen DNA/RNA sequence or antibodies (Quan *et al.*, 2008). Finding a pathogen nucleic acid sequence within host tissue does not necessarily prove that the pathogen is causing the plant disease symptoms (Studholme *et al.*, 2011). However, several published cases present concrete evidence for a relationship between the presence of disease and the pathogen found. For example, Adams and colleagues found that the presence of *Carrot yellow leaf virus* (CYLV) is strongly related to the occurrence of internal necrosis of carrots by using a metagenomic approach, which consisted of constructing an RNA-seq library and sequencing it on the Illumina Miseq instrument (Adams *et al.*, 2014).

At present, most of plant RNA viruses identified by NGS were obtained from either total nucleic acid or total double-stranded RNA extracted from infected plant tissue (Barba *et al.*, 2014). In another strategy, host nucleic acid was eliminated by hybridization to nucleic acid isolated from healthy plants in order to enrich for virus sequences prior to sequencing. Plant viruses can also be identified indirectly by detecting specific RNA molecules, called short interfering RNAs (siRNA), that the host plant generates in response to infection by RNA viruses (Mlotshwa *et al.*, 2008). NGS of siRNAs is a good approach to identify viruses infecting plants, even when they are present at very low titres and in symptomless infections.

Microbial metagenomic studies can be performed on environmental samples or individual plants (Dutta *et al.*, 2014; Roossinck *et al.*, 2015). In general, most viruses identified in wild plants cannot be associated with pathologies, although some of them can be pathogenic when infecting domesticated plants. Thus, the use of metagenomics data provides advance knowledge of viruses that could cause threats if they jump from wild species into domestic hosts.

Using high-throughput transcriptomic technologies, such as microarrays or RNA-seq, makes it possible to obtain gene expression profiles from the host plants and to detect gene expression changes in the host-associated pathogen. On the other hand, the use of different proteomics approaches allows for the identification of many resistance proteins that plants express in the presence of specific pathogen effectors that trigger its defence mechanisms. Using this methodology, several antimicrobial proteins expressed during phytopathogenic interactions have been identified, mainly by high resolution of two-dimensional polyacrylamide gel electrophoresis coupled with mass spectrometry (Gomez-Casati *et al.*, 2016; Mehta *et al.*, 2008).

All of the omics technologies discussed here can provide a wealth of information about the pathogen life cycle, and enable the discovery of novel virulence factors in pathogens as well as their host targets (Gomez-Casati *et al.*, 2016). Both approaches are important for developing strategies to improve disease resistance in plants.

Traditionally, the control of pathogen invasion and spread has been achieved through breeding programmes that focus on increasing host resistance to particular pathogens, or by the application of chemical pesticides and fungicides. A relatively straightforward approach to generate pathogen resistant plants is through introduction of genes from other organisms to develop new cellular metabolic pathways.

Several genes involved in specific immune responses to different phytopathogens were identified using RNA-seq technology; these include genes that encode a tyrosine protein kinase (Epk1) (Pombo *et al.*, 2014), a cell wall-associated kinase (SlWAK1) (Rosli *et al.*, 2013), and a nucleotide binding leucine-rich repeat receptor (NB-LRR) (Bernoux *et al.*, 2011).

Whereas some attempts to engineer disease resistance in economically important crop plants have failed, several have been successful. Cao and colleagues reported transgenic maize lines that constitutively express a mutant *E. coli* dsRNA-specific endoribonuclease gene and show increased resistance to the dsRNA virus that causes Maize rough dwarf disease (Cao *et al.*, 2013). Mysore and collaborators showed that the overexpression of a serine/threonine kinase gene (Pto) in tomato confers protection against *P. syringae* (Mysore *et al.*, 2003). In addition, the overexpression of NPR1, a gene involved in systemic acquired resistance in plants, resulted in increased disease resistance in apple, *Malus domestica*, to two important fungal pathogens, *Venturia inaequalis* and *Gymnosporangium juniperi-virginianae* (Gomez-Casati *et al.*, 2016; Malnoy *et al.*, 2007).

On the other side, the most usual way of obtaining virus resistant plants is by expressing in them a viral gene encoding the coat protein. So, the plant produces this viral protein before the virus attacks, and when the virus tries to infect the plant, the host gene silencing pathway is already activated and degrades the virus RNA early in the infection process. All genetically modified virus resistant plants existing on the market (e.g. squash and papayas) have this resistance mechanism (Ratcliff *et al.*, 1999; Voinnet, 2001).

## Conclusions

The application of bioinformatics to the study and characterization of biological phenomena represents a fundamental shift in the way that scientists study living organisms. As mentioned above, the many genomic, transcriptomic, proteomic, and metabolite databases available have the potential to accelerate the rate of functional discoveries in plant biology. Although these databases provide access to gene information, gene expression data, and metabolite profiles, as well as the experimental conditions used to generate them, additional measures are needed to complement genome annotation efforts effectively and to increase discovery of gene function and regulation. One possible course of action would be the generation of reference expression datasets for plant cells and/or tissues from specific stages of plant development. In addition, it would be highly desirable to develop databases which could contain increasingly integrated information about genomic, transcriptomic, metabolomic, as well as proteomic and phenomic data for each plant species.

## Acknowledgements

This work was supported by grants from ANPCyT (PICT 2013-2188 and 2014-2184).

## References

Adams, I.P., Skelton, A., Macarthur, R., Hodges, T., Hinds, H., Flint, L., Nath, P.D., Boonham, N., and Fox, A. (2014). Carrot yellow leaf virus is associated with carrot internal necrosis. PLOS ONE9, e109125. http://dx.doi.org/10.1371/journal.pone.0109125

Agarwal, P., Parida, S.K., Mahto, A., Das, S., Mathew, I.E., Malik, N., and Tyagi, A.K. (2014). Expanding frontiers in plant transcriptomics in aid of functional genomics and molecular breeding. Biotechnol. J. 9, 1480–1492. http://dx.doi.org/10.1002/biot.201400063

Aharoni, A., De Vos, C.H., Wein, M., Sun, Z., Greco, R., Kroon, A., Mol, J.N., and O'Connell, A.P. (2001). The strawberry FaMYB1 transcription factor suppresses anthocyanin and flavonol accumulation in transgenic tobacco. Plant J. *28*, 319–332. http://dx.doi.org/10.1046/j.1365-313X.2001.01154.x

Asif, M.H., Lakhwani, D., Pathak, S., Gupta, P., Bag, S.K., Nath, P., and Trivedi, P.K. (2014). Transcriptome analysis of ripe and unripe fruit tissue of banana identifies major metabolic networks involved in fruit ripening process. BMC Plant Biol. *14*, 316. http://dx.doi.org/10.1186/s12870-014-0316-1

Azim, M.K., Khan, I.A., and Zhang, Y. (2014). Characterization of mango (Mangifera indica L.) transcriptome and chloroplast genome. Plant Mol. Biol.*85*, 193–208. http://dx.doi.org/10.1007/s11103-014-0179-8

Barba, M., Czosnek, H., and Hadidi, A. (2014). Historical perspective, development and applications of next-generation sequencing in plant virology. Viruses *6*, 106–136. http://dx.doi.org/10.1016/j.pbi.2011.05.005

Bernoux, M., Ellis, J.G., and Dodds, P.N. (2011). New insights in plant immunity signaling activation. Curr. Opin. Plant Biol. *14*, 512–518. http://dx.doi.org/10.1016/j.pbi.2011.05.005

Bhardwaj, A.R., Joshi, G., Kukreja, B., Malik, V., Arora, P., Pandey, R., Shukla, R.N., Bankar, K.G., Katiyar-Agarwal, S., Goel, S., *etal.* (2015). Global insights into high temperature and drought stress regulated genes by RNA-Seq in economically important oilseed crop *Brassica juncea*. BMC Plant Biol. *15*, 9. http://dx.doi.org/10.1186/s12870-014-0405-1

Brown, P.O., and Botstein, D. (1999). Exploring the new world of the genome with DNA microarrays. Nat. Genet.*21* (Suppl. 1),33–37. http://dx.doi.org/10.1038/4462

Cantor, R.M., Lange, K., and Sinsheimer, J.S. (2010). Prioritizing GWAS results: A review of statistical methods and recommendations for their application. Am. J. Hum. Genet. *86*, 6–22. http://dx.doi.org/10.1016/j.ajhg.2009.11.017

Cao, X., Lu, Y., Di, D., Zhang, Z., Liu, H., Tian, L., Zhang, A., Zhang, Y., Shi, L., Guo, B., *et al.* (2013). Enhanced virus resistance in transgenic maize expressing a dsRNA-specific endoribonuclease gene from *E. coli*. PLOS ONE *8*, e60829. http://dx.doi.org/10.1371/journal.pone.0060829

D'Hont, A., Denoeud, F., Aury, J.M., Baurens, F.C., Carreel, F., Garsmeur, O., Noel, B., Bocs, S., Droc, G., Rouard, M., *et al.* (2012). The banana (*Musa acuminata*) genome and the evolution of monocotyledonous plants. Nature *488*, 213–217. http://dx.doi.org/10.1038/nature11241

Dai, Z., Wu, H., Baldazzi, V., van Leeuwen, C., Bertin, N., Gautier, H., Wu, B., Duchêne, E., Gomès, E., Delrot, S., *et al.* (2016). Inter-species comparative analysis of components of soluble sugar concentration in fleshy fruits. Front. Plant Sci. *7*, 649. http://dx.doi.org/10.3389/fpls.2016.00649

Debnath, M., Pandey, M., and Bisen, P.S. (2011). An omics approach to understand the plant abiotic stress. OMICS *15*, 739–762. http://dx.doi.org/10.1089/omi.2010.0146

Denoeud, F., Carretero-Paulet, L., Dereeper, A., Droc, G., Guyot, R., Pietrella, M., Zheng, C., Alberti, A., Anthony, F., Aprea, G., *etal.* (2014). The coffee genome provides insight into the convergent evolution of caffeine biosynthesis. Science *345*, 1181–1184. http://dx.doi.org/10.1126/science.1255274

Dutta, M., Sokhandan Bashir, N., Palmer, M.W., and Melcher, U. (2014). Genomic characterization of Ambrosia asymptomatic virus 1 and evidence of other Tymovirales members in the Oklahoma tallgrass prairie revealed by sequence analysis. Arch. Virol. *159*, 1755–1764. http://dx.doi.org/10.1016/j.tibtech.2004.03.002

Edwards, D., and Batley, J. (2004). Plant bioinformatics: from genome to phenome. Trends Biotechnol. *22*, 232–237. http://dx.doi.org/10.1016/j.tibtech.2004.03.002

Elshire, R.J., Glaubitz, J.C., Sun, Q., Poland, J.A., Kawamoto, K., Buckler, E.S., and Mitchell, S.E. (2011). A robust, simple genotyping-by-sequencing (GBS) approach for high diversity species. PLOS ONE*6*, e19379. http://dx.doi.org/10.1371/journal.pone.0019379

Evers, D., Overney, S., Simon, P., Greppin, H., and Hausman, J.F. (1999). Salt tolerance of *Solanum tuberosum* L. overexpressing an heterologous osmotin-like protein. Biol. Plant. *42*, 105–112. http://dx.doi.org/10.1023/a:1002131812340

Feuillet, C., Leach, J.E., Rogers, J., Schnable, P.S., and Eversole, K. (2011). Crop genome sequencing: lessons and rationales. Trends Plant Sci.*16*, 77–88. http://dx.doi.org/10.1016/j.tplants.2010.10.005

Findeis en, P., and Neumaier, M. (2009). Mass spectrometry-based clinical proteomics profiling: current status and future directions. Expert. Rev. Proteomics*6*, 457–459. http://dx.doi.org/10.1586/epr.09.67

Gallusci, P., Hodgman, C., Teyssier, E., and Seymour, G.B. (2016). DNA Methylation and Chromatin Regulation during Fleshy Fruit Development and Ripening. Front. Plant Sci.*7*, 807. http://dx.doi.org/10.3389/fpls.2016.00807

Gapper, N.E., Giovannoni, J.J., and Watkins, C.B. (2014). Understanding development and ripening of fruit crops in an 'omics' era. Hortic. Res.*1*, 14034. http://dx.doi.org/10.1038/hortres.2014.34

Garcia-Mas, J., Benjak, A., Sanseverino, W., Bourgeois, M., Mir, G., González, V.M., Hénaff, E., Câmara, F., Cozzuto, L., Lowy, E., *et al.* (2012). The genome of melon (Cucumis melo L.). Proc. Natl. Acad. Sci. U.S.A.*109*, 11872–11877. http://dx.doi.org/10.1073/pnas.1205415109

Glaubitz, J.C., Casstevens, T.M., Lu, F., Harriman, J., Elshire, R.J., Sun, Q., and Buckler, E.S. (2014). TASSEL-GBS: a high capacity genotyping by sequencing analysis pipeline. PLOS ONE*9*, e90346. http://dx.doi.org/10.1371/journal.pone.0090346

Gomez-Casati, D.F., Pagani, M.A., Busi, M.V., and Bhadauria, V. (2016). Omics approaches for the engineering of pathogen resistant plants. Curr. Issues. Mol. Biol.*19*, 89–98.

Gomez-Casati, D.F., Zanor, M.I., and Busi, M.V. (2013). Metabolomics in plants and humans: applications in the prevention and diagnosis of diseases. Biomed Res. Int. *2013*, 792527. http://dx.doi.org/10.1155/2013/792527

Guo, S., Zhang, J., Sun, H., Salse, J., Lucas, W.J., Zhang, H., Zheng, Y., Mao, L., Ren, Y., Wang, Z., *et al.* (2013). The draft genome of watermelon (Citrullus lanatus) and resequencing of 20 diverse accessions. Nat. Genet. *45*, 51–58. http://dx.doi.org/10.1038/ng.2470

Gupta, V., Estrada, A.D., Blakley, I., Reid, R., Patel, K., Meyer, M.D., Andersen, S.U., Brown, A.F., Lila, M.A., and Loraine, A.E. (2015). RNA-Seq analysis and annotation of a draft blueberry genome assembly identifies candidate genes involved in fruit ripening, biosynthesis of bioactive compounds, and stage-specific alternative splicing. Gigascience *4*, 5. http://dx.doi.org/10.1186/s13742-015-0046-9

Hamilton, J.P., and Buell, C.R. (2012). Advances in plant genome sequencing. Plant J. *70*, 177–190. http://dx.doi.org/10.1111/j.1365-313X.2012.04894.x

Hampson, C.R., Coleman, G.D., and Azarenko, A.N. (1996). Does the genome of *Corylus avellana* L. contain sequences homologous to the self-incompatibility gene of Brassica? Theor. Appl. Genet. *93*, 759–764. http://dx.doi.org/10.1007/BF00224073

Harrison, J., Grant, M.R., and Studholme, D.J. (2016). Draft Genome Sequences of Two Strains of Xanthomonas arboricola pv. celebensis Isolated from Banana Plants. Genome Announc. *4*, e01705-15. http://dx.doi.org/10.1128/genomeA.01705-15

He, N., Zhang, C., Qi, X., Zhao, S., Tao, Y., Yang, G., Lee, T.H., Wang, X., Cai, Q., Li, D., *et al.* (2013). Draft genome sequence of the mulberry tree Morus notabilis. Nat. Commun. *4*, 2445. http://dx.doi.org/10.1038/ncomms3445

Hirakawa, H., Shirasawa, K., Miyatake, K., Nunome, T., Negoro, S., Ohyama, A., Yamaguchi, H., Sato, S., Isobe, S., Tabata, S., *etal.* (2014). Draft genome sequence of eggplant (*Solanum melongena* L.): the representative solanum species indigenous to the old world. DNA Res. *21*, 649–660. http://dx.doi.org/10.1093/dnares/dsu027

Huang, S., Ding, J., Deng, D., Tang, W., Sun, H., Liu, D., Zhang, L., Niu, X., Zhang, X., Meng, M., *et al.* (2013). Draft genome of the kiwifruit Actinidia chinensis. Nat. Commun. *4*, 2640. http://dx.doi.org/10.1038/ncomms3640

Huang, S., Li, R., Zhang, Z., Li, L., Gu, X., Fan, W., Lucas, W.J., Wang, X., Xie, B., Ni, P., *etal.* (2009). The genome of the cucumber, Cucumis sativus L. Nat. Genet. *41*, 1275–1281. http://dx.doi.org/10.1038/ng.475

Jaillon, O., Aury, J.M., Noel, B., Policriti, A., Clepet, C., Casagrande, A., Choisne, N., Aubourg, S., Vitulo, N., Jubin, C., *et al.* (2007). The grapevine genome sequence suggests ancestral hexaploidization in major angiosperm phyla. Nature *449*, 463–467.

Jeck, W.R., Reinhardt, J.A., Baltrus, D.A., Hickenbotham, M.T., Magrini, V., Mardis, E.R., Dangl, J.L., and Jones, C.D. (2007). Extending assembly of short DNA sequences to handle error. Bioinformatics *23*, 2942–2944.

Jeong, M.J., Park, S.C., and Byun, M.O. (2001). Improvement of salt tolerance in transgenic potato plants by glyceraldehyde-3 phosphate dehydrogenase gene transfer. Mol. Cells *12*, 185–189.

Khan, M.S., Ahmad, D., and Khan, M.A. (2015). Utilization of genes encoding osmoprotectants in transgenic plants for enhanced abiotic stress tolerance. Electronic J. Biotechnol. *18*, 257–266. http://dx.doi.org/10.1016/j.ejbt.2015.04.002

Li, J.M., Huang, X.S., Li, L.T., Zheng, D.M., Xue, C., Zhang, S.L., and Wu, J. (2015). Proteome analysis of pear reveals key genes associated with fruit development and quality. Planta *241*, 1363–1379. http://dx.doi.org/10.1007/s00425-015-2263-y

Liu, M.J., Zhao, J., Cai, Q.L., Liu, G.C., Wang, J.R., Zhao, Z.H., Liu, P., Dai, L., Yan, G., Wang, W.J., *et al.* (2014). The complex jujube genome provides insights into fruit tree biology. Nat. Commun. *5*, 5315. http://dx.doi.org/10.1038/ncomms6315

Liu, R., How-Kit, A., Stammitti, L., Teyssier, E., Rolin, D., Mortain-Bertrand, A., Halle, S., Liu, M., Kong, J., Wu, C., *etal.* (2015). A DEMETER-like DNA demethylase governs tomato fruit ripening. Proc. Natl. Acad. Sci. U.S.A. *112*, 10804–10809. http://dx.doi.org/10.1038/ncomms6315

Lombardo, V.A., Osorio, S., Borsani, J., Lauxmann, M.A., Bustamante, C.A., Budde, C.O., Andreo, C.S., Lara, M.V., Fernie, A.R., and Drincovich, M.F. (2011). Metabolic profiling during peach fruit development and ripening reveals the metabolic networks that underpin each developmental stage. Plant Physiol. *157*, 1696–1710. http://dx.doi.org/10.1104/pp.111.186064

Malnoy, M., Jin, Q., Borejsza-Wysocka, E.E., He, S.Y., and Aldwinckle, H.S. (2007). Overexpression of the apple MpNPR1 gene confers increased disease resistance in Malus x domestica. Mol. Plant Microbe Interact.*20*, 1568–1580. http://dx.doi.org/10.1094/MPMI-20-12-1568

Mehta, A., Brasileiro, A.C., Souza, D.S., Romano, E., Campos, M.A., Grossi-de-Sá, M.F., Silva, M.S., Franco, O.L., Fragoso, R.R., Bevitori, R., *et al.* (2008). Plant-pathogen interactions: what is proteomics telling us? FEBS J.*275*, 3731–3746. http://dx.doi.org/10.1111/j.1742-4658.2008.06528.x

Ming, R., Hou, S., Feng, Y., Yu, Q., Dionne-Laporte, A., Saw, J.H., Senin, P., Wang, W., Ly, B.V., Lewis, K.L., *et al.* (2008). The draft genome of the transgenic tropical fruit tree papaya (Carica papaya Linnaeus). Nature *452*, 991–996. http://dx.doi.org/10.1038/nature06856

Mlotshwa, S., Pruss, G.J., and Vance, V. (2008). Small RNAs in viral infection and host defense. Trends Plant Sci. *13*, 375–382. http://dx.doi.org/10.1016/j.tplants.2008.04.009

Moing, A., Aharoni, A., Biais, B., Rogachev, I., Meir, S., Brodsky, L., Allwood, J.W., Erban, A., Dunn, W.B., Kay, L., *etal.* (2011). Extensive metabolic cross-talk in melon fruit revealed by spatial and developmental combinatorial metabolomics. New Phytol. *190*, 683–696. http://dx.doi.org/10.1111/j.1469-8137.2010.03626.x

Morin, R., Bainbridge, M., Fejes, A., Hirst, M., Krzywinski, M., Pugh, T., McDonald, H., Varhol, R., Jones, S., and Marra, M. (2008). Profiling the HeLa S3 transcriptome using randomly primed cDNA and massively parallel short-read sequencing. BioTechniques *45*, 81–94. http://dx.doi.org/10.2144/000112900

Motamayor, J.C., Mockaitis, K., Schmutz, J., Haiminen, N., Livingstone, D., Cornejo, O., Findley, S.D., Zheng, P., Utro, F., Royaert, S., *et al.* (2013). The genome sequence of the most widely cultivated cacao type and its use to identify candidate genes regulating pod color. Genome Biol. *14*, r53. http://dx.doi.org/10.1186/gb-2013-14-6-r53

Mysore, K.S., D'Ascenzo, M.D., He, X., and Martin, G.B. (2003). Overexpression of the disease resistance gene Pto in tomato induces gene expression changes similar to immune responses in human and fruitfly. Plant Physiol. *132*, 1901–1912. http://dx.doi.org/10.1104/pp.103.022731

Nguyen, C.V., Vrebalov, J.T., Gapper, N.E., Zheng, Y., Zhong, S., Fei, Z., and Giovannoni, J.J. (2014). Tomato

GOLDEN2-LIKE transcription factors reveal molecular gradients that function during fruit development and ripening. Plant Cell *26*, 585–601. http://dx.doi.org/10.1105/tpc.113.118794
Nock, C.J., Baten, A., and King, G.J. (2014). Complete chloroplast genome of Macadamia integrifolia confirms the position of the Gondwanan early-diverging eudicot family Proteaceae. BMC Genomics *15 Suppl 9*, S13. http://dx.doi.org/10.1186/1471-2164-15-S9-S13
Oms-Oliu, G., Hertog, M.L.A.T.M., Van de Poel, B., Ampofo-Asiama, J., Geeraerd, A.H., and Nicolaï, B.M. (2011). Metabolic characterization of tomato fruit during preharvest development, ripening, and postharvest shelf-life. Postharv Biol. Technol. *62*, 7–16. http://dx.doi.org/10.1016/j.postharvbio.2011.04.010
Osorio, S., Alba, R., Nikoloski, Z., Kochevenko, A., Fernie, A.R., and Giovannoni, J.J. (2012). Integrative comparative analyses of transcript and metabolite profiles from pepper and tomato ripening and development stages uncovers species-specific patterns of network regulatory behavior. Plant Physiol. *159*, 1713–1729. http://dx.doi.org/10.1104/pp.112.199711
Pedreschi, R., Hertog, M., Robben, J., Lilley, K.S., Karp, N.A., Baggerman, G., Vanderleyden, J., and Nicolaï, B. (2009). Gel-based proteomics approach to the study of metabolic changes in pear tissue during storage. J. Agric. Food Chem. *57*, 6997–7004. http://dx.doi.org/10.1021/jf901432h
Pedreschi, R., Muñoz, P., Robledo, P., Becerra, C., Defilippi, B.G., van Eekelen, H., Mumm, R., Westra, E., and de Vos, R.C.H. (2014). Metabolomics analysis of postharvest ripening heterogeneity of 'Hass' avocadoes. Postharv. Biol. Technol. *92*, 172–179. http://dx.doi.org/10.1021/jf901432h
Pombo, M.A., Zheng, Y., Fernandez-Pozo, N., Dunham, D.M., Fei, Z., and Martin, G.B. (2014). Transcriptomic analysis reveals tomato genes whose expression is induced specifically during effector-triggered immunity and identifies the Epk1 protein kinase which is required for the host response to three bacterial effector proteins. Genome Biol. *15*, 492. http://dx.doi.org/10.1186/s13059-014-0492-1
Powell, A.L., Nguyen, C.V., Hill, T., Cheng, K.L., Figueroa-Balderas, R., Aktas, H., Ashrafi, H., Pons, C., Fernandez-Munoz, R., Vicente, A., *etal.* (2012). Uniform ripening encodes a Golden 2-like transcription factor regulating tomato fruit chloroplast development. Science *336*, 1711–1715. http://dx.doi.org/10.1126/science.1222218
Qin, C., Yu, C., Shen, Y., Fang, X., Chen, L., Min, J., Cheng, J., Zhao, S., Xu, M., Luo, Y., *etal.* (2014). Whole-genome sequencing of cultivated and wild peppers provides insights into Capsicum domestication and specialization. Proc. Natl. Acad. Sci. U.S.A. *111*, 5135–5140. http://dx.doi.org/10.1073/pnas.1400975111
Quan, P.L., Briese, T., Palacios, G., and Ian Lipkin, W. (2008). Rapid sequence-based diagnosis of viral infection. Antiviral Res. *79*, 1–5. http://dx.doi.org/10.1016/j.antiviral.2008.02.002
Ratcliff, F.G., MacFarlane, S.A., and Baulcombe, D.C. (1999). Gene silencing without DNA. rna-mediated cross-protection between viruses Plant. Cell. *11*, 1207–1216.
Ray, S., and Satya, P. (2014). Next generation sequencing technologies for next generation plant breeding. Front. Plant Sci. *5*, 367. http://dx.doi.org/10.3389/fpls.2014.00367
Reinhardt, J.A., Baltrus, D.A., Nishimura, M.T., Jeck, W.R., Jones, C.D., and Dangl, J.L. (2009). De novo assembly using low-coverage short read sequence data from the rice pathogen *Pseudomonas syringae* pv. *oryzae*. Genome Res. *19*, 294–305. http://dx.doi.org/10.1101/gr.083311.108
Rhee, S.Y., Dickerson, J., and Xu, D. (2006). Bioinformatics and its applications in plant biology. Annu. Rev. Plant Biol. *57*, 335–360. http://dx.doi.org/10.1146/annurev.arplant.56.032604.144103
Roossinck, M.J., Martin, D.P., and Roumagnac, P. (2015). Plant virus metagenomics: advances in virus discovery. Phytopathology *105*, 716–727. http://dx.doi.org/10.1094/PHYTO-12-14-0356-RVW
Rosli, H.G., Zheng, Y., Pombo, M.A., Zhong, S., Bombarely, A., Fei, Z., Collmer, A., and Martin, G.B. (2013). Transcriptomics-based screen for genes induced by flagellin and repressed by pathogen effectors identifies a cell wall-associated kinase involved in plant immunity. Genome Biol. *14*, R139. http://dx.doi.org/10.1186/gb-2013-14-12-r139
Rudell, D.R., Mattheis, J.P., and Hertog, M.L. (2009). Metabolomic change precedes apple superficial scald symptoms. J. Agric. Food Chem. *57*, 8459–8466. http://dx.doi.org/10.1021/jf901571g
Sanger, F., Nicklen, S., and Coulson, A.R. (1977). DNA sequencing with chain-terminating inhibitors. Proc. Natl. Acad. Sci. U.S.A. *74*, 5463–5467.
Seymour, G.B., Østergaard, L., Chapman, N.H., Knapp, S., and Martin, C. (2013). Fruit development and ripening. Annu. Rev. Plant Biol. *64*, 219–241. http://dx.doi.org/10.1146/annurev-arplant-050312-120057
Shulaev, V., Sargent, D.J., Crowhurst, R.N., Mockler, T.C., Folkerts, O., Delcher, A.L., Jaiswal, P., Mockaitis, K., Liston, A., Mane, S.P., *et al.* (2011). The genome of woodland strawberry (*Fragaria vesca*). Nat. Genet. *43*, 109–116. http://dx.doi.org/10.1038/ng.740
Solis, J., Baisakh, N., Brandt, S.R., Villordon, A., and La Bonte, D. (2016). Transcriptome profiling of beach morning glory (*Ipomoea imperati*) under salinity and its comparative analysis with sweetpotato. PLOS ONE *11*, e0147398. http://dx.doi.org/10.1371/journal.pone.0147398
Stapley, J., Reger, J., Feulner, P.G., Smadja, C., Galindo, J., Ekblom, R., Bennison, C., Ball, A.D., Beckerman, A.P., and Slate, J. (2010). Adaptation genomics: the next generation. Trends Ecol. Evol.*25*, 705–712. http://dx.doi.org/10.1016/j.tree.2010.09.002
Studholme, D.J., Glover, R.H., and Boonham, N. (2011). Application of high-throughput DNA sequencing in phytopathology. Annu. Rev. Phytopathol. *49*, 87–105. http://dx.doi.org/10.1146/annurev-phyto-072910-095408
Studholme, D.J., Ibanez, S.G., MacLean, D., Dangl, J.L., Chang, J.H., and Rathjen, J.P. (2009). A draft genome sequence and functional screen reveals the repertoire of type III secreted proteins of *Pseudomonas syringae* pathovar *tabaci* 11528. BMC Genomics *10*, 395. http://dx.doi.org/10.1146/annurev-phyto-072910-095408

Tang, L., Kim, M.D., Yang, K.S., Kwon, S.Y., Kim, S.H., Kim, J.S., Yun, D.J., Kwak, S.S., and Lee, H.S. (2008). Enhanced tolerance of transgenic potato plants overexpressing nucleoside diphosphate kinase 2 against multiple environmental stresses. Transgenic Res. *17*, 705–715. http://dx.doi.org/10.1007/s11248-007-9155-2

Sikhakhane, T.N., Figlan, S., Mwadzingeni, L., Ortiz, R., and Tsilo, T.J. (2016). Integration of Next-generation Sequencing Technologies with Comparative Genomics in Cereals. In Agricultural and Biological Sciences 'Plant Genomics', InTech, ed. http://dx.doi.org/10.5772/61763

The Arabidopsis Genome Initiative (2000). Analysis of the genome sequence of the flowering plant *Arabidopsis thaliana*. Nature *408*, 796–815. http://dx.doi.org/10.1038/35048692

Tomato Genome Consortium. (2012). The tomato genome sequence provides insights into fleshy fruit evolution. Nature *485*, 635–641. http://dx.doi.org/10.1038/nature11119

Unamba, C.I., Nag, A., and Sharma, R.K. (2015). Next generation sequencing technologies: the doorway to the unexplored genomics of non-model plants. Front. Plant Sci. *6*, 1074. http://dx.doi.org/10.3389/fpls.2015.01074

VanBuren, R., Bryant, D., Bushakra, J.M., Vining, K.J., Edger, P.P., Rowley, E.R., Priest, H.D., Michael, T.P., Lyons, E., Filichkin, S.A., *et al.* (2016). The genome of black raspberry (Rubus occidentalis). Plant J. *87*, 535–547. http://dx.doi.org/10.1111/tpj.13215

Velasco, R., Zharkikh, A., Affourtit, J., Dhingra, A., Cestaro, A., Kalyanaraman, A., Fontana, P., Bhatnagar, S.K., Troggio, M., Pruss, D., *et al.* (2010). The genome of the domesticated apple (Malus × domestica Borkh.). Nat. Genet.*42*, 833–839. http://dx.doi.org/10.1038/ng.654

Verde, I., Abbott, A.G., Scalabrin, S., Jung, S., Shu, S., Marroni, F., Zhebentyayeva, T., Dettori, M.T., Grimwood, J., Cattonaro, F., *etal.* (2013). The high-quality draft genome of peach (*Prunus persica*) identifies unique patterns of genetic diversity, domestication and genome evolution. Nature Genet. *45*, 487–494. http://dx.doi.org/10.1038/ng.2586

Visscher, P.M., Brown, M.A., McCarthy, M.I., and Yang, J. (2012). Five years of GWAS discovery. Am. J. Hum. Genet. *90*, 7–24. http://dx.doi.org/10.1016/j.ajhg.2011.11.029

Voinnet, O. (2001). RNA silencing as a plant immune system against viruses. Trends Genet. *17*, 449–459.

Wang, Y., Zhou, L., Li, D., Dai, L., Lawton-Rauh, A., Srimani, P.K., Duan, Y., and Luo, F. (2015). Genome-wide comparative analysis reveals similar types of NBS genes in hybrid *Citrus sinensis* genome and original *Citrus clementine* genome and provides new insights into non-TIR NBS genes. PLOS ONE*10*, e0121893. http://dx.doi.org/10.1371/journal.pone.0121893

Wasukira, A., Coulter, M., Al-Sowayeh, N., Thwaites, R., Paszkiewicz, K., Kubiriba, J., Smith, J., Grant, M., and Studholme, D.J. (2014). Genome sequencing of *Xanthomonas vasicola* Pathovar *vasculorum* reveals variation in plasmids and genes encoding lipopolysaccharide synthesis, Type-IV Pilus and Type-III secretion effectors. Pathogens 3, 211–237. http://dx.doi.org/10.3390/pathogens3010211

Weber, A.P. (2015). Discovering new biology through sequencing of RNA. Plant Physiol. *169*, 1524–1531. http://dx.doi.org/10.1104/pp.15.01081

Wu, J., Wang, Z., Shi, Z., Zhang, S., Ming, R., Zhu, S., Khan, M.A., Tao, S., Korban, S.S., Wang, H., *et al.* (2013). The genome of the pear (*Pyrus bretschneideri* Rehd.). Genome Res. 23, 396–408. http://dx.doi.org/10.1101/gr.144311.112

Xu, Q., Chen, L.L., Ruan, X., Chen, D., Zhu, A., Chen, C., Bertrand, D., Jiao, W.B., Hao, B.H., Lyon, M.P., *et al.* (2013a). The draft genome of sweet orange (Citrus sinensis). Nat. Genet. *45*, 59–66. http://dx.doi.org/10.1038/ng.2472

Xu, Y., Gao, S., Yang, Y., Huang, M., Cheng, L., Wei, Q., Fei, Z., Gao, J., and Hong, B. (2013b). Transcriptome sequencing and whole genome expression profiling of chrysanthemum under dehydration stress. BMC Genomics *14*, 662. http://dx.doi.org/10.1186/1471-2164-14-662

Zhang, J., Liu, J., and Ming, R. (2014). Genomic analyses of the CAM plant pineapple. J. Exp. Bot. *65*, 3395–3404. http://dx.doi.org/10.1093/jxb/eru101

Zhang, Q., Chen, W., Sun, L., Zhao, F., Huang, B., Yang, W., Tao, Y., Wang, J., Yuan, Z., Fan, G., *et al.* (2012). The genome of Prunus mume. Nat. Commun. 3, 1318. http://dx.doi.org/10.1038/ncomms2290

Zhu, B., Yang, Y., Li, R., Fu, D., Wen, L., Luo, Y., and Zhu, H. (2015). RNA sequencing and functional analysis implicate the regulatory role of long non-coding RNAs in tomato fruit ripening. J. Exp. Bot. *66*, 4483–4495. http://dx.doi.org/10.1093/jxb/erv203

# Quantitative Genetics of Disease Resistance in Wheat

6

Vijai Bhadauria[1]* and Lucia Popescu[2]*

[1]Crop Development Centre and Department of Plant Sciences, University of Saskatchewan, Saskatoon, SK, Canada.
[2]Department of Soil Science, University of Saskatchewan, Saskatoon, SK, Canada.

*Correspondence: vijai.bhadauria@canada.ca and lucia.popescu@usask.ca
The authors contributed equally

https://doi.org/10.21775/9781910190654.07

## Abstract

Wheat (*Triticum aestivum* L.) is one of the top three global food security crops. Fusarium head blight is one of the major constraints in sustainable wheat production and resistance to the disease is polygenic. This review provides an overview of recent efforts in mapping these genes/loci with the objective to aid marker-assisted selection breeding.

Bread wheat (*Triticum aestivum* L.) is one of the top three stable food crops and accounts for 20% of calories consumed by people (Brenchley *et al.*, 2012). In terms of production, wheat ranked third with a global production of 729 million tonnes (Fig. 6.1A). Yield, maturity, grain quality and stress (biotic and abiotic) tolerance are the top four agronomic traits. The majority of these are polygenic and controlled by quantitative trait loci (QTLs). Wheat is an allohexaploid ($2n = 6x = 42$, AABBDD) containing three subgenomes: A-genome from *T. urartu*, B-genome from *Aegilops speltoides* and D-genome from *A. tauschii*.

Fusarium head blight (Fhb) is one of the most devastating diseases of wheat. It is caused by *Fusarium* species complex. The predominant species in North America is *Fusarium graminearum* Shwabe. The pathogen secretes the secondary metabolites deoxynivalenol and nivalenol, and the estrogenic metabolite zearalenone, which accumulate in seeds during infection. These mycotoxins constitute a major concern for human and animal health (Champeil *et al.*, 2004).

Inducible disease resistance in plants against pathogens is bilayered (Dangl and Jones, 2001). The first layer recognizes conserved microbial signatures, such as fungal chitin, using cell surface receptors and is known as microbe-associated molecular pattern-triggered immunity (MTI). MTI, also known as basal resistance, is incredibly effective against non-adapted pathogens, and therefore introgression of resistance from non-host of *F. graminearum* through bridge crossing and embryo rescue could provide durable resistance against this pathogen (lineage-exclusion breeding) (Bhadauria and Banniza, 2014). The second layer detects race-specific pathogen effector molecules using intracellular receptors. A majority of these receptors possess nucleotide binding and leucine rich repeat (LRR) domains. The LRR domain forms a horseshoe shaped structure, which recognizes isolate-specific effectors (avirulence proteins). The concave surface of the structure contains β strands arranged in a crescent shape and surrounded by α-helices (Fig. 6.1B).

Recent developments in next-generation sequencing have accelerated the *de novo* genome sequencing, re-sequencing and discovery of high-density molecular markers, especially single nucleotide polymorphism (SNP) markers. Despite this, the resolution of QTLs remains challenging for crops with complex genomes, such as bread wheat (~17 GB). Several genome complexity reduction-based genotyping approaches have been developed to overcome this bottleneck and used in

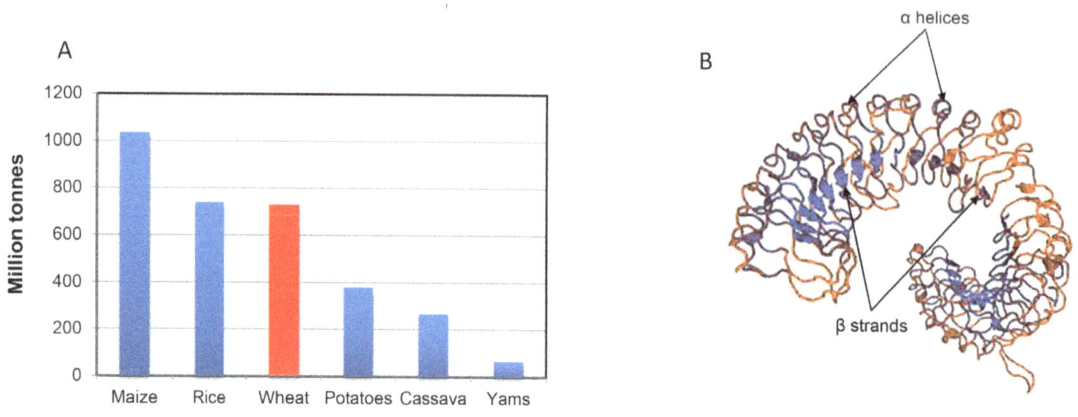


**Figure 6.1** (A) Global production of staple crops in 2014 (FAOSTAT). (B) Molecular structure of *Aegilops tauschii* RGA2 (GenBank EMT06480, NBLR).

the discovery of high-density SNP markers in maize, wheat and barley. These require restriction enzymes to reduce genome representation and multiplexing prior to massively parallel sequencing. This was first demonstrated in stickleback fish and *Neurospora crassa* through restriction enzyme-associated genomic DNA (RAD) sequencing (Baird *et al.*, 2008). More simplified and efficient versions of RAD, such as one and two restriction enzyme-based genotyping-by-sequencing (GBS) were developed. *Ape*KI-based GBS identified 25,185 and 24,186 biallelic SNPs in maize and barley recombinant inbred line (RIL) populations, respectively (Elshire *et al.*, 2011). Poland *et al.* (2012) used *Pst*I-*Msp*I-based GBS to map 34,000 and 20,000 SNPs onto the barley and wheat reference maps, respectively. The GBS approach has the advantage over other approaches, such as microarrays in which cross-hybridization between homologous genes causes false positive signals, which leads to false QTL discovery. Arruda *et al.* (2016) genotyped a panel of 273 elite winter wheat lines using three sets of restriction enzymes *Pst*I-*Msp*I, *Pst*I-*Hin*P1I and *Pst*I-*Bfa*I for reducing genome complexity and mapped 19,992 high-quality SNP across 21 chromosomes. Using a compressed mixed linear model, the authors identified two SNP markers IWGSC_CSS_3B_scaff_10352272_5482 and IWGSC_CSS_3B_scaff_10699215_3620, which were in the same LD block as *Fhb1*, a major effect QTL controlling Fusarium head blight located on chromosome 3B. Landraces are an excellent source of genetic diversity for agronomic traits. Cai *et al.* (2016) phenotyped a recombinant inbred population originated from a cross between Fhb-resistant *T. aestivum* landrace Haiyanzhong and susceptible cultivar Wheaton for *F. graminearum*-infected spikelets within spikes. The authors genotyped the population using *Pst*I-*Msp*I-based genotyping approach and identified 6,232 SNPs mapped across 21 chromosomes. Composite interval mapping revealed six QTLs in the landrace linked to Fhb resistance. The SNPs including the one linked to the QTL on chromosome 5A explaining ~16% of the variance in Fhb resistance response across the population was converted to KASP (Kompetitive allele-specific PCR) marker to pyramid Fhb-resistant QTLs in wheat through marker-assisted selection breeding. So far, over 100 QTLs controlling resistance to *F. graminearum* have been identified (Buerstmayr *et al.*, 2009) though candidate genes underlying these remain unidentified. Recently, Dhokane *et al.* (2016) identified six candidate genes underlying the QTL *qFhb6B-2* located on the short arm of chromosome 6B through comparative metabolomics and transcriptomics. These include 4-coumarate:CoA ligase (4CL), callose synthase (CS), basic helix–loop–helix (bHLH) transcription factor, glutathione S-transferase (GST), ABC transporter-4 (ABC4) and cinnamyl alcohol dehydrogenase (CAD). 4CL, CS and CAD confer resistance by reinforcing the cell wall (cell wall apposition). ABC4 and GST may have a role in reducing DON accumulation. Functional characterization of candidate genes through over-expression, silencing or knock-out is required to confirm their role in triggering Fhb-resistance.

## References


Arruda, M.P., Brown, P., Brown-Guedira, G., Krill, A.M., Thurber, C., Merrill, K.R., Foresman, B.J., and Kolb, F.L. (2016). Genome-wide association mapping of *Fusarium* head blight resistance in wheat using genotyping-by-sequencing. Plant Genome *9*. http://dx.doi.org/10.3835/plantgenome2015.04.0028

Baird, N.A., Etter, P.D., Atwood, T.S., Currey, M.C., Shiver, A.L., Lewis, Z.A., Selker, E.U., Cresko, W.A., and Johnson, E.A. (2008). Rapid SNP discovery and genetic mapping using sequenced RAD markers. PLOS ONE *3*, e3376. http://dx.doi.org/10.1371/journal.pone.0003376

Bhadauria, V., and Banniza, S. (2014). What lies ahead in post-genomics era: A perspective on genetic improvement of crops for fungal disease resistance. Plant Signal. Behav. *9*, e28503

Brenchley, R., Spannagl, M., Pfeifer, M., Barker, G.L., D'Amore, R., Allen, A.M., McKenzie, N., Kramer, M., Kerhornou, A., Bolser, D., *et al.* (2012). Analysis of the bread wheat genome using whole-genome shotgun sequencing. Nature *491*, 705–710. http://dx.doi.org/10.1038/nature11650

Buerstmayr, H., Ban, T., and Anderson, J.A. (2009). QTL mapping and marker-assisted selection for Fusarium head blight resistance in wheat: a review. Plant Breeding *128*, 1–26.

Cai, J., Wang, S., Li, T., Zhang, G., and Bai, G. (2016). Multiple Minor QTLs Are Responsible for Fusarium Head Blight Resistance in Chinese Wheat Landrace Haiyanzhong. PLOS ONE *11*, e0163292. http://dx.doi.org/10.1371/journal.pone.0163292

Champeil, A., Doré, T., and Fourbet, J.F. (2004). Fusarium head blight: Epidemiological origin of the effects of cultural practices on head blight attacks and the production of mycotoxins by *Fusarium* in wheat grains. Plant Sci. *166*, 1389–1415.

Dangl, J.L., and Jones, J.D. (2001). Plant pathogens and integrated defence responses to infection. Nature *411*, 826–833. http://dx.doi.org/10.1038/35081161

Dhokane, D., Karre, S., Kushalappa, A.C., McCartney. C. (2016). Integrated metabolo-transcriptomics reveals fusarium head blight candidate resistance genes in Wheat QTL-Fhb2. PLOS ONE *11*, e0155851.

Elshire, R.J., Glaubitz, J.C., Sun, Q., Poland, J.A., Kawamoto, K., Buckler, E.S., and Mitchell, S.E. (2011). A robust, simple genotyping-by-sequencing (GBS) approach for high diversity species. PLOS ONE *6*, e19379. http://dx.doi.org/10.1371/journal.pone.0019379

FAOSTAT. (2014). Food Agriculture Organization of The United Nations. Available at http://faostat.fao.org.

Poland, J.A., Brown, P.J., Sorrells, M.E., and Jannink, J.L. (2012). Development of high-density genetic maps for barley and wheat using a novel two-enzyme genotyping-by-sequencing approach. PLOS ONE 7, e32253. http://dx.doi.org/10.1371/journal.pone.0032253

# A Rice Genetic Improvement Boom by Next-generation Sequencing

7

Xiangchun Zhou†, Xufeng Bai† and Yongzhong Xing*

National Key Laboratory of Crop Genetic Improvement and National Center of Plant Gene Research, Huazhong Agricultural University, Wuhan, China.

*Correspondence: yzxing@mail.hzau.edu.cn
†These authors contributed equally

https://doi.org/10.21775/9781910190654.08

## Abstract

Rice (*Oryza sativa*. L) is a staple food crop for people worldwide, and a key goal has been to increase its grain yield. An increasing population that relies on a decreasing level of farmland has rendered the traditional method for the isolation and use of genetic loci in rice breeding unsatisfactory. Recently, the rapid development in next-generation sequencing (NGS) has boosted the number of genome sequences for hundreds to thousands of rice varieties. A MutMap strategy and bulk segregation analysis (BSA) has been developed to directly identify candidate genes based on NGS. The genome-wide association analysis (GWAS) has become a commonly used approach towards identifying the genetic loci and candidate genes for several traits that are closely associated with grain yield. The Multi-parent Advanced Generation Inter-Cross population (MAGIC) is introduced here to discuss potential applications for mapping QTLs for rice varietal development. These strategies broaden the capacity of functional gene identification and its application as a complementary method to insert mutants that comprise T-DNA and transposons. High-throughput SNP analysis platforms, such as the SNP array, provide novel strategies for genomic-assisted selections (GAS) for rice genetic improvements. Moreover, accurate genome sequence information enables genome editing for the utilization of key recessive genes that control important agronomic traits. This review summarizes how NGS accelerates rice genetic improvements through the identification and utilization of key functional genes that regulate agronomic traits.

## Introduction

The world population will reach 9 billion in 2050, and a 70–100% increase in food production will be required to meet the demand if the population continues to increase at the current rate (Godfray *et al.*, 2010). Rice is a vitally important food crop worldwide (Xing and Zhang, 2010). The limited area of farmland, the occurrence of extreme climates, and emerging biotic and abiotic stresses have driven the need for rice genetic improvements. Additional knowledge of functional genes for agronomic traits is expected to inform the breeding design for rice. Next-generation sequencing (NGS) provides numerous genome resources by re-sequencing diverse accessions, including landraces and varieties. Millions of single nucleotide polymorphisms (SNPs) are available, and they facilitate the genome-wide association study (GWAS). Recently, NGS has enabled rice genetic improvements by identifying hundreds of key functional genes or by establishing associations between SNPs and agronomic traits.

## The available methods for identification of functional genes

There are over 40,000 putative genes in rice (Feng *et al.*, 2002; Goff *et al.*, 2002; Sasaki *et al.*, 2002; Yu *et al.*, 2002). Most genes are quantitative trait loci (QTL), each with a minor effect. Currently, a small group of genes that control agronomic traits, such as the yield traits and heading date (which are all controlled by multiple QTLs), have been identified by a forward genetics strategy. In the past two decades, thousands of QTLs have been located (www.gramene.org/qtl), but only approximately 100 of the QTLs were cloned by map-based cloning. Map-based cloning is time-consuming and laborious. Therefore, several alternative approaches have been developed to efficiently discover functional genes, including gene knockdown and gene knockout.

Antisense and RNA interference (RNAi) technologies (Chuang and Meyerowitz, 2000) decrease the expression of target genes, which can lead to phenotypic changes. This method enables a direct mutation of a target gene. Direct gene knockout is accomplished using engineering techniques, such as Zinc finger nuclease (ZFN), transcription activator-like effector nuclease (TALEN), and clustered regularly interspaced short palindromic repeats-associated protein 9 (CRISPR-Cas9). Random gene knockout can cause functional mutations, which are randomly produced by a transposon insertion, a T-DNA insertion or treatments with physical factors and chemical mutagens.

## Mutant library for functional genomics in rice

### T-DNA insertion library

The isolation of genes using T-DNA mutants based on reverse genetics is common, and a series of T-DNA tagging populations have been developed. For example, approximately 100,000 fertile rice T-DNA lines have been tagged in Korea (Jeon *et al.*, 2000; Jeong *et al.*, 2002), and over 30,000 T-DNA insertion lines have been resulted (Wu *et al.*, 2003). Several other groups have reported T-DNA insertion lines in the genetic background of japonica rice (Chen *et al.*, 2003; Sallaud *et al.*, 2003; Sha *et al.*, 2004; Yin and Wang, 2000). No matter how large the T-DNA library is, it does not ensure that all genes will be inserted because the T-DNA insertion preferentially occurs in gene-rich regions (An *et al.*, 2005; An *et al.*, 2003; Barakat *et al.*, 2000; Chen *et al.*, 2003; Wu *et al.*, 2003). Moreover, T-DNA insertions are non-uniformly distributed on each chromosome; they are higher at the distal ends and lower near the centromeres. Additionally, several regions show extreme peaks and valleys of insertion frequency, which suggests hot and cold spots for T-DNA integration (Jeong *et al.*, 2006). Applying this strategy to *indica* rice has been particularly difficult because the transformation efficiency is not high enough to produce the thousands of requisite transgenic plants. Therefore, other mutant library types are needed to complement the T-DNA library vacancy.

### Transposon-derived library

A two-element *Activator/Dissociation* (*Ac/Ds*) transposon system has been extended to rice from maize (Enoki *et al.*, 1999; Greco *et al.*, 2001, 2003; Kolesnik *et al.*, 2004). The generation of a *Ds* transposon population has been achieved using a regeneration procedure that involves culturing seed-derived calli that carry *Ac* and non-autonomous *Ds* elements. *Ds* insertions are randomly distributed throughout the genome, but a hot spot for *Ds* insertions has been identified on chromosome 7 within a 40-kb region (Kolesnik *et al.*, 2004). However, its insertion preference and unstable insertion across different generations have also been reported (An *et al.*, 2005; Greco *et al.*, 2003).

*Tos17* is a retrotransposon in rice. Although its copy number is low in plants (one to five, depending on the cultivar), it can increase up to 30 copies during tissue culture (Hirochika *et al.*, 1996). A previous report describes the generation of 47,196 *Tos17*-induced insertion mutants in rice (Miyao *et al.*, 2003). Although insertion targets are distributed throughout the chromosomes, they tend to cluster, and 76% of the clusters are located in genic regions. Moreover, *Tos* populations can only play minor roles in genome-wide mutagenesis because the number of original donor sites is limited (Kim *et al.*, 2004).

### EMS-induced mutant library

The establishment of a large cohort of insertion mutants will accelerate investigations of rice gene functions via reverse genetics approaches.

However, the preferred mutagenesis locations of the *Ac/Ds*-, *Tos17*- and T-DNA-derived mutant libraries make them incapable of covering all genes. Therefore, physical and chemical mutagenesis strategies are strong candidates to complement their shortcomings. Fast neutrons, gamma rays, and diepoxybutane induce deletions or translocations; ethyl methanesulfonate (EMS) always induces point mutations. IR64, an *indica* rice, possesses multiple favourable agronomic characteristics (e.g. wide adaptability, high yield potential, tolerance to multiple diseases and pests, and a suitable quality for eating) and is an ideal genotype for identifying mutational phenotype changes (Wu *et al.*, 2005). Approximately 60,000 IR64 mutants have been generated using chemicals [diepoxybutane (DEB) and EMS] and irradiation (fast neutron and gamma ray). Targeting Induced Local Lesions in Genomes (TILLING) is an efficient way to identify point mutations (Comai and Henikoff, 2006). However, TILLING has largely been avoided as a first option towards identifying mutations because the procedure, which includes the specific gene primer design, PCR application, digestion with a crude celery juice extract that contains the CELI nuclease, and electrophoresis with a denaturing polyacrylamide gel, is lengthy.

## Next-generation sequencing-based MutMap strategy for gene mutation

### MutMap strategies

It is easy to obtain a large number of mutants using the transposon technique and chemical and physical mutagens, but it is difficult to target the mutation that is responsible for the phenotype change. The rapid development of NGS has illuminated and improved the ability to identify nucleotide variations at the genome level. MutMap has been used for gene isolation in rice (Abe *et al.*, 2012). The MutMap strategy frequently involves multiple steps (Fig. 7.1). A rice elite cultivar with

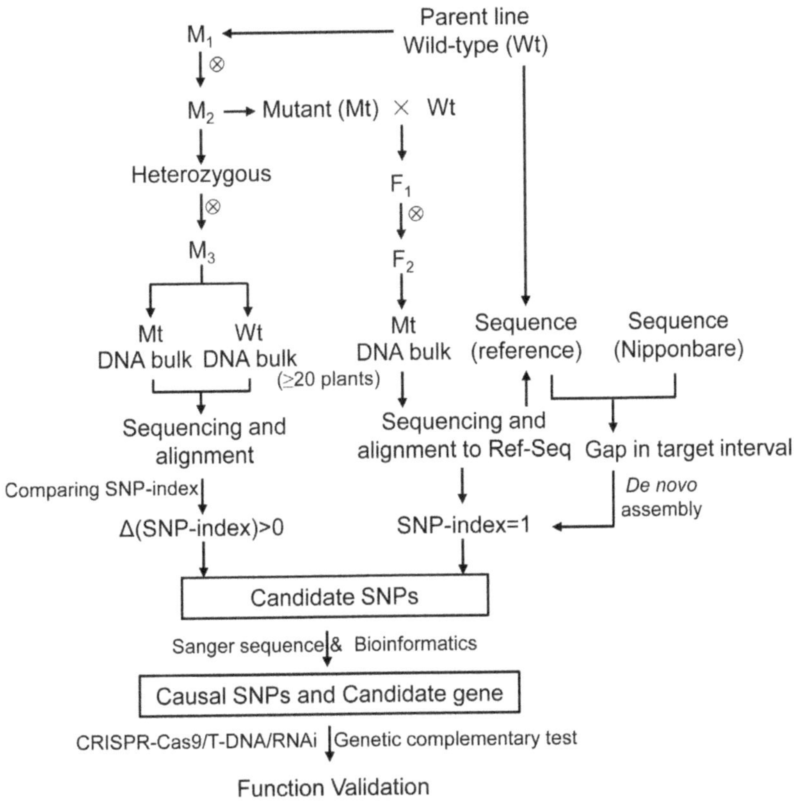


**Figure 7.1** The combined scheme of the different strategies of MutMap.

a reference genome sequence, such as Hitomebore from Northern Japan, has been mutagenized with the EMS mutagen to yield the $M_1$ generation, which self-pollinated to yield $M_2$; the $M_2$ segregation population yielded the recessive homozygous mutant with an altered agronomic trait, such as heading date, plant height, panicle architecture or grain size. The mutant was crossed with the Hitomebore wild-type to produce the heterozygous $F_1$, and the $F_2$ progeny were developed by self-pollination of $F_1$. The $F_2$ segregation population comprised over 100 plants. The DNA from 20 mutants was equally bulked and subjected to whole genome sequencing. The depth of the sequence was typically above 10-fold coverage of the rice genome (370 Mb). These mutant sequences were aligned to the reference sequence of the wild-type plants. The SNP-index was calculated, and the candidate SNPs were identified with a SNP-index = 1. The causal SNPs and the candidate genes were determined by further analysis of the Sanger sequence and with bioinformatics. Finally, the candidate gene was verified by overexpression, RNAi or knockout. *OsCOA1* and *OsRR22*, which are two genes that control pale green leaf and salinity tolerance, were identified by the MutMap strategy (Table 7.1) and validated with mutants and transgenic plants (Abe *et al.*, 2012; Takagi *et al.*, 2015).

The mutants with early development lethality or sterility cannot be crossed with the wild-type; this makes them unsuitable for MutMap applications. MutMap+ has been developed by a versatile extension of MutMap to address this problem (Fekih *et al.*, 2013). MutMap+ selects a heterozygous plant from the $M_2$ generation to give rise to the $M_3$ segregating population by selfing, which contains the wild-type and homozygous mutants. The DNA bulks (≥ 20 plants) for the wild-type and mutants have been constructed for whole-genome sequencing. The sequences from both bulks were aligned to the reference sequence of the cultivar that was used for mutagenesis. SNP-index plots were calculated for both the mutant and wild-type $M_3$ bulks. The two SNP-index plots were compared to identify the region with SNP-index = 1 that was specific to the mutant bulk. In other words, the Δ (SNP-index) plot that was obtained by subtracting the wild-type bulk SNP-index value from that of the mutant bulk was a positive value [Δ (SNP-index) > 0]. The candidate SNPs were likely harboured in the genomic region of the peak value of the Δ (SNP-index). The causal SNPs and the candidate genes were then determined by further analysis of the Sanger sequence and with bioinformatics. Finally, the candidate gene was verified by overexpression, RNAi or gene knockout (Fig. 7.1). *OsNAP6*, which encodes a SufBD protein that regulates dwarfism and pale green leaf, has been identified by the MutMap+ strategy, and its function has been validated by RNAi (Takagi *et al.*, 2015). A gene (Os08g0139100) for albino seeding, which is incapable of growing to maturity, has also been isolated with the MutMap+ method (Takagi *et al.*, 2015). In specific cases where the re-sequenced cultivar/line displays a significant structural variation from the reference genome (Nipponbare), mutations in genome regions that are missing from the reference genome (gaps) cannot be identified by simple alignment. A method called 'MutMap-Gap' has been developed (Takagi *et al.*, 2013b), which involves delineating a candidate region that harbours a mutation of interest using the MutMap method, followed by de novo assembly, alignment, and identification of the mutation within the genome gaps. The candidate SNPs are further identified based on SNP-index = 1, and the causal SNPs and candidate gene are determined by the analysis of the Sanger sequence and with bioinformatics. Finally, they are validated with mutants and/or transgenic plants (Fig. 7.1). *HIT7*, which encodes a NBS-LRR type R protein that is resistant to rice blast fungus, has been isolated using this method (Takagi *et al.*, 2013b).

## Options for further analysis of MutMap detection

MutMap is an efficient approach for isolating mutated genes, but the prerequisite for this strategy's use is the easy identification of distinct phenotypic types in a $F_2$ population that is derived from a cross between a wild-type and mutant. Specifically, a single Mendelian segregation ratio of 3:1 should be observed before MutMap is used to target the mutation. There is always a large candidate region when the MutMap strategy is used for gene identification. For example, the candidate regions of the five genes that were isolated by the MutMap strategy ranged from 2.1 to 5.8 Mb; the number of SNPs in the target region ranged from two to ten (Table 7.1). The SNPs of the coding

**Table 7.1** The summary information of the cloned genes using the different MutMap strategies

| Mutants | Sequence depth | Candidate region (Mb) | SNPs | Causal SNP | Validation | Gene | Function annotation | Strategy | Reference |
|---|---|---|---|---|---|---|---|---|---|
| Pale green leaf | >12× | 2.1 | 7 | Leu to Phe | Mutant, RNAi and complementary test | *OsCAO1* (Os10g0567400) | Chlorophyll a oxygenase | MutMap | Abe *et al*., 2012 |
| Salinity tolerance | 20× | 5.8 | 2 | Premature stop codon | Tos17 mutant | *OsRR22* (Os06g0183100) | B-type response regulator | MutMap | Takagi *et al*., 2015 |
| Dwarfism and pale green leaf | Mt:19.4 × Wt:11.7× | 5.2 | 8 | Ala to Thr | RNAi | *OsNAP6* (Os01g0127300) | SufBD protein | MutMap+ | Fekih *et al*., 2013 |
| Albino seeding | Mt:12 × Wt:11.4× | 4.3 | 10 | Premature stop codon | | Os08g0139100 | Homologues of chloroplast precursor differentiation and greening | MutMap+ | |
| Blast fungus resistance | Mt1:6.6× | 4.1 | 4 | Premature stop codon | Association analysis | *HIT7* (Os09t0327600–01) | NBS-LRR type R protein | MutMap-Gap | Takagi *et al*., 2013b |
| | Mt2:7.7× | 5.8 | 10 | Splicing error | | | | | |

region induced a premature stop codon, amino acid change and splicing error that frequently caused varied phenotypes and were commonly considered the most likely causal SNPs (Table 7.1). The corresponding gene that carries the causal SNP may be the candidate gene. This has been further validated by another method, which produced mutants or transgenic plants (Fig. 7.1). Notably, the regulatory region is also important for gene function, and mutations in the regulatory region may result in a mutant phenotype. Attention should be directed towards its candidate. Moreover, when the candidate SNPs are located on other genes without the annotation of function (unknown protein), it is difficult to predict the causal SNP and candidate gene. All the genes that harbour the candidate SNPs should be verified with other resource mutants and transgenic plants. Alternatively, a large $F_2$ population should be used to isolate the mutation by map-based cloning.

## Genome-wide quantitative trait loci mapping

Most agronomic traits are quantitatively inherited in rice, and the natural variations in a segregating population frequently exhibit continuous extremes in high and low phenotypic values. These natural variations provide us with the opportunity to breed super varieties. Generating favourable variations from this diversity is the most important task towards genetic improvements. NGS has more advantages towards QTL mapping than traditional molecular marker techniques. It is ideally suited for a natural collection or for a bi- or multi-parental population.

### Genome-wide association study

The GWAS is a routinely used method for identifying the association between markers and traits in humans and plants. Recently, several NGS projects have been widely applied to rice genomic studies, which have provided millions of SNPs. Because of their abundance and uniform distributions throughout the genome, SNPs are becoming the most desirable molecular markers for QTL scanning. Thus, GWAS has been regularly applied for different traits in rice. The agronomic traits have been investigated by several research groups using GWAS. A mixed linear model (MLM) has been introduced as an improved method to simultaneously account for the population structure and unequal relatedness among individuals, which should decrease false-positive associations (Zhang *et al.*, 2010). A report by Huang *et al.* (2010) describes the identification of approximately 3.6 million SNPs through the sequencing of 517 rice landraces; it also details the construction of a high-density haplotype map of the rice genome. GWAS was performed for 14 agronomic traits in a population of the *Oryza sativa indica* subspecies. The association population was extended to a larger and more diverse sample of 950 worldwide rice varieties, and GWAS was performed for 11 grain-related traits (Huang *et al.*, 2011). GWAS was performed for agronomic traits and metabolites by other groups (Bai *et al.*, 2016; Chen *et al.*, 2014b; Crowell *et al.*, 2016; Magwa *et al.*, 2016a,b; Zhao *et al.*, 2011). Huang *et al.* (2015) developed an integrated genomic approach using the SNP data that was obtained by NGS to construct a genome map for 1495 elite hybrid rice varieties and their inbred parental lines. In total, 38 agronomic traits were investigated, and 130 associated loci were identified. Based on the in-depth analyses of the effects of the heterozygous genotypes of these loci, they found that high-yielding hybrid varieties had several superior alleles, but most parental inbred lines had a small number. Therefore, they concluded that the accumulation of numerous rare superior alleles with positive dominance was an important contributor to the heterotic phenomena. Huang *et al.* (2016) reported the genomic sequencing of 10,074 $F_2$ lines from 17 representative hybrid rice crosses and constructed a dense genotype map with 347,803 recombination events using the NGS-based genotyping method. They classified modern hybrid rice varieties into three groups, which represented different hybrid breeding systems, and obtained support from several loci for a partial dominance of the heterozygous locus for yield-related traits. These results address the genomic architecture of heterosis and rice hybrid breeding. Han *et al.* (2016) performed an association analysis of 11 known flowering genes at the SNP and haplotype levels. The haplotype-level association analysis detected 7 of the 11 known genes, including the only four detected at the SNP level. Notably, the haplotype-level association strongly increased the power of QTL mapping. Moreover, the haplotype-level association mapping suggested favourable haplotypes for agronomic traits. However, genome-wide haplotype-level association mapping continues to be a difficult procedure.

The resolution of GWAS is dependent on the SNP local linkage disequilibrium (LD). The average LD decay in cultivated rice has been extended from 100 kb to 200 kb (Huang *et al.*, 2010; Mather *et al.*, 2007; McNally *et al.*, 2009). Although it is difficult to map a QTL to a single gene, this resolution can provide valuable information for marker-selected breeding.

Recently, a Multi-parent Advanced Generation Inter-Cross population (MAGIC) was introduced to discuss potential applications for mapping QTLs and for rice varietal development. MAGIC that is derived from a cross between multiple parents has a genotypic diversity and reduced linkage drag (Leung *et al.*, 2015). There is a clear genealogy among the lines and between the generations, which can effectively avoid the population structure. Several MAGIC populations that were derived from diverse elite parental lines have been used to detect QTLs for different traits in rice. The MAGIC genetic map was constructed using the SNP markers based on NGS and/or SNP array (Bandillo *et al.*, 2013; Meng *et al.*, 2016a,b). A GWAS with these MAGIC populations had a higher detection power than with the assembled populations/germplasm collection and had a higher resolution than the bi-parental populations (Meng *et al.*, 2016a). Moreover, the Hidden Markov Model (HMM) was used for a QTL analysis in MAGIC, which led to a low false positive rate and a low bias of the estimated QTL effect sizes. An implementation of the approach is available as an R package (Verbyla *et al.*, 2014). Currently, GWAS is primarily performed based on the SNPs. Here, we strongly suggest the use of bin-based GWAS for QTL mapping; this may improve the precision of the QTL location and reduce the false positive associations compared with those of the SNP-based GWAS.

### Quick QTL mapping via bulk segregation analysis

$F_2$ and/or $BC_1F_2$ are generally used for QTL mapping. To further isolate the candidate genes that underlie the QTLs that are mapped in the $F_2$, RIL or $BC_1F_2$ primary populations, advanced populations, such as near isogenic lines (NILs), are required. However, five to six cropping seasons are necessary to develop the advanced population, which is a lengthy time period. Fortunately, the BSA strategy (Takagi *et al.*, 2013a) has become more vigorous towards identifying QTLs with the NGS application in rice, particularly for major QTLs. Two DNA bulks are constructed by selecting extremely different phenotypic groups from the mapping populations. The QTL region is quickly mapped with an in-depth comparative analysis of the NGS-based sequence. Specifically, if the allele frequency in one region is very different between these two bulks, the region is most likely to contribute to the phenotypic difference. Applications of this strategy will quickly provide valuable information for breeders and will shorten the required times for major QTL isolations in rice.

## Genomic selection for breeding

### Genetic diagnosis

The selection of parental lines is a key step in the genetic research and breeding processes of rice (Chen *et al.*, 2013). Breeders often choose varieties with diverse genetic backgrounds for their breeding programs. For example, to screen QTLs for different agronomic traits, a population developed from the varieties with dramatically different backgrounds will enhance the mapping power and will map numerous QTLs. Phylogenetic analysis is useful with this breeding protocol. Rice varieties can be clustered into several subgroups based on the SNP analysis. The distance in the neighbour-joining tree indicates the genetic similarities among these varieties. The cluster information can be used to guide the parental line selections for genetic studies and rice breeding, such as in the construction of MAGIC populations for genetic studies. Parent founders with higher diversity may provide more opportunities for recombination, which enables efficient and precise explorations of multiple desirable traits (Bandillo *et al.*, 2013).

On the other hand, for hybrid breeding, we can perform genetic diagnosis through the SNP analysis to select the parental lines that have superior alleles to control yield-related traits, such as the plant height, heading date, panicle or grain shape. Exploitation of hybrid vigour (heterosis) has been accomplished in rice production. Several yield-related genes, such as *sd1, Ghd7, Ghd8, gs3, dep1, ipa1,* and so on (Table 7.2), have been identified as being critical for hybrids and as having partial heterotic effects (Huang *et al.*, 2015, 2016). Huang *et al.* (2016) found that genes for yield-related traits were enriched for positive dominance effects, which indicated that superior alleles could be optimally utilized by complementing parents. Furthermore, most of these superior alleles were contributed by the male parents. These findings will be helpful towards rational genetic designs for producing rice hybrids. For example, using SNP analysis, we can choose a more suitable male parent with the desired alleles for *sd1, gs3, tac1* and *Ghd8* and a more suitable female parent with superior alleles for *nal1, ipa1,* and *dep1*. With these favourable alleles, the expected hybrid would exhibit a high-yield potential with an ideal plant architecture.

## Genomic-assisted selection

Several NGS projects have been widely applied to rice genomic studies (Huang *et al.*, 2010, 2011, 2012; McNally *et al.*, 2009; Xu *et al.*, 2011), which have generated millions of SNPs. Benefiting from NGS, high-throughput SNP arrays have been developed to genotype numerous samples in a short timeframe by identifying thousands of SNPs that are distributed in the whole genome. Various SNP array platforms based on different principles are available in rice. An Affymetrix 44K SNP array has been developed and successfully used in GWAS (Famoso *et al.*, 2011; McCouch *et al.*, 2010; McNally *et al.*, 2009; Zhao *et al.*, 2011). However, most SNPs on the array are derived from the *Oryza*SNP project with 20 varieties (McNally *et al.*, 2009), which may not adequately represent the rice species. Various SNP chips based on the

**Table 7.2** Collection of key genes controlling yield-related traits

| Gene | Chromosome | Trait | Yield-related phenotype | Heterosis | Reference |
|---|---|---|---|---|---|
| *sd1** | 1 | PH | High YD, reduced PH | + | Spielmeyer *et al.*, 2002 |
| *hd3a* | 6 | HD | High YD, delayed HD | + | Kojima *et al.*, 2002 |
| *Hd1*** | 6 | HD | High YD, delayed HD | + | Yano *et al.*, 2000 |
| *Ghd7* | 7 | HD | High YD, delayed HD | + | Xue *et al.*, 2008 |
| *Ghd8* | 8 | HD | High YD, delayed HD | + | Yan *et al.*, 2011 |
| *Ghd7.1* | 7 | HD | High YD, delayed HD | | Yan *et al.*, 2013 |
| *gs3* | 3 | GL and KGW | Increased GL and KGW | + | Fan *et al.*, 2006 |
| *GS5* | 5 | GW and KGW | Increased GW and KGW | | Li *et al.*, 2011 |
| *GW2* | 2 | GW and KGW | Increased GW and KGW | + | Song *et al.*, 2007 |
| *GW5* | 5 | GW and KGW | Increased GW and KGW | + | Weng *et al.*, 2008 |
| *TGW6* | 6 | KGW | Increased YD and KGW | + | Ishimaru *et al.*, 2013 |
| *gn1a* | 1 | TGN | Increased YD and KGW | + | Ashikari *et al.*, 2005 |
| *osgi* | 1 | GN | Increased YD and GN | | Izawa *et al.*, 2011 |
| *dep1* | 9 | GN | Increased YD, SPP and GN | + | Huang *et al.*, 2009 |
| *nal1* | 4 | GN | Increased YD, SPP and GN | + | Fujita *et al.*, 2013 |
| *MOC1* | 6 | Tillering | Increased YD and tillers | | Li *et al.*, 2003 |
| *LAX1* | 1 | Tillering | Increased YD and tillers | + | Komatsu *et al.*, 2001 |
| *htd1/ osccd7* | 4 | Tillering | Increased YD and tillers | + | Kulkarni *et al.*, 2014 |
| *ipa1/ osspl14* | 8 | Ideal architecture | Increased YD, an ideal architecture | + | Miura *et al.*, 2010 |
| *tac1* | 9 | Tiller angle | increased YD, a compact plant status | + | Yu *et al.*, 2007 |

*Gene in lower case means the negative regulation of rice yield; **Gene in capital format means the positive controlling. +, means that partial dominance or overdominance effect has been reported within the gene (Huang *et al.*, 2015, 2016). YD, yield; PH, plant height; HD, heading date; GL, grain length; GW, grain width; KGW, kilo-grain weight; TGN, total grain number; GN, grain number; SPP, spikelets per panicle.

Illumina GoldenGate SNP Chip have also been developed for genetic analysis (Chen *et al.*, 2011; Yamamoto *et al.*, 2010; Zhao *et al.*, 2010). SNP arrays that are specific for rice genotyping have recently been designed based on the Illumina Infinium technology; they include RICE6K (Yu *et al.*, 2014) and RiceSNP50 (Chen *et al.*, 2014). RICE6K is based on the low-coverage genome sequences (~1×) of the 520 rice germplasm collections (Huang *et al.*, 2010). It contains 5556 SNP sites and accommodates 80 functional markers that cover 40 rice genes controlling traits such as grain yield, grain quality, heading date, hybrid fertility and stress resistance (Yu *et al.*, 2014). RiceSNP50, a new high-density array, was developed after the 520 accessions (~1×) (Huang *et al.*, 2010) and 281 accessions were sequenced (2.5–15×) (Chen *et al.*, 2014). RiceSNP50 has been synthesized with 51,478 SNPs that are evenly distributed along the whole rice genome. These SNP arrays have become useful for rice genetic research and breeding (Chen *et al.*, 2014; Hu *et al.*, 2013; Tan *et al.*, 2013; Yu *et al.*, 2014).

Marker-assisted selection (MAS) is a valuable tool for improving target traits with DNA markers that are linked to the genes/QTLs underlying the traits (Xu and Crouch, 2008). A promising MAS approach, known as genomic-assisted selection (GAS), has been widely used for genetic improvements in rice (Heffner *et al.*, 2009; Nakaya and Isobe, 2012). GAS is a high-throughput and efficient approach that uses genome-wide SNP markers to improve the agronomic traits; GAS completes the foreground and background examination in a single SNP array assay. During the breeding process with backcross (BC), breeders aim to introduce favourable alleles from a donor parent into the genome of a desired variety without changing the genetic background. The progeny that carry the target genes would then be selected during the early generations of the BC populations (positive selection), while the genetic background of the progeny that carry the donor alleles of interest would be screened to minimize the donor background in advanced BC populations, such as $BC_4F_1$. Rice SNP arrays, such as RICE6K and RiceSNP50, are able to achieve the goals of foreground and background selection. Closely linked low-density SSR markers or functional markers are typically used for positive selection. For GS3 (Fan *et al.*, 2006, 2009; Mao *et al.*, 2010), a major gene that controls grain length, a SNP between C and A in the second exon induces the differentiation of the rice grain length. A total of 180 varieties with genetic diversity have been used for association analysis with the C/A mutation and grain length; they have confirmed the causal mutation to be a SNP (Fan *et al.*, 2009). Based on the causal SNP, a CAPS marker, SF28, has been developed, and it works well in the MAS of the rice grain length (Fan *et al.*, 2009; Lu *et al.*, 2013; Wang *et al.*, 2011). We can also use the RICE6K SNP array for this selection because it accommodates 80 functional probes for 40 genes, including *GS3*. For an advanced BC population, the individuals that possess the target alleles are required for the genetic background selection. For example, 29 rice individuals with introgressions of *Pi1* (Hua *et al.*, 2012) and *Pi2* (Zhou *et al.*, 2006) into Kongyu 131, a japonica cultivar that is popular in north-east China, were examined by RICE6K (Yu *et al.*, 2014). The results showed that several improved lines with minimal donor parental backgrounds could be selected, but the target genomic regions also contained large, dragged fragments from the donor parent. These results suggest that closely linked markers or gene functional markers should be used for selection and should be performed on early generations. As another example, the RiceSNP50 array was used in rice BC breeding to assist a selection of the fragment with *Pigm(t)* (Deng *et al.*, 2006) from Gumei 4 (a rice variety that possesses broad-spectrum resistance) into an elite parental line, R608. The results showed that two out of four $BC_4F_1$ individuals were identified as possessing the *Pigm(t)* allele, and one improved line that contained one additional fragment from the donor parent was selected for further breeding (Chen *et al.*, 2014). These application examples indicate that the high efficiencies of Rice SNP arrays make them applicable to rice GAS breeding programmes.

## Genome editing

Genome editing is a precise approach for targeting genetic manipulation and is becoming a popular tool for rice genetic improvements. Genome editing relies on accurate genome sequence information to target DNA fragments for mutations, which can result in a DNA insertion, deletion, replacement or reverse replication. ZFN,

TALEN and CRISPR-Cas9 are the three preferred genome-editing tools (Barrangou, 2012; Bibikova *et al.*, 2003; Bogdanove and Voytas, 2011; Pennisi, 2013).

In rice, several important genes that negatively control agronomic traits have been targeted through genome editing (Table 7.3). In 2012, TALEN was used to improve bacterial blight resistance by mutating the *OsSWEET14* promoter region, which harbours the cis-elements that interact with pathogens (Li *et al.*, 2012). This mutation decreased the binding effect between the promoter and the effectors that were secreted by the bacterial blight pathogen. The following year, the genes that showed negative effects on the rice architecture, yield, and quality were targeted by TALEN and CRISPR (Shan *et al.*, 2013a,b), with the aim of improving the rice yield and quality. The examples of genome editing applications to rice indicate that the induction-specific mutations in the negative regulatory genes can rapidly produce desirable agronomic traits. There are numerous, well-performing lines that contain one or more undesirable traits for the rice breeding process. There are also dozens of accessible genes that can be edited to improve these traits. Heading date is a vital trait for rice diversification and domestication (Izawa, 2007; Meyer and Purugganan, 2013). The molecular regulatory network for flowering is becoming clearer and more complete. Several important flowering genes have been extensively studied (Xue *et al.*, 2008; Yan *et al.*, 2011; Yano *et al.*, 2000); they provide flexibility towards manipulating the heading date by providing different gene combinations for selection. For example, we previously observed that different allele-combinations of the major heading date genes [*Ghd7* (Xue *et al.*, 2008), *Ghd8* (Yan *et al.*, 2011) and *Hd1* (Yano *et al.*, 2000)] largely defined eco-geographical adaptation in rice cultivars (Zhang *et al.*, 2015). Therefore, we suppose that by targeting *Ghd7*, *Ghd8*, or *Hd1*, we can rapidly design the flowering time of the desired varieties to adapt to different ecological regions based on the combinations of these genes. On the other hand, modifications of the *R* genes, such as *Xa5* (Iyer-Pascuzzi *et al.*, 2008; Iyer and McCouch, 2004) and *Xa13* (Antony *et al.*, 2010; Chu *et al.*, 2006; Yuan *et al.*, 2009), would cause the *xa5* and *xa13* recessive genes to function particularly well. In the rice bacterial blight system, the dominant allele, *Xa5*, is necessary and sufficient for pathogen susceptibility, but *xa5* is a completely recessive gene that contributes to bacterial blight resistance (Iyer-Pascuzzi *et al.*, 2008). Hence, we believe that through genome editing, we can efficiently obtain improved lines with the *xa5xa5* or *xa13xa13* homozygous alleles, which would enhance bacterial blight resistance.

Gene replacements and insertions have recently been achieved in rice (Li *et al.*, 2016). Using a pair of single guide RNAs (sgRNAs) that targeted adjacent introns and a donor DNA

**Table 7.3** Application examples of genome editing in rice genetic improvements

| Mutation types | Genes | Aims | Tool | Repair | Reference |
|---|---|---|---|---|---|
| Gene disruption | *OsSWEET14* | Bacterial blight resistance | TALEN/CRISPR | NHEJ | Li *et al.*, 2012 |
| | *OsBADH2* | Rice with fragrance | TALEN/CRISPR | NHEJ | Shan *et al.*, 2013a,b, 2015 |
| | *Waxy* | Lower levels of amylose | CRISPR | NHEJ | Ma *et al.*, 2015 |
| | *ROC5* | Rolling leaves rice | CRISPR | NHEJ | Feng *et al.*, 2013 |
| | *SD1* | Semi-dwarfing breeding | TALEN | NHEJ | Shan *et al.*, 2013a |
| | *DEP1* | Erect panicle breeding | TALEN | NHEJ | Li *et al.*, 2016; Shan *et al.*, 2013a |
| | *Gn1a* | High yield | TALEN | NHEJ | Li *et al.*, 2016; Shan *et al.*, 2013a |
| | *GS3* | Bacterial blight resistance | CRISPR | NHEJ | Li *et al.*, 2016 |
| | *IPA1* | Bacterial blight resistance | CRISPR | NHEJ | Li *et al.*, 2016 |
| Gene replacement | *EPSPS* | Glyphosate resistance | CRISPR | NHEJ | Li *et al.*, 2016 |
| Gene insertion | *EPSPS* | Glyphosate resistance | CRISPR | NHEJ | Li *et al.*, 2016 |

template with the same sgRNAs sites, gene replacements were achieved for an endogenous gene, 5-enolpyruvylshikimate-3-phosphate synthase (*EPSPS*), in rice at a frequency of 2.0%. Moreover, the gene insertion frequency was 2.2% with this intron targeting method. Genetically modified plants with the *OsEPSPS* gene substitution were glyphosate resistant. These achievements for target insertions in rice genes may assist breeders with the introduction useful gene clusters as breeding modules, which could accelerate genetic improvements for the utilization of multiple genes.

The promise of genome editing in rice is its fast and accurate modifications/introductions of ideal agronomic traits for genetic improvements. However, there are always critical limitations, such as off-target effects. Procedures to minimize the off-target effects have been developed and include the double nickase system (Mali *et al.*, 2013) and an inactivated Cas9 (dCas9) that is fused to FokI (Guilinger *et al.*, 2014). Nevertheless, off-target editing can still occur. Another limitation is food safety. Whether the improved lines with the targeted DNA modifications/introductions can be described as genetically modified organisms (GMOs) remains a critical question (Hartung and Schiemann, 2014; Lusser and Davies, 2013; Pauwels *et al.*, 2014). Araki and Ishii (2015) suggest that the genome editing events that result in small insertions/deletions through the non-homologous end joining (NHEJ) repair pathway should not be regulated as GMOs, while genome editing that induces a foreign template by homologous recombination should be considered as a GMO. This issue remains controversial (Schaart *et al.*, 2016; Wolt *et al.*, 2016).

## Omics integration will push rice genetic improvement forward

Several NGS-based omics systems have been established, and high-throughput omics measurements are routinely explored in rice, including genomics, transcriptomics, and epigenomics. Additionally, proteomics (Chen *et al.*, 2016; Das *et al.*, 2016), metabolomics (Chen *et al.*, 2014) and phenomics (McCouch *et al.*, 2016; Yang *et al.*, 2014, 2015) have been established for high-throughput measurements of phenotypic values, including biochemical, physiological and agronomic traits. Determining the association between the genome variation [SNP, Insertion/deletion (InDel) and copy number variation (CNV)] and trait performance is the crucial step towards estimating genetic effects for rice breeding. GWAS for metabolites has detected more loci with higher resolution than agronomical traits, and 36 candidate genes have been identified, five of which are functionally validated in rice (Chen *et al.*, 2014). However, the relationship between the metabolites and agronomical traits is unclear for most cases. Comparison of the GWAS for agronomics with that for the metabolites, such as assimilates and phytohormones, would address uncertainties by providing compensatory or supporting information for one dataset. False positives are less likely when multiple sources of evidence point to the same gene or pathway. A complete biological model is likely to be discovered only when omics at different levels are jointly considered in an analysis. Integration of these omics datasets could categorize them into two clades – a meta-dimensional analysis and a multistaged analysis (Ritchie *et al.*, 2015). Meta-dimensional analysis is a method where different scales of data are simultaneously combined to identify a complex. Meta-dimensional models involve multiple variables from different data types, but the models for a multistaged analysis are constructed using two different data scales at a time, in a stepwise or linear manner. The main objective of this method is to divide the analysis into multiple steps, where the associations are initially determined between the different data types and subsequently between the data type and trait phenotype of interest (Ritchie *et al.*, 2015). Developing powerful statistical methods will bridge the gap between our ability to generate vast amounts of data and our biological understanding to uncover the genetic basis that underlies the large datasets. A systems genomics approach can provide a more thorough and informative interrogation of genotype-to-phenotype associations.

## Acknowledgements

This work was partially supported by natural science foundation of China (31571751) and the Natural Science Foundation of Hubei province, China (2015CFA006).

## References


Abe, A., Kosugi, S., Yoshida, K., Natsume, S., Takagi, H., Kanzaki, H., Matsumura, H., Yoshida, K., Mitsuoka, C., Tamiru, M., *et al.* (2012). Genome sequencing reveals agronomically important loci in rice using MutMap. Nat. Biotechnol. *30*, 174–178. http://dx.doi.org/10.1038/nbt.2095

An, G., Lee, S., Kim, S.H., and Kim, S.R. (2005). Molecular genetics using T-DNA in rice. Plant Cell Physiol. *46*, 14–22. https://dx.doi.org/10.1093/pcp/pci502

An, S., Park, S., Jeong, D.H., Lee, D.Y., Kang, H.G., Yu, J.H., Hur, J., Kim, S.R., Kim, Y.H., Lee, M., *et al.* (2003). Generation and analysis of end sequence database for T-DNA tagging lines in rice. Plant Physiol. *133*, 2040–2047. http://dx.doi.org/10.1104/pp.103.030478

Antony, G., Zhou, J., Huang, S., Li, T., Liu, B., White, F., and Yang, B. (2010). Rice *xa13* recessive resistance to bacterial blight is defeated by induction of the disease susceptibility gene Os-*11N3*. Plant Cell *22*, 3864–3876. http://dx.doi.org/10.1105/tpc.110.078964

Araki, M., and Ishii, T. (2015). Towards social acceptance of plant breeding by genome editing. Trends Plant Sci. *20*, 145–149. http://dx.doi.org/10.1016/j.tplants.2015.01.010

Ashikari, M., Sakakibara, H., Lin, S., Yamamoto, T., Takashi, T., Nishimura, A., Angeles, E.R., Qian, Q., Kitano, H., and Matsuoka, M. (2005). *Cytokinin oxidase* regulates rice grain production. Science *309*, 741–745. http://dx.doi.org/10.1126/science.1113373

Bai, X.F., Zhao, H., Huang, Y., Xie, W.B., Han, Z.M., Zhang, B., Guo, Z.L., Yang, L., Dong, H.J., Xue, W.Y., *et al.* (2016). Genome-wide association analysis reveals different genetic control in panicle architecture between *indica* and *japonica* Rice. Plant Genome *9*, 2. https://dx.doi.org/10.3835/plantgenome2015.11.0115

Bandillo, N., Raghavan, C., Muyco, P.A., Sevilla, M.A., Lobina, I.T., Dilla-Ermita, C.J., Tung, C.W., McCouch, S., Thomson, M., Mauleon, R., *et al.* (2013). Multi-parent advanced generation inter-cross (MAGIC) populations in rice: progress and potential for genetics research and breeding. Rice *6*, 11. http://dx.doi.org/10.1186/1939-8433-6-11

Barakat, A., Gallois, P., Raynal, M., Mestre-Ortega, D., Sallaud, C., Guiderdoni, E., Delseny, M., and Bernardi, G. (2000). The distribution of T-DNA in the genomes of transgenic *Arabidopsis* and rice. FEBS Lett. *471*, 161–164. http://dx.doi.org/10.1016/S0014-5793(00)01393-4

Barrangou, R. (2012). RNA-mediated programmable DNA cleavage. Nat. Biotechnol. *30*, 836–838. http://dx.doi.org/10.1038/nbt.2357

Bibikova, M., Beumer, K., Trautman, J.K., and Carroll, D. (2003). Enhancing gene targeting with designed zinc finger nucleases. Science *300*, 764. http://dx.doi.org/10.1126/science.1079512

Bogdanove, A.J., and Voytas, D.F. (2011). TAL effectors: customizable proteins for DNA targeting. Science *333*, 1843–1846. http://dx.doi.org/10.1126/science.1204094

Chen, H., He, H., Zhou, F., Yu, H., and Deng, X.W. (2013). Development of genomics-based genotyping platforms and their applications in rice breeding. Curr. Opin. Plant Biol. *16*, 247–254. http://dx.doi.org/10.1016/j.pbi.2013.04.002

Chen, H., He, H., Zou, Y., Chen, W., Yu, R., Liu, X., Yang, Y., Gao, Y.M., Xu, J.L., Fan, L.M., *et al.* (2011). Development and application of a set of breeder-friendly SNP markers for genetic analyses and molecular breeding of rice (*Oryza sativa* L.). Theor. Appl. Genet. *123*, 869–879. http://dx.doi.org/10.1007/s00122-011-1633-5

Chen, H., Xie, W., He, H., Yu, H., Chen, W., Li, J., Yu, R., Yao, Y., Zhang, W., He, Y., *et al.* (2014). A high-density SNP genotyping array for rice biology and molecular breeding. Mol. Plant *7*, 541–553. http://dx.doi.org/10.1093/mp/sst135

Chen, L., Huang, Y., Xu, M., Cheng, Z., Zhang, D., and Zheng, J. (2016). iTRAQ-based quantitative proteomics analysis of black rice grain development reveals metabolic pathways associated with anthocyanin biosynthesis. PLOS ONE *11*, e0159238. http://dx.doi.org/10.1371/journal.pone.0159238

Chen, S., Jin, W., Wang, M., Zhang, F., Zhou, J., Jia, Q., Wu, Y., Liu, F., and Wu, P. (2003). Distribution and characterization of over 1000 T-DNA tags in rice genome. Plant J. *36*, 105–113. http://dx.doi.org/10.1046/j.1365-313X.2003.01860.x

Chen, W., Gao, Y., Xie, W., Gong, L., Lu, K., Wang, W., Li, Y., Liu, X., Zhang, H., Dong, H., *et al.* (2014). Genome-wide association analyses provide genetic and biochemical insights into natural variation in rice metabolism. Nat. Genet. *46*, 714–721. http://dx.doi.org/10.1038/ng.3007

Chu, Z., Fu, B., Yang, H., Xu, C., Li, Z., Sanchez, A., Park, Y.J., Bennetzen, J.L., Zhang, Q., and Wang, S. (2006). Targeting *xa13*, a recessive gene for bacterial blight resistance in rice. Theor. Appl. Genet. *112*, 455–461. http://dx.doi.org/10.1007/s00122-005-0145-6

Chuang, C.F., and Meyerowitz, E.M. (2000). Specific and heritable genetic interference by double-stranded RNA in *Arabidopsis thaliana*. Proc. Natl. Acad. Sci. U.S.A. *97*, 4985–4990. http://dx.doi.org/10.1073/pnas.060034297

Comai, L., and Henikoff, S. (2006). TILLING: practical single-nucleotide mutation discovery. Plant J. *45*, 684–694. http://dx.doi.org/10.1111/j.1365-313X.2006.02670.x

Crowell, S., Korniliev, P., Falcão, A., Ismail, A., Gregorio, G., Mezey, J., and McCouch, S. (2016). Genome-wide association and high-resolution phenotyping link *Oryza sativa* panicle traits to numerous trait-specific QTL clusters. Nat. Commun. 7, 10527. http://dx.doi.org/10.1038/ncomms10527

Das, K., Panda, B.B., Sekhar, S., Kariali, E., Mohapatra, P.K., and Shaw, B.P. (2016). Comparative proteomics of the superior and inferior spikelets at the early grain filling stage in rice cultivars contrast for panicle compactness and ethylene evolution. J. Plant Physiol. *202*, 65–74. http://dx.doi.org/10.1016/j.jplph.2016.07.008

Deng, Y., Zhu, X., Shen, Y., and He, Z. (2006). Genetic characterization and fine mapping of the blast resistance locus *Pigm*(*t*) tightly linked to *Pi2* and *Pi9* in a broad-spectrum resistant Chinese variety. Theor. Appl. Genet. *113*, 705–713. http://dx.doi.org/10.1007/s00122-006-0338-7

Enoki, H., Izawa, T., Kawahara, M., Komatsu, M., Koh, S., Kyozuka, J., and Shimamoto, K. (1999). Ac as a tool for the functional genomics of rice. Plant J. *19*, 605–613. http://dx.doi.org/10.1046/j.1365-313X.1999.00549.x

Famoso, A.N., Zhao, K., Clark, R.T., Tung, C.W., Wright, M.H., Bustamante, C., Kochian, L.V., and McCouch, S.R. (2011). Genetic architecture of aluminum tolerance in rice (*Oryza sativa*) determined through genome-wide association analysis and QTL mapping. PLOS Genet. *7*, e1002221. http://dx.doi.org/10.1371/journal.pgen.1002221

Fan, C., Xing, Y., Mao, H., Lu, T., Han, B., Xu, C., Li, X., and Zhang, Q. (2006). *GS3*, a major QTL for grain length and weight and minor QTL for grain width and thickness in rice, encodes a putative transmembrane protein. Theor. Appl. Genet. *112*, 1164–1171. http://dx.doi.org/10.1007/s00122-006-0218-1

Fan, C., Yu, S., Wang, C., and Xing, Y. (2009). A causal C-A mutation in the second exon of *GS3* highly associated with rice grain length and validated as a functional marker. Theor. Appl. Genet. *118*, 465–472. http://dx.doi.org/10.1007/s00122-008-0913-1

Fekih, R., Takagi, H., Tamiru, M., Abe, A., Natsume, S., Yaegashi, H., Sharma, S., Sharma, S., Kanzaki, H., Matsumura, H., *et al.* (2013). MutMap+: genetic mapping and mutant identification without crossing in rice. PLOS ONE *8*, e68529. http://dx.doi.org/10.1371/journal.pone.0068529

Feng, Q., Zhang, Y., Hao, P., Wang, S., Fu, G., Huang, Y., Li, Y., Zhu, J., Liu, Y., Hu, X., *et al.* (2002). Sequence and analysis of rice chromosome 4. Nature *420*, 316–320. http://dx.doi.org/10.1038/nature01183

Feng, Z., Zhang, B., Ding, W., Liu, X., Yang, D.L., Wei, P., Cao, F., Zhu, S., Zhang, F., Mao, Y., *et al.* (2013). Efficient genome editing in plants using a CRISPR/Cas system. Cell Res. *23*, 1229–1232. http://dx.doi.org/10.1038/cr.2013.114

Fujita, D., Trijatmiko, K.R., Tagle, A.G., Sapasap, M.V., Koide, Y., Sasaki, K., Tsakirpaloglou, N., Gannaban, R.B., Nishimura, T., Yanagihara, S., *et al.* (2013). *NAL1* allele from a rice landrace greatly increases yield in modern *indica* cultivars. Proc. Natl. Acad. Sci. U.S.A. *110*, 20431–20436. http://dx.doi.org/10.1073/pnas.1310790110

Godfray, H.C., Beddington, J.R., Crute, I.R., Haddad, L., Lawrence, D., Muir, J.F., Pretty, J., Robinson, S., Thomas, S.M., and Toulmin, C. (2010). Food security: the challenge of feeding 9 billion people. Science *327*, 812–818. http://dx.doi.org/10.1126/science.1185383

Goff, S.A., Ricke, D., Lan, T.H., Presting, G., Wang, R., Dunn, M., Glazebrook, J., Sessions, A., Oeller, P., Varma, H., *et al.* (2002). A draft sequence of the rice genome (*Oryza sativa* L. ssp. *japonica*). Science *296*, 92–100. http://dx.doi.org/10.1126/science.1068275

Greco, R., Ouwerkerk, P.B., De Kam, R.J., Sallaud, C., Favalli, C., Colombo, L., Guiderdoni, E., Meijer, A.H., Hoge Dagger, J.H., and Pereira, A. (2003). Transpositional behaviour of an *Ac/Ds* system for reverse genetics in rice. Theor. Appl. Genet. *108*, 10–24. http://dx.doi.org/10.1007/s00122-003-1416-8

Greco, R., Ouwerkerk, P.B., Taal, A.J., Favalli, C., Beguiristain, T., Puigdomènech, P., Colombo, L., Hoge, J.H., and Pereira, A. (2001). Early and multiple *Ac* transpositions in rice suitable for efficient insertional mutagenesis. Plant Mol. Biol. *46*, 215–227. http://dx.doi.org/10.1023/A:1010607318694

Guilinger, J.P., Thompson, D.B., and Liu, D.R. (2014). Fusion of catalytically inactive Cas9 to FokI nuclease improves the specificity of genome modification. Nat. Biotechnol. *32*, 577–582. http://dx.doi.org/10.1038/nbt.2909

Han, Z., Zhang, B., Zhao, H., Ayaad, M., and Xing, Y. (2016). Genome-wide association studies reveal that diverse heading date genes respond to short and long day lengths between *indica* and *japonica* rice. Front. Plant Sci. *7*, 1270. http://dx.doi.org/10.3389/fpls.2016.01270

Hartung, F., and Schiemann, J. (2014). Precise plant breeding using new genome editing techniques: opportunities, safety and regulation in the EU. Plant J. *78*, 742–752. http://dx.doi.org/10.1111/tpj.12413

Heffner, E.L., Sorrells, M.E., and Jannink, J.L. (2009). Genomic Selection for Crop Improvement. Crop Sci. *49*, 1–12. http://dx.doi.org/10.2135/cropsci2008.08.0512

Hirochika, H., Sugimoto, K., Otsuki, Y., Tsugawa, H., and Kanda, M. (1996). Retrotransposons of rice involved in mutations induced by tissue culture. Proc. Natl. Acad. Sci. U.S.A. *93*, 7783–7788. http://dx.doi.org/10.1073/pnas.93.15.7783

Hu, W., Wen, M., Han, Z., Tan, C., and Xing, Y. (2013). Scanning QTLs for grain shape using a whole genome SNP array in rice. J. Plant Biochem. Physiol. *1*, 104. http://dx.doi.org/10.4172/2329-9029.1000104

Hua, L., Wu, J., Chen, C., Wu, W., He, X., Lin, F., Wang, L., Ashikawa, I., Matsumoto, T., Wang, L., *et al.* (2012). The isolation of *Pi1*, an allele at the *Pik* locus which confers broad spectrum resistance to rice blast. Theor. Appl. Genet. *125*, 1047–1055. http://dx.doi.org/10.1007/s00122-012-1894-7

Huang, X., Kurata, N., Wei, X., Wang, Z.X., Wang, A., Zhao, Q., Zhao, Y., Liu, K., Lu, H., Li, W., *et al.* (2012). A map of rice genome variation reveals the origin of cultivated rice. Nature *490*, 497–501. http://dx.doi.org/10.1038/nature11532

Huang, X., Qian, Q., Liu, Z., Sun, H., He, S., Luo, D., Xia, G., Chu, C., Li, J., and Fu, X. (2009). Natural variation at the *DEP1* locus enhances grain yield in rice. Nat. Genet. *41*, 494–497. http://dx.doi.org/10.1038/ng.352

Huang, X., Wei, X., Sang, T., Zhao, Q., Feng, Q., Zhao, Y., Li, C., Zhu, C., Lu, T., Zhang, Z., *et al.* (2010). Genome-wide association studies of 14 agronomic traits in rice landraces. Nat. Genet. *42*, 961–967. http://dx.doi.org/10.1038/ng.695

Huang, X., Yang, S., Gong, J., Zhao, Q., Feng, Q., Zhan, Q., Zhao, Y., Li, W., Cheng, B., Xia, J., *et al.* (2016). Genomic architecture of heterosis for yield traits in rice. Nature *537*, 629–633. http://dx.doi.org/10.1038/nature19760

Huang, X., Yang, S., Gong, J., Zhao, Y., Feng, Q., Gong, H., Li, W., Zhan, Q., Cheng, B., Xia, J., *et al.* (2015). Genomic analysis of hybrid rice varieties reveals numerous superior alleles that contribute to heterosis. Nat. Commun. *6*, 6258. http://dx.doi.org/10.1038/ncomms7258

Huang, X., Zhao, Y., Wei, X., Li, C., Wang, A., Zhao, Q., Li, W., Guo, Y., Deng, L., Zhu, C., *et al.* (2011). Genome-wide association study of flowering time and grain yield traits in a worldwide collection of rice germplasm. Nat. Genet. *44*, 32–39. http://dx.doi.org/10.1038/ng.1018

Ishimaru, K., Hirotsu, N., Madoka, Y., Murakami, N., Hara, N., Onodera, H., Kashiwagi, T., Ujiie, K., Shimizu, B., Onishi, A., *et al.* (2013). Loss of function of the IAA-glucose hydrolase gene *TGW6* enhances rice grain weight and increases yield. Nat. Genet. *45*, 707–711. http://dx.doi.org/10.1038/ng.2612

Iyer-Pascuzzi, A.S., Jiang, H., Huang, L., and McCouch, S.R. (2008). Genetic and functional characterization of the rice bacterial blight disease resistance gene *xa5*. Phytopathology *98*, 289–295. http://dx.doi.org/10.1094/PHYTO-98-3-0289

Iyer, A.S., and McCouch, S.R. (2004). The rice bacterial blight resistance gene *xa5* encodes a novel form of disease resistance. Mol. Plant Microbe Interact. *17*, 1348–1354. http://dx.doi.org/10.1094/MPMI.2004.17.12.1348

Izawa, T. (2007). Adaptation of flowering-time by natural and artificial selection in *Arabidopsis* and rice. J. Exp. Bot. *58*, 3091–3097. http://dx.doi.org/10.1093/jxb/erm159

Izawa, T., Mihara, M., Suzuki, Y., Gupta, M., Itoh, H., Nagano, A.J., Motoyama, R., Sawada, Y., Yano, M., Hirai, M.Y., *et al.* (2011). Os-*GIGANTEA* confers robust diurnal rhythms on the global transcriptome of rice in the field. Plant Cell *23*, 1741–1755. http://dx.doi.org/10.1105/tpc.111.083238

Jeon, J.S., Lee, S., Jung, K.H., Jun, S.H., Jeong, D.H., Lee, J., Kim, C., Jang, S., Yang, K., Nam, J., *et al.* (2000). T-DNA insertional mutagenesis for functional genomics in rice. Plant J. *22*, 561–570. http://dx.doi.org/10.1046/j.1365-313x.2000.00767.x

Jeong, D.H., An, S., Kang, H.G., Moon, S., Han, J.J., Park, S., Lee, H.S., An, K., and An, G. (2002). T-DNA insertional mutagenesis for activation tagging in rice. Plant Physiol. *130*, 1636–1644. http://dx.doi.org/10.1104/pp.014357

Jeong, D.H., An, S., Park, S., Kang, H.G., Park, G.G., Kim, S.R., Sim, J., Kim, Y.O., Kim, M.K., Kim, S.R., *et al.* (2006). Generation of a flanking sequence-tag database for activation-tagging lines in *japonica* rice. Plant J. *45*, 123–132. http://dx.doi.org/10.1111/j.1365-313X.2005.02610.x

Kim, C.M., Piao, H.L., Park, S.J., Chon, N.S., Je, B.I., Sun, B., Park, S.H., Park, J.Y., Lee, E.J., Kim, M.J., *et al.* (2004). Rapid, large-scale generation of *Ds* transposant lines and analysis of the *Ds* insertion sites in rice. Plant J. *39*, 252–263. http://dx.doi.org/10.1111/j.1365-313X.2004.02116.x

Kojima, S., Takahashi, Y., Kobayashi, Y., Monna, L., Sasaki, T., Araki, T., and Yano, M. (2002). *Hd3a*, a rice ortholog of the *Arabidopsis FT* gene, promotes transition to flowering downstream of *Hd1* under short-day conditions. Plant Cell Physiol. *43*, 1096–1105. http://dx.doi.org/10.1093/pcp/pcf156

Kolesnik, T., Szeverenyi, I., Bachmann, D., Kumar, C.S., Jiang, S., Ramamoorthy, R., Cai, M., Ma, Z.G., Sundaresan, V., and Ramachandran, S. (2004). Establishing an efficient *Ac/Ds* tagging system in rice: large-scale analysis of *Ds* flanking sequences. Plant J. *37*, 301–314. http://dx.doi.org/10.1046/j.1365-313X.2003.01948.x

Komatsu, M., Maekawa, M., Shimamoto, K., and Kyozuka, J. (2001). The *LAX1* and *FRIZZY PANICLE 2* genes determine the inflorescence architecture of rice by controlling rachis-branch and spikelet development. Dev. Biol. *231*, 364–373. http://dx.doi.org/10.1006/dbio.2000.9988

Kulkarni, K.P., Vishwakarma, C., Sahoo, S.P., Lima, J.M., Nath, M., Dokku, P., Gacche, R.N., Mohapatra, T., Robin, S., Sarla, N., *et al.* (2014). A substitution mutation in *OsCCD7* cosegregates with dwarf and increased tillering phenotype in rice. J. Genet. *93*, 389–401. http://dx.doi.org/10.1007/s12041-014-0389-5

Leung, H., Raghavan, C., Zhou, B., Oliva, R., Choi, I.R., Lacorte, V., Jubay, M.L., Cruz, C.V., Gregorio, G., Singh, R.K., *et al.* (2015). Allele mining and enhanced genetic recombination for rice breeding. Rice *8*, 34. http://dx.doi.org/10.1186/s12284-015-0069-y

Li, J., Meng, X., Zong, Y., Chen, K., Zhang, H., Liu, J., Li, J., and Gao, C. (2016). Gene replacements and insertions in rice by intron targeting using CRISPR-Cas9. Nat. Plants *2*, 16139. http://dx.doi.org/10.1038/nplants.2016.139

Li, M., Li, X., Zhou, Z., Wu, P., Fang, M., Pan, X., Lin, Q., Luo, W., Wu, G., and Li, H. (2016). Reassessment of the Four Yield-related Genes *Gn1a*, *DEP1*, *GS3*, and *IPA1* in Rice Using a CRISPR/Cas9 System. Front. Plant Sci. *7*, 377. http://dx.doi.org/10.3389/fpls.2016.00377

Li, T., Liu, B., Spalding, M.H., Weeks, D.P., and Yang, B. (2012). High-efficiency TALEN-based gene editing produces disease-resistant rice. Nat. Biotechnol. *30*, 390–392. http://dx.doi.org/10.1038/nbt.2199

Li, X., Qian, Q., Fu, Z., Wang, Y., Xiong, G., Zeng, D., Wang, X., Liu, X., Teng, S., Hiroshi, F., *et al.* (2003). Control of tillering in rice. Nature *422*, 618–621. http://dx.doi.org/10.1038/nature01518

Li, Y., Fan, C., Xing, Y., Jiang, Y., Luo, L., Sun, L., Shao, D., Xu, C., Li, X., Xiao, J., *et al.* (2011). Natural variation in *GS5* plays an important role in regulating grain size and yield in rice. Nat. Genet. *43*, 1266–1269. http://dx.doi.org/10.1038/ng.977

Lu, L., Shao, D., Qiu, X., Sun, L., Yan, W., Zhou, X., Yang, L., He, Y., Yu, S., and Xing, Y. (2013). Natural variation and artificial selection in four genes determine grain shape in rice. New Phytol. *200*, 1269–1280. http://dx.doi.org/10.1111/nph.12430

Lusser, M., and Davies, H.V. (2013). Comparative regulatory approaches for groups of new plant breeding techniques. New Biotechnol. *30*, 437–446. http://dx.doi.org/10.1016/j.nbt.2013.02.004

Ma, X., Zhang, Q., Zhu, Q., Liu, W., Chen, Y., Qiu, R., Wang, B., Yang, Z., Li, H., Lin, Y., *et al.* (2015). A robust CRISPR/Cas9 system for convenient, high-efficiency multiplex genome editing in monocot and dicot Plants. Mol. Plant *8*, 1274–1284. http://dx.doi.org/10.1016/j.molp.2015.04.007

Magwa, R.A., Zhao, H., and Xing, Y. (2016a). Genome-wide association mapping revealed a diverse genetic basis of seed dormancy across subpopulations in rice (*Oryza sativa* L.). BMC Genet. *17*, 28. http://dx.doi.org/10.1186/s12863-016-0340-2

Magwa, R.A., Zhao, H., Yao, W., Xie, W., Yang, L., Xing, Y., and Bai, X. (2016b). Genomewide association analysis for awn length linked to the seed shattering gene *qSH1* in rice. J. Genet. *95*, 639–646. http://dx.doi.org/10.1007/s12041-016-0679-1

Mali, P., Aach, J., Stranges, P.B., Esvelt, K.M., Moosburner, M., Kosuri, S., Yang, L., and Church, G.M. (2013). CAS9 transcriptional activators for target specificity screening and paired nickases for cooperative genome engineering. Nat. Biotechnol. *31*, 833–838. http://dx.doi.org/10.1038/nbt.2675

Mao, H., Sun, S., Yao, J., Wang, C., Yu, S., Xu, C., Li, X., and Zhang, Q. (2010). Linking differential domain functions of the GS3 protein to natural variation of grain size in rice. Proc. Natl. Acad. Sci. U.S.A. *107*, 19579–19584. http://dx.doi.org/10.1073/pnas.1014419107

Mather, K.A., Caicedo, A.L., Polato, N.R., Olsen, K.M., McCouch, S., and Purugganan, M.D. (2007). The extent of linkage disequilibrium in rice (*Oryza sativa* L.). Genetics *177*, 2223–2232. http://dx.doi.org/10.1534/genetics.107.079616

McCouch, S.R., Wright, M.H., Tung, C.W., Maron, L.G., McNally, K.L., Fitzgerald, M., Singh, N., DeClerck, G., Agosto-Perez, F., Korniliev, P., *et al.* (2016). Open access resources for genome-wide association mapping in rice. Nat. Commun. *7*, 10532. http://dx.doi.org/10.1038/ncomms10532

McCouch, S.R., Zhao, K.Y., Wright, M., Tung, C.W., Ebana, K., Thomson, M., Reynolds, A., Wang, D., DeClerck, G., Ali, M.L., *et al.* (2010). Development of genome-wide SNP assays for rice. Breeding Science *60*, 524–535. http://dx.doi.org/10.1270/jsbbs.60.524

McNally, K.L., Childs, K.L., Bohnert, R., Davidson, R.M., Zhao, K., Ulat, V.J., Zeller, G., Clark, R.M., Hoen, D.R., Bureau, T.E., *et al.* (2009). Genomewide SNP variation reveals relationships among landraces and modern varieties of rice. Proc. Natl. Acad. Sci. U.S.A. *106*, 12273–12278. http://dx.doi.org/10.1073/pnas.0900992106

Meng, L.J., Guo, L.B., Ponce, K., Zhao, X.Q., and Ye, G.Y. (2016a). Characterization of Three Indica Rice Multiparent Advanced Generation Intercross (MAGIC) Populations for Quantitative Trait Loci Identification. Plant Genome *9*, 0. http://dx.doi.org/10.3835/plantgenome2015.10.0109

Meng, L.J., Zhao, X.Q., Ponce, K., Ye, G.Y., and Leung, H. (2016b). QTL mapping for agronomic traits using multi-parent advanced generation inter-cross (MAGIC) populations derived from diverse elite *indica* rice lines. Field Crops Res. *189*, 19–42. http://dx.doi.org/10.1016/j.fcr.2016.02.004

Meyer, R.S., and Purugganan, M.D. (2013). Evolution of crop species: genetics of domestication and diversification. Nat. Rev. Genet. *14*, 840–852. http://dx.doi.org/10.1038/nrg3605

Miura, K., Ikeda, M., Matsubara, A., Song, X.J., Ito, M., Asano, K., Matsuoka, M., Kitano, H., and Ashikari, M. (2010). *OsSPL14* promotes panicle branching and higher grain productivity in rice. Nat. Genet. *42*, 545–549. http://dx.doi.org/10.1038/ng.592

Miyao, A., Tanaka, K., Murata, K., Sawaki, H., Takeda, S., Abe, K., Shinozuka, Y., Onosato, K., and Hirochika, H. (2003). Target site specificity of the *Tos17* retrotransposon shows a preference for insertion within genes and against insertion in retrotransposon-rich regions of the genome. Plant Cell *15*, 1771–1780. http://dx.doi.org/10.1105/tpc.012559

Nakaya, A., and Isobe, S.N. (2012). Will genomic selection be a practical method for plant breeding? Ann. Bot. *110*, 1303–1316. http://dx.doi.org/10.1093/aob/mcs109

Pauwels, K., Podevin, N., Breyer, D., Carroll, D., and Herman, P. (2014). Engineering nucleases for gene targeting: safety and regulatory considerations. New Biotechnol. *31*, 18–27. http://dx.doi.org/10.1016/j.nbt.2013.07.001

Pennisi, E. (2013). The CRISPR craze. Science *341*, 833–836. http://dx.doi.org/10.1126/science.341.6148.833

Ritchie, M.D., Holzinger, E.R., Li, R., Pendergrass, S.A., and Kim, D. (2015). Methods of integrating data to uncover genotype-phenotype interactions. Nat. Rev. Genet. *16*, 85–97. http://dx.doi.org/10.1038/nrg3868

Sallaud, C., Meynard, D., van Boxtel, J., Gay, C., Bès, M., Brizard, J.P., Larmande, P., Ortega, D., Raynal, M., Portefaix, M., *et al.* (2003). Highly efficient production and characterization of T-DNA plants for rice (*Oryza sativa* L.) functional genomics. Theor. Appl. Genet. *106*, 1396–1408. http://dx.doi.org/10.1007/s00122-002-1184-x

Sasaki, T., Matsumoto, T., Yamamoto, K., Sakata, K., Baba, T., Katayose, Y., Wu, J., Niimura, Y., Cheng, Z., Nagamura, Y., *et al.* (2002). The genome sequence and structure of rice chromosome 1. Nature *420*, 312–316. http://dx.doi.org/10.1038/nature01184

Schaart, J.G., van de Wiel, C.C., Lotz, L.A., and Smulders, M.J. (2016). Opportunities for Products of New Plant Breeding Techniques. Trends Plant Sci. *21*, 438–449. http://dx.doi.org/10.1016/j.tplants.2015.11.006

Sha, Y., Li, S., Pei, Z., Luo, L., Tian, Y., and He, C. (2004). Generation and flanking sequence analysis of a rice T-DNA tagged population. Theor. Appl. Genet. *108*, 306–314. http://dx.doi.org/10.1007/s00122-003-1423-9

Shan, Q., Wang, Y., Chen, K., Liang, Z., Li, J., Zhang, Y., Zhang, K., Liu, J., Voytas, D.F., Zheng, X., *et al.* (2013a). Rapid and efficient gene modification in rice and *Brachypodium* using TALENs. Mol. Plant *6*, 1365–1368. http://dx.doi.org/10.1093/mp/sss162

Shan, Q., Wang, Y., Li, J., Zhang, Y., Chen, K., Liang, Z., Zhang, K., Liu, J., Xi, J.J., Qiu, J.L., *et al.* (2013b). Targeted genome modification of crop plants using a CRISPR-Cas system. Nat. Biotechnol. *31*, 686–688. http://dx.doi.org/10.1038/nbt.2650

Shan, Q., Zhang, Y., Chen, K., Zhang, K., and Gao, C. (2015). Creation of fragrant rice by targeted knockout of the *OsBADH2* gene using TALEN technology. Plant Biotechnol. J. *13*, 791–800. http://dx.doi.org/10.1111/pbi.12312

Song, X.J., Huang, W., Shi, M., Zhu, M.Z., and Lin, H.X. (2007). A QTL for rice grain width and weight encodes a previously unknown RING-type E3 ubiquitin ligase. Nat. Genet. *39*, 623–630. http://dx.doi.org/10.1038/ng2014

Spielmeyer, W., Ellis, M.H., and Chandler, P.M. (2002). Semidwarf (*sd-1*), "green revolution" rice, contains a defective gibberellin 20-oxidase gene. Proc. Natl. Acad. Sci. U.S.A. *99*, 9043–9048. http://dx.doi.org/10.1073/pnas.132266399

Takagi, H., Abe, A., Yoshida, K., Kosugi, S., Natsume, S., Mitsuoka, C., Uemura, A., Utsushi, H., Tamiru, M., Takuno, S., *et al.* (2013a). QTL-seq: rapid mapping of quantitative trait loci in rice by whole genome resequencing of DNA from two bulked populations. Plant J. *74*, 174–183. http://dx.doi.org/10.1111/tpj.12105

Takagi, H., Tamiru, M., Abe, A., Yoshida, K., Uemura, A., Yaegashi, H., Obara, T., Oikawa, K., Utsushi, H., Kanzaki, E., *et al.* (2015). MutMap accelerates breeding of a salt-tolerant rice cultivar. Nat. Biotechnol. *33,* 445–449. http://dx.doi.org/10.1038/nbt.3188

Takagi, H., Uemura, A., Yaegashi, H., Tamiru, M., Abe, A., Mitsuoka, C., Utsushi, H., Natsume, S., Kanzaki, H., Matsumura, H., *et al.* (2013b). MutMap-Gap: whole-genome resequencing of mutant F2 progeny bulk combined with de novo assembly of gap regions identifies the rice blast resistance gene *Pii.* New Phytol. *200,* 276–283. http://dx.doi.org/10.1111/nph.12369

Tan, C., Han, Z., Yu, H., Zhan, W., Xie, W., Chen, X., Zhao, H., Zhou, F., and Xing, Y. (2013). QTL scanning for rice yield using a whole genome SNP array. J. Genet. Genomics *40,* 629–638. http://dx.doi.org/10.1016/j.jgg.2013.06.009

Verbyla, A.P., George, A.W., Cavanagh, C.R., and Verbyla, K.L. (2014). Whole-genome QTL analysis for MAGIC. Theor. Appl. Genet. *127,* 1753–1770. http://dx.doi.org/10.1007/s00122-014-2337-4

Wang, C., Chen, S., and Yu, S. (2011). Functional markers developed from multiple loci in *GS3* for fine marker-assisted selection of grain length in rice. Theor. Appl. Genet. *122,* 905–913. http://dx.doi.org/10.1007/s00122-010-1497-0

Weng, J., Gu, S., Wan, X., Gao, H., Guo, T., Su, N., Lei, C., Zhang, X., Cheng, Z., Guo, X., *et al.* (2008). Isolation and initial characterization of *GW5*, a major QTL associated with rice grain width and weight. Cell Res. *18,* 1199–1209. http://dx.doi.org/10.1038/cr.2008.307

Wolt, J.D., Wang, K., and Yang, B. (2016). The regulatory status of genome-edited crops. Plant Biotechnol. J. *14,* 510–518. http://dx.doi.org/10.1111/pbi.12444

Wu, C., Li, X., Yuan, W., Chen, G., Kilian, A., Li, J., Xu, C., Li, X., Zhou, D.X., Wang, S., *et al.* (2003). Development of enhancer trap lines for functional analysis of the rice genome. Plant J. *35,* 418–427. http://dx.doi.org/10.1046/j.1365-313X.2003.01808.x

Wu, J.L., Wu, C., Lei, C., Baraoidan, M., Bordeos, A., Madamba, M.R., Ramos-Pamplona, M., Mauleon, R., Portugal, A., Ulat, V.J., *et al.* (2005). Chemical- and irradiation-induced mutants of *indica* rice IR64 for forward and reverse genetics. Plant Mol. Biol. *59,* 85–97. http://dx.doi.org/10.1007/s11103-004-5112-0

Xing, Y., and Zhang, Q. (2010). Genetic and molecular bases of rice yield. Annu. Rev. Plant Biol. *61,* 421–442. http://dx.doi.org/10.1146/annurev-arplant-042809-112209

Xu, X., Liu, X., Ge, S., Jensen, J.D., Hu, F., Li, X., Dong, Y., Gutenkunst, R.N., Fang, L., Huang, L., *et al.* (2011). Resequencing 50 accessions of cultivated and wild rice yields markers for identifying agronomically important genes. Nat. Biotechnol. *30,* 105–111. http://dx.doi.org/10.1038/nbt.2050

Xu, Y.B., and Crouch, J.H. (2008). Marker-assisted selection in plant breeding: From publications to practice. Crop Sci. *48,* 391–407. http://dx.doi.org/10.2135/cropsci2007.04.0191

Xue, W., Xing, Y., Weng, X., Zhao, Y., Tang, W., Wang, L., Zhou, H., Yu, S., Xu, C., Li, X., *et al.* (2008). Natural variation in *Ghd7* is an important regulator of heading date and yield potential in rice. Nat. Genet. *40,* 761–767. http://dx.doi.org/10.1038/ng.143

Yamamoto, T., Nagasaki, H., Yonemaru, J., Ebana, K., Nakajima, M., Shibaya, T., and Yano, M. (2010). Fine definition of the pedigree haplotypes of closely related rice cultivars by means of genome-wide discovery of single-nucleotide polymorphisms. BMC Genomics *11,* 267. http://dx.doi.org/10.1186/1471-2164-11-267

Yan, W., Liu, H., Zhou, X., Li, Q., Zhang, J., Lu, L., Liu, T., Liu, H., Zhang, C., Zhang, Z., *et al.* (2013). Natural variation in *Ghd7.1* plays an important role in grain yield and adaptation in rice. Cell Res. *23,* 969–971. http://dx.doi.org/10.1038/cr.2013.43

Yan, W.H., Wang, P., Chen, H.X., Zhou, H.J., Li, Q.P., Wang, C.R., Ding, Z.H., Zhang, Y.S., Yu, S.B., Xing, Y.Z., *et al.* (2011). A major QTL, *Ghd8,* plays pleiotropic roles in regulating grain productivity, plant height, and heading date in rice. Mol. Plant *4,* 319–330. http://dx.doi.org/10.1093/mp/ssq070

Yang, W., Guo, Z., Huang, C., Duan, L., Chen, G., Jiang, N., Fang, W., Feng, H., Xie, W., Lian, X., *et al.* (2014). Combining high-throughput phenotyping and genome-wide association studies to reveal natural genetic variation in rice. Nat. Commun. *5,* 5087. http://dx.doi.org/10.1038/ncomms6087

Yang, W., Guo, Z., Huang, C., Wang, K., Jiang, N., Feng, H., Chen, G., Liu, Q., and Xiong, L. (2015). Genome-wide association study of rice (*Oryza sativa* L.) leaf traits with a high-throughput leaf scorer. J. Exp. Bot. *66,* 5605–5615. http://dx.doi.org/10.1093/jxb/erv100

Yano, M., Katayose, Y., Ashikari, M., Yamanouchi, U., Monna, L., Fuse, T., Baba, T., Yamamoto, K., Umehara, Y., Nagamura, Y., *et al.* (2000). *Hd1,* a major photoperiod sensitivity quantitative trait locus in rice, is closely related to the *Arabidopsis* flowering time gene *CONSTANS.* Plant Cell *12,* 2473–2484. http://dx.doi.org/10.1105/tpc.12.12.2473

Yin, Z., and Wang, G. (2000). Evidence of multiple complex patterns of T-DNA integration into the rice genome. Theor. Appl. Genet. *100,* 461–470. http://dx.doi.org/10.1007/s001220050060

Yu, B., Lin, Z., Li, H., Li, X., Li, J., Wang, Y., Zhang, X., Zhu, Z., Zhai, W., Wang, X., *et al.* (2007). *TAC1,* a major quantitative trait locus controlling tiller angle in rice. Plant J. *52,* 891–898. http://dx.doi.org/10.1111/j.1365-313X.2007.03284.x

Yu, H., Xie, W., Li, J., Zhou, F., and Zhang, Q. (2014). A whole-genome SNP array (RICE6K) for genomic breeding in rice. Plant Biotechnol. J. *12,* 28–37. http://dx.doi.org/10.1111/pbi.12113

Yu, J., Hu, S., Wang, J., Wong, G.K., Li, S., Liu, B., Deng, Y., Dai, L., Zhou, Y., Zhang, X., *et al.* (2002). A draft sequence of the rice genome (*Oryza sativa* L. ssp. *indica*). Science *296,* 79–92. http://dx.doi.org/10.1126/science.1068037

Yuan, M., Chu, Z., Li, X., Xu, C., and Wang, S. (2009). Pathogen-induced expressional loss of function is the key factor in race-specific bacterial resistance conferred by a recessive *R* gene *xa13* in rice. Plant Cell Physiol. *50,* 947–955. http://dx.doi.org/10.1093/pcp/pcp046

Zhang, J., Zhou, X., Yan, W., Zhang, Z., Lu, L., Han, Z., Zhao, H., Liu, H., Song, P., Hu, Y., *et al.* (2015). Combinations of the *Ghd7, Ghd8* and *Hd1* genes largely define the ecogeographical adaptation and yield potential of cultivated rice. New Phytol. *208*, 1056–1066. http://dx.doi.org/10.1111/nph.13538

Zhang, Z., Ersoz, E., Lai, C.Q., Todhunter, R.J., Tiwari, H.K., Gore, M.A., Bradbury, P.J., Yu, J., Arnett, D.K., Ordovas, J.M., *et al.* (2010). Mixed linear model approach adapted for genome-wide association studies. Nat. Genet. *42*, 355–360. http://dx.doi.org/10.1038/ng.546

Zhao, K., Tung, C.W., Eizenga, G.C., Wright, M.H., Ali, M.L., Price, A.H., Norton, G.J., Islam, M.R., Reynolds, A., Mezey, J., *et al.* (2011). Genome-wide association mapping reveals a rich genetic architecture of complex traits in *Oryza sativa*. Nat. Commun. *2*, 467. http://dx.doi.org/10.1038/ncomms1467

Zhao, K., Wright, M., Kimball, J., Eizenga, G., McClung, A., Kovach, M., Tyagi, W., Ali, M.L., Tung, C.W., Reynolds, A., *et al.* (2010). Genomic diversity and introgression in *O. sativa* reveal the impact of domestication and breeding on the rice genome. PLOS ONE *5*, e10780. http://dx.doi.org/10.1371/journal.pone.0010780

Zhou, B., Qu, S., Liu, G., Dolan, M., Sakai, H., Lu, G., Bellizzi, M., and Wang, G.L. (2006). The eight amino-acid differences within three leucine-rich repeats between *Pi2* and *Piz-t* resistance proteins determine the resistance specificity to *Magnaporthe grisea*. Mol. Plant Microbe In. *19*, 1216–1228. http://dx.doi.org/10.1094/MPMI-19-1216

# Dual RNA-seq to Elucidate the Plant–Pathogen Duel

8

Sanushka Naidoo[1]*, Erik Andrei Visser[1], Lizahn Zwart[1], Yves du Toit[1], Vijai Bhadauria[2] and Louise Simone Shuey[1]*

[1]Department of Genetics, Forestry and Agricultural Biotechnology Institute, Genomics Research Institute, University of Pretoria, Pretoria, South Africa.
[2]Crop Development Centre and Department of Plant Sciences, University of Saskatchewan, Saskatoon, SK, Canada.

*Correspondence: Sanushka.Naidoo@fabi.up.ac.za and Louise.Shuey@fabi.up.ac.za

https://doi.org/10.21775/9781910190654.09

## Abstract
RNA-sequencing technology has been widely adopted to investigate host responses during infection with pathogens. Dual RNA-sequencing (RNA-seq) allows the simultaneous capture of pathogen-specific transcripts during infection, providing a more complete view of the interaction. In this review, we focus on the design of dual RNA-seq experiments and the application of downstream data analysis to gain biological insight into both sides of the interaction. Recent literature in this area demonstrates the power of the dual RNA-seq approach and shows that it is not limited to model systems where genomic resources are available. Sequencing costs continue to decrease and single cell transcriptomics is becoming more feasible. In combination with proteomics and metabolomics studies, these technological advances are likely to contribute to our understanding of the temporal and spatial aspects of dynamic plant–pathogen interactions.

## A dual approach *in planta*
The interaction between plants and pathogens is an active and dynamic process that can be likened to a duel. Plants have complex defence mechanisms that can be rendered ineffective when pathogens interfere with one of the various processes required for host defence. These processes include penetration resistance, recognition by Pattern Recognition Receptors (PRRs), phytohormone signalling pathways, secretory pathways, secondary metabolite production, and plant cell death (Dou and Zhou, 2012). Until recently, transcriptomic approaches have been applied in the host and pathogen separately to obtain the gene expression profile of each organism and gain insight into infection biology or host defence mechanisms.

RNA sequencing (RNA-seq) is a powerful technology that does not rely on any prior knowledge of transcripts and can generate vast quantities of data with much smaller costs involved than for older techniques such as microarrays (Pareek *et al.*, 2011; Wilhelm and Landry, 2009). An advantage of RNA-seq in the field of plant–pathogen interactions is that both plant and pathogen transcripts can be detected simultaneously and accurately in the same sample. This tactic, known as dual RNA-seq, *in planta* RNA-seq, simultaneous RNA-seq, or comparative RNA-seq, is a relatively new technique both in the plant and medical fields. In plants, it allows for the study of plant–pathogen interactions in herbaceous crops (Chen *et al.*, 2013; Kunjeti *et al.*, 2012; Lowe *et al.*, 2014) as well as trees (Hayden *et al.*, 2014; Liang *et al.*, 2014; Teixeira *et al.*, 2014). This review outlines technical considerations for dual RNA-seq experiments, summarizes recent insights drawn from such approaches in plant–pathogen interactions, and provides an overview of the next generation of dual approaches. Since this technique is most useful to study interactions with pathogens with complex prokaryotic and eukaryotic genomes, viral pathogens have not been included in this review.

## It's all in the design

Experimental design considerations for a dual RNA-seq experiment can be divided into three broad categories: sample generation, data generation and data analysis. An overview of the process can be found in Fig. 8.1.

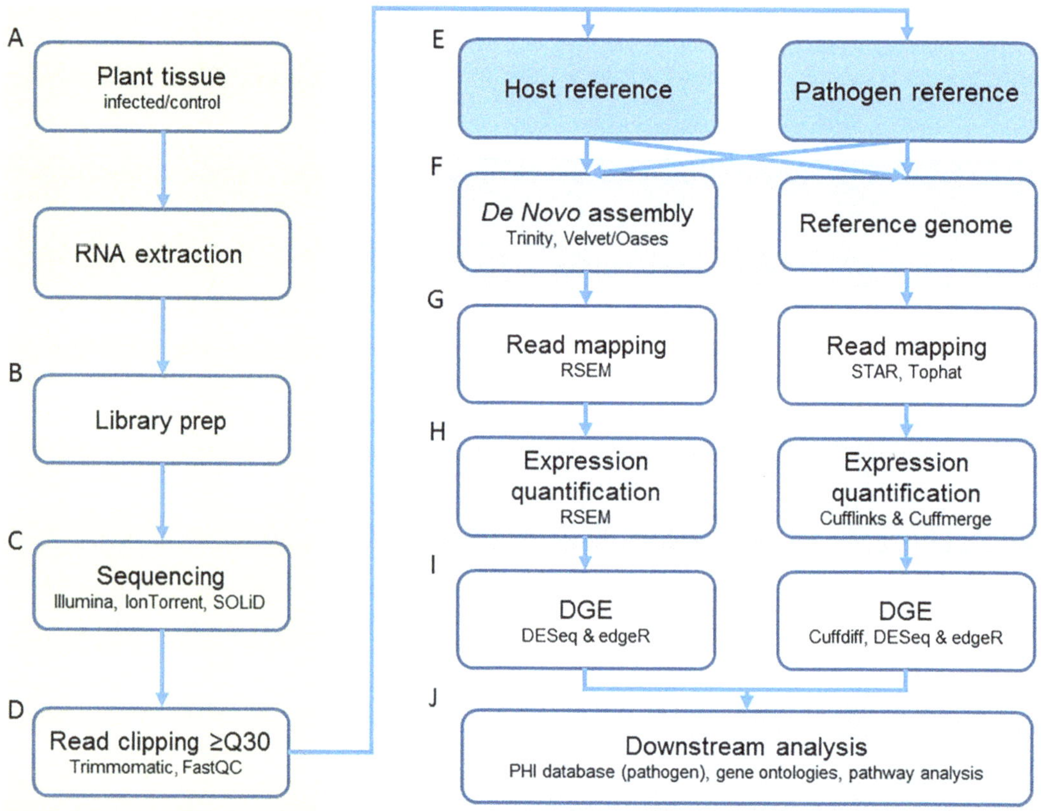


**Figure 8.1** Flow chart of a dual RNA-seq experiment with example software programs. (A) Experimental design considerations can include comparisons between resistant and susceptible interactions, normalized to a mock inoculated control. (B) Different library preparation options include enrichments for mRNA, small RNAs, stranded RNA and total RNA. (C) The sequencing platform can vary based on availability and aim of the study. Paired-end sequencing on the Illumina platform is a common approach for RNA-seq. Deep sequencing is required for dual RNA-seq approaches. Read length can vary depending on the application and sequencing platform used. (D) Downstream read quality control can be implemented. Filtering for a minimum Phred quality score of Q30 is generally optimal, but the threshold is data dependent. (E) Dual RNA-seq is performed by mapping to both the host and pathogen reference genome sequences, or to the host first with the remaining reads mapped to the pathogen reference. Endophyte contamination can be removed by mapping to common contaminant sequences obtained from databases such as Refseq. (F) If a reference is not available, a *de novo* transcriptome can be assembled from the reads. (G) The mapping approach will differ based on the type of reference used and different methods can be used when mapping to the host or the pathogen. (H) Other programs commonly used for expression quantification include HTSeq, featureCounts (Liao *et al.*, 2014) and Limma (Ritchie *et al.*, 2015). (I) Differential Gene Expression (DGE) analysis can be performed using a number of methods. The two examples listed here can be used for both transcriptome and genome-based DGE analysis. (J) Genes identified as differentially expressed can be used in programs and databases such as BinGO (Maere *et al.*, 2005), MapMan (Thimm *et al.*, 2004) and Kegg (Ogata *et al.*, 1999) to predict biological significance (Maere *et al.*, 2005).

## Sample generation

When considering experimental design for sample generation, the main factors include trial design, sample harvesting approach and sample handling (reviewed in Yang and Wei, 2015).

Two important trial design and sample harvesting considerations for dual RNA-seq experiments are the predicted gene number in the pathogen and host genomes, and the relative amounts of pathogen and host cells within a given sample (Westermann *et al.*, 2012). Both of these factors influence the amount of pathogen RNA relative to host RNA within a sample. A lower ratio of pathogen to host RNA requires greater sequencing depth to capture the full extent of biological variation within the pathogen.

A dual RNA-seq experiment considering an interaction between a eukaryotic host and prokaryotic pathogen requires approximately 10 to 20 times as many reads than would usually be required. This is partly due to the smaller amount of cellular RNA within prokaryotic cells relative to eukaryotes (Westermann *et al.*, 2012). While the relative amounts of cellular RNA between host and pathogen are more similar in eukaryote–eukaryote interactions, the higher read coverage is still necessary due to the lower quantity of pathogen versus host cells, which results in less pathogen RNA per sample.

An important trial design consideration specific to dual RNA-seq experiments is the inclusion of a control for pathogen gene expression. This can be done by comparing *in planta* expression of pathogen genes to *in planta* gene expression of a non-pathogenic strain and/or pathogen gene expression in an agar culture or spore suspension (Kawahara *et al.*, 2012). Synthetic RNA spike-ins can also be included to quantify both pathogen and host RNA (Box 8.1).

## Data generation

The main experimental design considerations for data generation include the level of sample replication, library construction and sequencing (Liu *et al.*, 2014).

### Sample replication

One of the first factors to consider in experimental design is the level of sample replication (Auer and Doerge, 2010). Sample replication is divided into technical replication, which is defined as performing the same analysis multiple times on the same sample, and biological replication, a study dependent term that can be loosely defined as harvesting the same type of sample from the same type of organism from the same conditions.

Technical variation arises when errors occur in the experimental procedure and can be accounted for through technical replication. Illumina sequencing produces negligible technical variability, removing the need for technical replication in RNA-seq experiments (Marioni *et al.*, 2008). However, when coverage is low for certain transcripts, technical variation can still arise (McIntyre *et al.*, 2011). Thus, technical replication should be considered for dual RNA-seq experiments where there is low representation of pathogen RNA within a sample, resulting in low coverage of pathogen transcripts.

**Box 8.1 Total RNA quantification**

It is not always possible to accurately predict the amounts of host and pathogen RNA that will be present in a sample. While it is possible to measure the amount of host and pathogen DNA in a sample using qRT-PCR, this is not always an accurate reflection of total host and pathogen RNA. This problem can be circumvented by the addition of RNA spike-ins to samples. An RNA spike-in for RNA-seq is a mixture of synthesized RNA transcripts of known sequence, concentration and abundance. While inclusion of an RNA spike-in could increase the cost of sequencing due to increased coverage requirements, it can be used to measure sensitivity and accuracy of sequencing as well as to detect biases that can occur during RNA-seq (Jiang *et al.*, 2011). Furthermore, standard curves can be generated from RNA spike-ins. This allows for more accurate quantification of transcript abundance (Jiang *et al.*, 2011). In dual RNA-seq experiments, it is possible to use these standard curves to estimate host and pathogen RNA levels within a sample. However, it is important to ensure that none of the spike-in sequences are present in the genome of either host or pathogen, as this could preclude accurate quantification of genes containing similar sequences and the use of those spike-in sequences (Jiang *et al.*, 2011).

While technical replication can be excluded due to reliability of the technology, biological replication remains crucial to all RNA-seq experiments. Besides accounting for biological variation (Hansen *et al.*, 2011; Nettleton, 2014), biological replication significantly affects the power and accuracy of differential expression analyses. Liu *et al.* (2014) showed that increasing the number of biological replicates sequenced increased the number of accurately identified differentially expressed genes, whereas increased read depth produced diminishing returns for both statistical power and the precision with which differential expression is detected. This is especially important in dual RNA-seq experiments where biological variation is introduced from both pathogen and host.

## Library construction and sequencing

The main factors to consider during library construction and sequencing are depletion methods, strandedness, insert size, read length, and read depth. The use of strand-specific rather than non-strand-specific libraries [reviewed in Levin *et al.* (2010)], allows the accurate detection of anti-sense transcription and can allow accurate expression quantification of overlapping transcripts. Thus, strand-specific sequencing in dual RNA-seq experiments could enable detection of evidence suggesting host–pathogen interaction through anti-sense transcription.

The choice of insert size is dependent on the complexity of the transcriptome and the target RNA species (reviewed in Head *et al.*, 2014). Insert size selection can be a limiting factor in which RNA species can be analysed because inclusion of a size selection step during library preparation results in loss of transcripts shorter than the selected insert size. Insert size selection also imposes an upper limit on read length, since reads longer than the insert size will sequence into adapters, providing no new information.

Apart from insert size, the choice of read length is dependent mainly on the objectives of the study and the quality of the reference sequence used for mapping. If a high quality and well annotated reference sequence is available, increasing read length above 50 bp is unnecessary for accurate detection of differential expression (Chhangawala *et al.*, 2015). Similarly, sequencing of paired-end instead of single-end reads does not significantly affect detection of differential expression in these cases (Chhangawala *et al.*, 2015). Conversely, when studying organisms with less well defined reference sequences, sequencing of longer paired-end reads increases the accuracy of splice junction detection (Chhangawala *et al.*, 2015). When no reference sequence is available, it is often assumed that longer reads equate to increased accuracy for *de novo* assembly. Similar to the detection of differential expression, however, there seems to be a species-specific threshold beyond which increasing read length becomes redundant (Chang *et al.*, 2014).

In cases where a high quality reference is available, less coverage is required for accurate transcript identification and quantification, compared to cases where a reference is missing. This is because gaps in an assembly arising from low coverage can be filled using the underlying reference sequence. For studies relying on *de novo* assembly, a predicted minimum of 30× total reference coverage is required (Martin and Wang, 2011), while genome-guided assembly can be accomplished with coverage below 10× (Denoeud *et al.*, 2008).

For an RNA-seq experiment to be representative, it is important to make sure that the number of reads is sufficient to account for the least represented RNA species. This is also referred to as sufficient sequencing depth. To obtain adequate depth for a dual RNA-seq experiment, enough reads need to be sequenced to have at least 1× coverage of the least represented pathogen transcript in the sample with the lowest level of pathogen to host RNA. However, it is almost impossible to know this information when performing *de novo* RNA-seq experiments.

Techniques to deplete or enrich certain RNA species, such as RNA fractionation and poly(A) selection, can enhance detection of transcripts with low expression in eukaryotes (Sims *et al.*, 2014). Depletion of the rRNA fraction can further reduce the required sequencing depth of an experiment and, unlike poly(A) selection, allow for detection of non-poly(A) transcripts. Although depletion-based methods allow for selection of non-poly(A) RNA species, these methods can bias quantification of abundant transcripts and decrease exon coverage and power to detect splice junctions due to the presence of sequenced introns from pre-mRNA in eukaryotes (Martin and Wang, 2011; Sims *et al.*, 2014).

## Data analysis

As with sample and data generation, data analysis considerations are dependent on the underlying biological questions. Data analysis for the majority of RNA-seq experiments follows three sequential steps: (1) quality control, (2) mapping, expression quantification and DE analysis, and (3) downstream analysis. Due to the variety of tools and platforms that can be used for RNA-seq data analysis (Grant *et al.*, 2011), programs typically used for RNA-seq data analysis may be created for specific analysis types and tested within a specific experimental context. Thus, it is often advisable to repeat an analysis using different programs and compare the outputs.

### Quality control

Quality control for dual RNA-seq studies is similar to that used for traditional RNA-seq studies. However, contaminant filtering becomes more complicated as reads from both host and pathogen need to be retained. While reads originating from the host and pathogen can be separated by mapping to the host and pathogen reference sequences (Schulze *et al.*, 2015; Westermann *et al.*, 2012), contamination of various forms should be considered in order to improve the accuracy and efficiency with which genes and transcripts are mapped and quantified. Contamination may occur in two main forms: non-mRNA species (which constitute the majority of the total RNA extracted) and reads representing mRNA extracted from organisms (saprophytes and endophytes) other than the organisms of interest. These forms of contamination may skew the quantification of genes and transcripts when assembling and mapping reads to the reference. Westermann *et al.* (2012) provide insight into dealing with contaminating RNA which is species and study dependent.

Contamination in the form of RNA extracted from endophytic or saprophytic organisms is especially important in plant–pathogen interaction studies. Saprophytes may be present at the sites of wounding due to the degradation of tissue that occurs, while endophytes colonize areas below the surface of the plant tissue without causing symptoms. Thus RNA from these types of organisms can be present in RNA-seq libraries. While surface sterilization could be used to decrease the presence of saprophytes, the process is time consuming and may result in damage to host RNA. Surface sterilization could also result in decreased pathogen representation, which is counterproductive for a dual RNA-seq experiment. Therefore, removal of these contaminating sequences requires bioinformatics intervention. This can be accomplished through stringent mapping of data to a database of common contaminant cDNA sequences constructed from databases such as RefSeq, UniRef100 and GenBank (Ikeue *et al.*, 2015). In cases where reference genomes are available for known endophytes and saprophytes, stringent alignment to these references could also be used to filter reads (Zuluaga *et al.*, 2015).

### Mapping, expression quantification and differential expression analysis

Mapping is the reconstruction of the transcriptome through alignment of reads to a reference sequence. In dual RNA-seq experiments, reads are mapped to the host reference sequence and the unaligned sequences are retained and mapped to the pathogen reference sequence (Teixeira *et al.*, 2014). A common program used for read alignment to a reference is the short read aligner Bowtie, which is part of the Tophat package of the Tuxedo pipeline (Trapnell *et al.*, 2012). Box 8.2 describes mapping and splice site determination. Bowtie allows the user to set the number of mismatches between the query and reference sequence, effectively setting a stringency threshold for the alignment (Langmead and Salzberg, 2012). This affects the stringency with which reads will be aligned and effectively assigned to the host or pathogen reference.

Once the reads have been assembled and filtered into host and pathogen libraries, transcript abundance quantification and differential expression analysis can be performed (Boxes 8.3 and 8.4, respectively). Expression levels are quantified by counting the number of reads mapped to each gene/transcript, normalized across the length of the gene/transcript to account for bias across abundant gene regions, relative to the number of reads in the original library. Programs like Cufflinks (Trapnell *et al.*, 2012) and RSEM (Li and Dewey, 2011) can be used to accurately quantify relative numbers of genes and transcripts. Differential expression analysis is commonly performed using packages such as Cuffdiff, DESeq and EdgeR (Anders and Huber, 2010; Robinson *et al.*, 2010; Trapnell *et al.*, 2013).

**Box 8.2 Mapping and splice site determination**

One of the most important aspects when mapping RNA-seq data to a reference genome is the detection of splice site junctions, and this should be considered when choosing an appropriate program. In many programs, the accuracy of estimating isoform abundances is still debatable. It is important to detect splice site junctions with as much accuracy as possible (Chandramohan *et al.*, 2013), particularly when *a priori* splice site annotations are unavailable. Programs discussed here include Tophat (Kim *et al.*, 2013), MapSplice (Wang *et al.*, 2010), SOAPsplice (Huang *et al.*, 2011), GSNAP (Wu and Nacu, 2010) and CRAC (Philippe *et al.*, 2013). CRAC is the only known splice junction mapper that includes the discovery of transcriptomic splice junctions and chimeras, single nucleotide variants, and indels in a single analysis step. MapSplice specializes in determining splice site junctions from empirical methods rather than from the canonical intronic dinucleotide (ITDN) detecting methods, while TopHat makes use of the ultrafast short read aligner Bowtie and performs canonical splice site detection. SOAPsplice is capable of detecting true splice site junctions when there is low coverage, and provides efficient use of computational memory by making use of the Burrows–Wheeler transformation. GSNAP makes use of both single-end and paired-end reads, and can tolerate a variety of read lengths from very short to long length reads. GSNAP detects splice junctions using a probabilistic model as well as user defined splice site databases. Once successful detection of splice site junctions is achieved, the quantification of gene/transcript expression is the next step.

**Box 8.3 Gene and transcript quantification**

Programs discussed here – Cufflinks (Trapnell *et al.*, 2012), IsoEM (Nicolae *et al.*, 2011), HTSeq (Anders *et al.*, 2015) and RSEM (Li and Dewey, 2011), have proven to estimate gene abundances with similar accuracy, however, their accuracy for transcript abundance estimation varies. Cufflinks is capable of calculating both gene and isoform level abundance estimates, although it has proven to be less efficient when considering computational memory, making it a time-consuming but accurate choice. HTSeq does not retain reads that map to multiple genes, which can prevent accurate expression quantification of overlapping genes. RSEM allows accurate quantification of gene and isoform abundances for organisms that do not have a reference genome. Isoform abundance estimations can be improved with paired-end read libraries when provided with an annotation of expected isoforms. IsoEM outperforms its competitors in its ability to estimate isoform abundances by making use of paired-end libraries and annotated lists of expected isoforms.

If a reference sequence is incomplete or unavailable, *de novo* assembly can be performed to create a reference sequence. Yazawa *et al.* (2013) describe a method for *de novo* assembly of the *in planta* pathogen transcriptome in the *Sorghum bicolor* and *Bipolaris sorghicola* interaction. Several putative defence and virulence related genes were identified. In this case, however, a reference sequence was available for the host, and host sequences could easily be removed from the assembled pathogen transcriptome by alignment to the host reference sequence, resulting in 232,760 non-host transcripts (Yazawa *et al.*, 2013). In cases where neither reference sequence is available, high quality annotation of *de novo* assembled sequences is crucial because separation of host and pathogen sequences will have to be done through alignment-based homology. When *de novo* assembly is required, it is important to consider the complexity of the transcriptome to be assembled. Increased transcriptome complexity, such as a large number of alternative splicing events, complicates the alignment of reads (Chang *et al.*, 2014). The accuracy of *de novo* assemblies increases with read length, which reduces mapping uncertainty. While there are algorithms that account for mapping uncertainty, thereby allowing the use of shorter read lengths (Li *et al.*, 2010),

**Box 8.4 Differential expression analysis**

When considering differential expression analysis, a number of confounding issues can arise, such as inherent errors in next-generation sequencing technologies, biases introduced due to length and composition of genes and transcripts, the existence of multiple gene isoforms, and the problem of overdispersion (Zhang *et al.*, 2014). Overdispersion can occur when biological replicates are increased to account for biological variation. Using a Poisson distribution to model this variation results in underestimates. When considering differential gene expression, using a variance stabilizing transformation decreases overdispersion (Soneson and Delorenzi, 2013). However, if a negative binomial (NB) distribution is used, it effectively eliminates overdispersion (Anders and Huber, 2010; Zhang *et al.*, 2014). Programs such as Cuffdiff, edgeR (Robinson *et al.*, 2010) and DESeq (Anders and Huber, 2010) use very different statistical approaches to account for these confounding issues. EdgeR employs the empirical Bayes and weighted likelihood methods which allow for gene-specific variation while generalized linear models are used to account for complex experimental designs (Chen *et al.*, 2014). DESeq has the option to use a variance stabilizing transformation, which results in differentially expressed gene candidates being balanced over the large dynamic range which is typical of RNA-seq experiments. Finally, Cuffdiff, a popular choice for differential gene expression analysis, estimates expression at transcript-level resolution and stabilizes variability across different biological replicates. Cuffdiff2 is capable of identifying differentially expressed transcripts and genes as well as promoter-preference changes where applicable (Trapnell *et al.*, 2013).

these algorithms assume the availability of a high quality reference sequence.

### Downstream analysis

Potential genes and pathways involved in the host–pathogen interaction can be identified during downstream analysis by comparing annotations of the identified differentially expressed genes to databases. These resources utilize controlled vocabularies provided by the Gene Ontology Consortium (GOC), which defines how genes function biologically within three categories: biological process, molecular function and cellular localization (Ashburner *et al.*, 2000). Once potential pathways or biological functions have been identified through GOC, a Gene Set Enrichment Analysis (GSEA) can be conducted. This may aid in identifying pathways that are perturbed in the host and pathogen during the interaction.

The pathogen–host interactions database (www.phi-base.org) provides curated information on genes that affect the interaction between host and pathogens based on functional genetics studies with links to experimental evidence (Urban *et al.*, 2015). A BLAST analysis against the nucleotide or predicted amino acid sequences in the database provides clues about whether pathogen genes expressed *in planta* are putative effectors, virulence factors or pathogenicity factors. *PathoPlant* (www.pathoplant.de) (Bülow *et al.*, 2004) is a plant–pathogen interactions database with information on 350 interactions from model plant–pathogen systems. The database also allows the identification of signal transduction pathway components potentially induced during pathogen challenge.

For insight into host responses, the TreeGenes database (Wegrzyn *et al.*, 2008), Phytozome database (http://phytozome.jgi.doe.gov) (Goodstein *et al.*, 2012), The Plant Genome Integrative Explorer (PlantGenIE; www.plantgenie.org) (Sundell *et al.*, 2015), and Gramene (www.gramene.org) (Liang *et al.*, 2008) provide tools for analyses and visualization. Gramene contains the Plant Reactome database, which can be used to identify plant metabolic and regulatory pathways (Tello-Ruiz *et al.*, 2016).

## Weapons in plant–pathogen duels

The ultimate goal of a dual RNA-seq experiment is to gain biological insight into the interaction that is being studied, specifically those factors that determine disease severity. Since the dual RNA-seq approach is relatively new, it has not yet been applied to many systems. Several recent examples, representing interactions between a variety of plant and pathogen species, are listed in Table 8.1. This

**Table 8.1** Summary of dual RNA-seq experiments showing genomic resources utilized, biological replicates (BR), mapping statistics and biological insight for host and pathogen

| Host | Pathogen | Pathogen type | Time-point(s) | BR | Host | Pathogen | % reads mapped to pathogen | Main finding: host | Main finding: pathogen | Reference |
|---|---|---|---|---|---|---|---|---|---|---|
| *Populus trichocarpa* × *deltoides* | *Melampsora larici-populina* (virulent and avirulent isolates) | Fungus, biotroph | 18, 21, 24 hpi (incompatible) 18, 24, 48 hpi (compatible) | 1 | *P. trichocarpa* genome v2.2 | Genome v1.0 | 1.30% | Host sulfate transporter with greatly increased transcription after infection | 19 candidate effectors identified | Petre *et al.*, 2012 |
| *Oryza sativa* | *Magnaporthe oryzae* (compatible, incompatible isolates) | Fungus, hemibiotroph | 24 hpi | 2 | *Oryza sativa* build 5 genome | *Magnaporthe grisea* release 6 genome | 0.1–0.2% | PR and phytoalexin genes in plant. More pronounced changes in incompatible interactions | Secreted protein genes were identified among the 240 up-regulated pathogen genes | Kawahara *et al.*, 2012 |
| *Triticum aestivum* | *Zymoseptoria tritici* | Fungus, hemibiotroph | 7, 13, 56 dpi | 3 | – | *Mycosphaerella graminicola* genome v2.0 | Not reported | – | PCWDEs had different expression profiles at the biotrophic, necrotrophic and saprotrophic stages | Brunner *et al.*, 2013 |
| *Sorghum bicolor* | *Bipolaris sorghicola* | Fungus, necrotroph | 12, 24 hpi | 4 | *Sorghum* genome | *De novo* assembled transcriptome | Not reported | Pathogen-induced receptor, signalling, transcription factor and redox genes were expressed | LysM domain containing protein gene, PCWDEs, TFs identified | Yazawa *et al.*, 2013 |
| *Theobroma cacao* | *Moniliophthora perniciosa* | Fungus, hemibiotroph | 30 dpi | 5 | *Theobroma cacao* genome v0.9 | *M. perniciosa* genome | 0.15–0.44% | Genes associated with carbohydrate degradation, secondary metabolism and cell wall modification were up-regulated, genes involved in starch biosynthesis and nitrogen assimilation were down-regulated | Putative effectors identified, model for biotrophic phase proposed | Teixeira *et al.*, 2014 |

| | | | | | | | | | | |
|---|---|---|---|---|---|---|---|---|---|---|
| *Notholithocarpus densiflorus* | *Phytophthora ramorum* | Oomycete, hemibiotroph | 1, 5 dpi | 3 | *De novo* transcriptome | *P. ramorum* genome | 0.1–4.5% | Many R genes and pathogenesis-related (PR) genes were transcribed in the host, as well as genes related to cell death and signalling | Pathogenicity factor genes with roles in necrosis and host cell damage were expressed by the pathogen | Hayden *et al.*, 2014 |
| *Brassica napus* | *Leptosphaeria biglobosa* | Fungus, necrotroph | 7, 14 dpi | 2 | *B. napus*, *B. rapa* and *B. oleracea* exon array | Genome | 5–44% | Jasmonic acid (JA) and salicylic acid (SA) pathway genes as well as cell wall remodelling genes were expressed | Expressed more cell wall degrading enzyme genes than the hemibiotroph | Lowe *et al.*, 2014 |
| | *Leptosphaeria maculans* | Fungus, hemibiotroph | 7, 14 dpi | 2 | *B. napus*, *B. rapa* and *B. oleracea* exon array | *Genome* | 6–61% | SA related gene was expressed | Might employ carbohydrate binding enzymes (CAZy) such as LysM to evade the PAMP-triggered immunity and also expressed avirulence genes | Lowe *et al.*, 2014 |
| *Arabidopsis thaliana* | *Hyaloperonospora arabidopsidis* (virulent, avirulent isolates) | Fungus, biotroph | 1, 3, 5 dpi | 3 | Genome | Genome | 0.022–1.452% | SA responses required for defence | Pathogen suppresses host SA-related defence | Asai *et al.*, 2014 |
| *Eucalyptus nitens* | *Phytophthora cinnamomi* | Oomycete, hemibiotroph | 5 dpi | 3 | *E. grandis* v1.1 | *P. cinnamomi* draft genome | 1% | PR-9 gene highly expressed | Crinkler family effector highly expressed | Meyer *et al.*, 2016 |

demonstrates the versatility of dual RNA-seq and the importance of experimental design in this type of study. Experimental design directly affects the kind of results that can be obtained from a dual RNA-seq experiment. The number of biological replicates among the studies in Table 8.1 ranges from one to five. With fewer biological replicates, very few differentially expressed genes were identified, but the information could still be used to guide functional characterization studies of the interaction. Knowledge of disease progression and the physical interaction between the host and pathogen is valuable when designing the study and interpreting the biological significance of the results. Examples include information about the timing of hyphal penetration of host cells or the switch between biotrophic and necrotrophic stages in hemibiotrophic pathogens, which can then be used to select the most prudent time points for RNA-seq.

Different sample groups can be chosen to understand those responses that are most likely to be involved in a particular phenotype, such as increased or reduced host resistance or pathogen virulence. The approaches listed in Table 8.1 included comparisons between resistant and susceptible hosts inoculated with a virulent pathogen, as well as comparisons between resistant hosts inoculated with virulent or avirulent pathogen strains.

The percentage of reads mapping to the pathogen reference sequence varies drastically, but it is typically very low compared to that of the host. Another technical aspect that becomes apparent is that several studies were possible without the use of an existing reference sequence, which expands the possible number of interactions that can be studied.

While the interactions are quite complex, researchers have been able to use the information to subsequently target genes for functional characterization. For example, Petre *et al.* (2016) and Yokotani *et al.* (2014) proceeded to characterize candidate effectors and a transcription factor, respectively. Many new approaches are becoming available for functional characterization. For example, the CRISPR-Cas9 system appears to be a good alternative to T-DNA knock-outs in fungi and its feasibility was recently demonstrated in *Trichoderma reesei* (Liu *et al.*, 2015) and *Pyricularia oryzae* (Arazoe *et al.*, 2015) following codon usage optimization of the Cas9 for use in filamentous fungi. This method may be favoured over conventional knock-out and knock-down systems due to its efficiency and the ability to produce knock-downs using a less efficient Cas9 nuclease (Doudna and Charpentier, 2014).

## Next generation of dual approaches

The decrease in sequencing costs will promote dual RNA-seq approaches for plant–pathogen interaction studies and could allow multiple time-points and replicates to be investigated. Recent multiplexing approaches in RNA-seq such as RNA-tagSeq (Shishkin *et al.*, 2015) and Ligation Mediated RNA sequencing (Hou *et al.*, 2015) are set to further reduce costs of sequencing. Additionally, the number of sequenced genomes of plant hosts and their pathogens is rapidly increasing, facilitating dual RNA-sequencing studies [e.g. Joint Genome Initiative: 1000 fungal genomes project, http://1000.fungalgenomes.org (Kovalchuk *et al.*, 2013; Spatafora *et al.*, 2013)]. The sequencing of new plant transcriptomes [1000 Plant Transcriptomes (1KP Project), www.iplantcollaborative.org] and improvement of annotations is constantly occurring, providing more informative resources for dual RNA-seq experiments.

Laser capture microdissection coupled with an expression profiling technique may become a valuable tool for understanding the temporal and spatial dynamics of plant–pathogen interactions (Podgorny and Lazarev, 2016). Laser capture microdissection was applied to stomatal and surrounding areas of grapevine leaves in contact with the downy mildew pathogen *Plasmopara viticola* during early stages of infection, allowing the researchers greater resolution of transcriptional responses compared to whole leaves (Lenzi *et al.*, 2016). Laser capture coupled with microarray expression profiling was also used to determine the expression of *Melampsora larici* genes in the palisade mesophyll of poplar leaves and in the uredinium, revealing genetic signatures indicative of biotrophy and the switch to sporulation, respectively (Hacquard *et al.*, 2010).

We are currently in an age where single cell RNA-seq transcriptomics is possible [reviewed in Saliba *et al.* (2014)]. There are two technologies which will allow direct RNA-sequencing using nanopores. One is where each nucleotide is read while the RNA molecule is threaded through the pore. The second strategy is RNA exosequencing where RNA is successively cleaved using polynucleotide phosphorylase, whereafter each nucleotide is read separately in the nanopore (Ayub *et al.*, 2013). Dual RNA-seq of infected single cells together with an intracellular bacterial pathogen was demonstrated with the latter approach (Hardwick *et al.*, 2011).

While we have reviewed a dual RNA-seq approach to dissect the interplay between plants and their pathogens, insight into the interaction is also possible at the level of proteomics and metabolomics (Allwood *et al.*, 2010; Rees *et al.*, 2015). Protein–protein interaction networks can reveal important triggers governing host resistance (Wang *et al.*, 2013). We predict that a combination of these approaches, together with gene function studies, will provide a systems level understanding of the interaction biology between the two organisms, which will allow more effective management of plant–pathogen duels for plant protection and improvement.

## Glossary

**Biological replicate:** Repetition of measurements using different samples obtained under the same experimental conditions to account for biological variance.

**Biological variance:** Variation between data sets obtained from different samples within an experimental group arising from genetic differences or environmental influences on gene expression.

**Coverage:** The average number of reads that represent any given nucleotide in a reference or *de novo* assembled genome sequence. For a transcriptome assembly or expression study, coverage should be calculated based on the least expressed transcript.

**Dual RNA-seq:** A technique that makes use of high-throughput next-generation sequencing platforms to assess gene expression changes in the pathogen and its host simultaneously.

**Fragment size:** The combined length of the insert and the adapters within a sequencing library.

**Insert size:** The length of the DNA (or RNA) target between adapters within a sequencing library.

**Library complexity:** The number of unique molecules present within a sequencing library.

**Overdispersion:** The presence of higher variability within a dataset than predicted/accounted for by a statistical model.

**Poly(A) selection:** Selective amplification of poly(A) RNA species during library preparation by using poly(dT) primers for cDNA synthesis.

**Read depth:** The average read depth of a sequencing experiment is the read length multiplied by the number of reads, divided by the sum of the size of all genes being considered.

**Read length:** The number of bases sequenced per read.

**rRNA depletion:** Selective removal of the majority of ribosomal RNA species during library preparation.

**Sequencing complexity:** The number of unique molecules sequenced relative to the expected number of unique molecules.

**Strand-specific library:** A double-stranded cDNA library where information about which strand represents the original mRNA has been preserved.

**Technical replicate:** Repetition of measurements using the same sample to account for technical variance.

**Technical variance:** Variation between data sets obtained from the same sample using the same procedure arising from artefacts introduced by the process.

**Unstranded library:** A double-stranded cDNA library where information regarding which strand represents the original mRNA was not preserved. This results from the standard library preparation approach where synthesis of double stranded cDNA is primed randomly before the addition of adaptors.

## References

Allwood, J.W., Clarke, A., Goodacre, R., and Mur, L.A. (2010). Dual metabolomics: a novel approach to understanding plant-pathogen interactions. Phytochemistry *71*, 590–597. http://dx.doi.org/10.1016/j.phytochem.2010.01.006

Anders, S., and Huber, W. (2010). Differential expression analysis for sequence count data. Genome Biol. *11*, R106. http://dx.doi.org/10.1186/gb-2010-11-10-r106

Anders, S., Pyl, P.T., and Huber, W. (2015). HTSeq – a Python framework to work with high-throughput sequencing data. Bioinformatics *31*, 166–169. http://dx.doi.org/10.1093/bioinformatics/btu638

Arazoe, T., Miyoshi, K., Yamato, T., Ogawa, T., Ohsato, S., Arie, T., and Kuwata, S. (2015). Tailor-made CRISPR/Cas system for highly efficient targeted gene replacement in the rice blast fungus. Biotechnol. Bioeng. *112*, 2543–2549. http://dx.doi.org/10.1002/bit.25662

Asai, S., Rallapalli, G., Piquerez, S.J., Caillaud, M.C., Furzer, O.J., Ishaque, N., Wirthmueller, L., Fabro, G., Shirasu, K., and Jones, J.D. (2014). Expression profiling during *Arabidopsis*/downy mildew interaction reveals a highly-expressed effector that attenuates responses to salicylic acid. PLOS Pathog. *10*, e1004443. http://dx.doi.org/10.1371/journal.ppat.1004443

Ashburner, M., Ball, C.A., Blake, J.A., Botstein, D., Butler, H., Cherry, J.M., Davis, A.P., Dolinski, K., Dwight, S.S., Eppig, J.T., *et al.* (2000). Gene Ontology: tool for the unification of biology. Nat. Gen. *25*, 25–29.

Auer, P.L., and Doerge, R.W. (2010). Statistical design and analysis of RNA sequencing data. Genetics *185*, 405–416. http://dx.doi.org/10.1534/genetics.110.114983

Ayub, M., Hardwick, S.W., Luisi, B.F., and Bayley, H. (2013). Nanopore-based identification of individual nucleotides for direct RNA sequencing. Nano Lett. *13*, 6144–6150. http://dx.doi.org/10.1021/nl403469r

Brunner, P.C., Torriani, S.F., Croll, D., Stukenbrock, E.H., and McDonald, B.A. (2013). Coevolution and life cycle specialization of plant cell wall degrading enzymes in a hemibiotrophic pathogen. Mol. Biol. Evol. *30*, 1337–1347. http://dx.doi.org/10.1093/molbev/mst041

Bülow, L., Schindler, M., Choi, C., and Hehl, R. (2004). PathoPlant: a database on plant-pathogen interactions. In Silico Biol. *4*, 529–536

Chandramohan, R., Wu, P.Y., Phan, J.H., and Wang, M.D. (2013). Benchmarking RNA-Seq quantification tools. Conf. Proc. IEEE Eng. Med. Biol. Soc. *2013*, 647–650. http://dx.doi.org/10.1109/EMBC.2013.6609583

Chang, Z., Wang, Z., and Li, G. (2014). The impacts of read length and transcriptome complexity for de novo assembly: a simulation study. PLOS ONE *9*, e94825. http://dx.doi.org/10.1371/journal.pone.0094825

Chen, X.R., Xing, Y.P., Li, Y.P., Tong, Y.H., and Xu, J.Y. (2013). RNA-seq reveals infection-related gene expression changes in *Phytophthora capsici*. PLOS ONE *8*. http://dx.doi.org/10.1371/journal.pone.0074588

Chen, Y., Lun, A.T.L., and Smyth, G.K. (2014). Diferential Expression Analysis of Complex RNA-seq Experiments Using edgeR. In Statistical Analysis of Next Generation Sequencing Data, S. Datta, and D. Nettleton, eds. (Cham, Switzerland: Springer International Publishing), pp. 51–74.

Chhangawala, S., Rudy, G., Mason, C.E., and Rosenfeld, J.A. (2015). The impact of read length on quantification of differentially expressed genes and splice junction detection. Genome Biol. *16*, 131. http://dx.doi.org/10.1186/s13059-015-0697-y

Denoeud, F., Aury, J.M., Da Silva, C., Noel, B., Rogier, O., Delledonne, M., Morgante, M., Valle, G., Wincker, P., Scarpelli, C., *et al.* (2008). Annotating genomes with massive-scale RNA sequencing. Genome Biol. *9*, R175. http://dx.doi.org/10.1186/gb-2008-9-12-r175

Dou, D., and Zhou, J.M. (2012). Phytopathogen effectors subverting host immunity: different foes, similar battleground. Cell Host Microbe *12*, 484–495. http://dx.doi.org/10.1016/j.chom.2012.09.003

Doudna, J.A., and Charpentier, E. (2014). Genome editing. The new frontier of genome engineering with CRISPR-Cas9. Science *346*, 1258096. http://dx.doi.org/10.1126/science.1258096

Goodstein, D.M., Shu, S., Howson, R., Neupane, R., Hayes, R.D., Fazo, J., Mitros, T., Dirks, W., Hellsten, U., Putnam, N., *et al.* (2012). Phytozome: a comparative platform for green plant genomics. Nucleic Acids Res. *40*, D1178–D1186. http://dx.doi.org/10.1093/nar/gkr944

Grant, G.R., Farkas, M.H., Pizarro, A.D., Lahens, N.F., Schug, J., Brunk, B.P., Stoeckert, C.J., Hogenesch, J.B., and Pierce, E.A. (2011). Comparative analysis of RNA-Seq alignment algorithms and the RNA-Seq unified mapper (RUM). Bioinformatics *27*, 2518–2528. http://dx.doi.org/10.1093/bioinformatics/btr427

Hacquard, S., Delaruelle, C., Legue, V., Tisserant, E., Kohler, A., Frey, P., Martin, F., and Duplessis, S. (2010). Laser capture microdissection of uredinia formed by *Melampsora larici-populina* revealed a transcriptional switch between biotrophy and sporulation. Mol. Plant-Microbe Interact. *23*, 1275–1286. http://dx.doi.org/10.1094/mpmi-05-10-0111

Hansen, K.D., Wu, Z., Irizarry, R.A., and Leek, J.T. (2011). Sequencing technology does not eliminate biological variability. Nat. Biotechnol. *29*, 572–573. http://dx.doi.org/10.1038/nbt.1910

Hardwick, S.W., Chan, V.S., Broadhurst, R.W., and Luisi, B.F. (2011). An RNA degradosome assembly in *Caulobacter crescentus*. Nucleic Acids Res. *39*, 1449–1459. http://dx.doi.org/10.1093/nar/gkq928

Hayden, K.J., Garbelotto, M., Knaus, B.J., Cronn, R.C., Rai, H., and Wright, J.W. (2014). Dual RNA-seq of the plant pathogen *Phytophthora ramorum* and its tanoak host. Tree Genet. Genomes *10*, 489–502. http://dx.doi.org/10.1007/s11295-014-0698-0

Head, S.R., Komori, H.K., LaMere, S.A., Whisenant, T., Van Nieuwerburgh, F., Salomon, D.R., and Ordoukhanian, P. (2014). Library construction for next-generation sequencing: overviews and challenges. BioTechniques *56*, 61–64, 66, 68, passim. http://dx.doi.org/10.2144/000114133

Hou, Z., Jiang, P., Swanson, S.A., Elwell, A.L., Nguyen, B.K., Bolin, J.M., Stewart, R., and Thomson, J.A. (2015). A cost-effective RNA sequencing protocol for large-scale gene expression studies. Sci. Rep. *5*, 9570. http://dx.doi.org/10.1038/srep09570

Huang, S., Zhang, J., Li, R., Zhang, W., He, Z., Lam, T.W., Peng, Z., and Yiu, S.M. (2011). SOAPsplice:genome-wide *ab initio* detection of splice junctions from RNA-Seq data. Front. Genet. *2*, 46. http://dx.doi.org/10.3389/fgene.2011.00046

Ikeue, D., Schudoma, C., Zhang, W., Ogata, Y., Sakamoto, T., Kurata, T., Furuhashi, T., Kragler, F., and Aoki, K. (2015). A bioinformatics approach to distinguish plant parasite and host transcriptomes in interface tissue by classifying RNA-Seq reads. Plant Methods *11*, 34. http://dx.doi.org/10.1186/s13007-015-0066-6

Jiang, L., Schlesinger, F., Davis, C.A., Zhang, Y., Li, R., Salit, M., Gingeras, T.R., and Oliver, B. (2011). Synthetic spike-in standards for RNA-seq experiments. Genome Res. *21*, 1543–1551. http://dx.doi.org/10.1101/gr.121095.111

Kawahara, Y., Oono, Y., Kanamori, H., Matsumoto, T., Itoh, T., and Minami, E. (2012). Simultaneous RNA-seq analysis of a mixed transcriptome of rice and blast fungus interaction. PLOS ONE *7*, e49423. http://dx.doi.org/10.1371/journal.pone.0049423

Kim, D., Pertea, G., Trapnell, C., Pimentel, H., Kelley, R., and Salzberg, S.L. (2013). TopHat2: accurate alignment of transcriptomes in the presence of insertions, deletions and gene fusions. Genome Biol. *14*, R36. http://dx.doi.org/10.1186/gb-2013-14-4-r36

Kovalchuk, A., Keriö, S., Oghenekaro, A.O., Jaber, E., Raffaello, T., and Asiegbu, F.O. (2013). Antimicrobial defenses and resistance in forest trees: challenges and perspectives in a genomic era. Annu. Rev. Phytopathol. *51*, 221–244. http://dx.doi.org/10.1146/annurev-phyto-082712-102307

Kunjeti, S.G., Evans, T.A., Marsh, A.G., Gregory, N.F., Kunjeti, S., Meyers, B.C., Kalavacharla, V.S., and Donofrio, N.M. (2012). RNA-Seq reveals infection-related global gene changes in *Phytophthora phaseoli*, the causal agent of lima bean downy mildew. Mol. Plant Pathol. *13*, 454–466. http://dx.doi.org/10.1111/j.1364-3703.2011.00761.x

Langmead, B., and Salzberg, S.L. (2012). Fast gapped-read alignment with Bowtie 2. Nat. Methods *9*, 357–359. http://dx.doi.org/10.1038/nmeth.1923

Lenzi, L., Caruso, C., Bianchedi, P.L., Pertot, I., and Perazzolli, M. (2016). Laser microdissection of grapevine leaves reveals site-specific regulation of transcriptional response to *Plasmopara viticola*. Plant Cell Physiol. 57. http://dx.doi.org/10.1093/pcp/pcv166

Levin, J.Z., Yassour, M., Adiconis, X., Nusbaum, C., Thompson, D.A., Friedman, N., Gnirke, A., and Regev, A. (2010). Comprehensive comparative analysis of strand-specific RNA sequencing methods. Nat. Methods *7*, 709–715. http://dx.doi.org/10.1038/nmeth.1491

Li, B., and Dewey, C.N. (2011). RSEM: accurate transcript quantification from RNA-Seq data with or without a reference genome. BMC Bioinf. *12*, 323. http://dx.doi.org/10.1186/1471-2105-12-323

Li, B., Ruotti, V., Stewart, R.M., Thomson, J.A., and Dewey, C.N. (2010). RNA-Seq gene expression estimation with read mapping uncertainty. Bioinformatics *26*, 493–500. http://dx.doi.org/10.1093/bioinformatics/btp692

Liang, C., Jaiswal, P., Hebbard, C., Avraham, S., Buckler, E.S., Casstevens, T., Hurwitz, B., McCouch, S., Ni, J., Pujar, A., *et al.* (2008). Gramene: a growing plant comparative genomics resource. Nucleic Acids Res. *36*, D947–D953. http://dx.doi.org/10.1093/nar/gkm968

Liang, H., Staton, M., Xu, Y., Xu, T., and LeBoldus, J. (2014). Comparative expression analysis of resistant and susceptible *Populus* clones inoculated with *Septoria musiva*. Plant Sci. *223*, 69–78. http://dx.doi.org/10.1016/j.plantsci.2014.03.004

Liao, Y., Smyth, G.K., and Shi, W. (2014). featureCounts: an efficient general purpose program for assigning sequence reads to genomic features. Bioinformatics *30*, 923–930. http://dx.doi.org/10.1093/bioinformatics/btt656

Liu, R., Chen, L., Jiang, Y., Zhou, Z., and Zou, G. (2015). Efficient genome editing in filamentous fungus *Trichoderma reesei* using the CRISPR/Cas9 system. Cell Discovery *1*, 15007. http://dx.doi.org/10.1038/celldisc.2015.7

Liu, Y., Zhou, J., and White, K.P. (2014). RNA-seq differential expression studies: more sequence or more replication? Bioinformatics *30*, 301–304. http://dx.doi.org/10.1093/bioinformatics/btt688

Lowe, R.G.T., Cassin, A., Grandaubert, J., Clark, B.L., Van de Wouw, A.P., Rouxel, T., and Howlett, B.J. (2014). Genomes and transcriptomes of partners in plant-fungal- interactions between canola (*Brassica napus*) and two *Leptosphaeria* species. PLOS ONE 9. http://dx.doi.org/10.1371/journal.pone.0103098

Maere, S., Heymans, K., and Kuiper, M. (2005). BiNGO: a Cytoscape plugin to assess overrepresentation of gene ontology categories in biological networks. Bioinformatics *21*, 3448–3449. http://dx.doi.org/10.1093/bioinformatics/bti551

Marioni, J.C., Mason, C.E., Mane, S.M., Stephens, M., and Gilad, Y. (2008). RNA-seq: an assessment of technical reproducibility and comparison with gene expression arrays. Genome Res. *18*, 1509–1517. http://dx.doi.org/10.1101/gr.079558.108

Martin, J.A., and Wang, Z. (2011). Next-generation transcriptome assembly. Nat. Rev. Genet. *12*, 671–682. http://dx.doi.org/10.1038/nrg3068

McIntyre, L.M., Lopiano, K.K., Morse, A.M., Amin, V., Oberg, A.L., Young, L.J., and Nuzhdin, S.V. (2011). RNA-seq: technical variability and sampling. BMC Bioinf. *12*, 293. http://dx.doi.org/10.1186/1471-2164-12-293

Meyer, F.E., Shuey, L.S., Naidoo, S., Mamni, T., Berger, D.K., Myburg, A.A., van den Berg, N., and Naidoo, S. (2016). Dual RNA-sequencing of *Eucalyptus nitens* during *Phytophthora cinnamomi* challenge reveals pathogen and host factors influencing compatibility. Front. Plant Sci. 7, 191. http://dx.doi.org/10.3389/fpls.2016.00191

Nettleton, D. (2014). Design of RNA Sequencing Experiments. In Statistical Analysis of Next Generation Sequencing Data, S. Datta, and D. Nettleton, eds. (Cham, Switzerland: Springer International Publishing), pp. 93–113.

Nicolae, M., Mangul, S., Măndoiu, I.I., and Zelikovsky, A. (2011). Estimation of alternative splicing isoform frequencies from RNA-Seq data. Algorithms Mol. Biol. *6*, 9. http://dx.doi.org/10.1186/1748-7188-6-9

Ogata, H., Goto, S., Sato, K., Fujibuchi, W., Bono, H., and Kanehisa, M. (1999). KEGG: Kyoto Encyclopedia of Genes and Genomes. Nucleic Acids Res. *27*, 29–34. http://dx.doi.org/10.1093/nar/27.1.29

Pareek, C.S., Smoczynski, R., and Tretyn, A. (2011). Sequencing technologies and genome sequencing. J. Appl. Genet. *52*, 413–435. http://dx.doi.org/10.1007/s13353-011-0057-x

Petre, B., Lorrain, C., Saunders, D., Win, J., Sklenar, J., Duplessis, S., and Kamoun, S. (2016). Rust fungal effectors mimic host transit peptides to translocate into chloroplasts Cell. Microbiol. *18*, 453–465. http://dx.doi.org/10.1111/cmi.12530

Petre, B., Morin, E., Tisserant, E., Hacquard, S., Da Silva, C., Poulain, J., Delaruelle, C., Martin, F., Rouhier, N., Kohler, A., *et al.* (2012). RNA-seq of early-infected poplar leaves by the rust pathogen *Melampsora larici-populina* uncovers PtSultr3;5, a fungal-induced host sulfate transporter. PLOS ONE 7. http://dx.doi.org/10.1371/journal.pone.0044408

Philippe, N., Salson, M., Commes, T., and Rivals, E. (2013). CRAC: an integrated approach to the analysis of RNA-seq reads. Genome Biol. *14*, R30. http://dx.doi.org/10.1186/gb-2013-14-3-r30

Podgorny, O.V., and Lazarev, V.N. (2016). Laser microdissection: A promising tool for exploring microorganisms and their interactions with hosts. J. Microbiol. Methods http://dx.doi.org/10.1016/j.mimet.2016.01.001

Rees, M.A., Kleifeld, O., Crellin, P.K., Ho, B., Stinear, T.P., Smith, A.I., and Coppel, R.L. (2015). Proteomic characterization of a natural host pathogen interaction: repertoire of *in vivo* expressed bacterial and host surface-associated proteins. J. Proteome Res. *14*, 120–132. http://dx.doi.org/10.1021/pr5010086

Ritchie, M.E., Phipson, B., Wu, D., Hu, Y.F., Law, C.W., Shi, W., and Smyth, G.K. (2015). Limma powers differential expression analyses for RNA-sequencing and microarray studies. Nucleic Acids Res. *43*. http://dx.doi.org/10.1093/nar/gkv007

Robinson, M.D., McCarthy, D.J., and Smyth, G.K. (2010). edgeR: a Bioconductor package for differential expression analysis of digital gene expression data. Bioinformatics *26*, 139–140. http://dx.doi.org/10.1093/bioinformatics/btp616

Saliba, A.E., Westermann, A.J., Gorski, S.A., and Vogel, J. (2014). Single-cell RNA-seq: advances and future challenges. Nucleic Acids Res. *42*, 8845–8860. http://dx.doi.org/10.1093/nar/gku555

Schulze, S., Henkel, S.G., Driesch, D., Guthke, R., and Linde, J. (2015). Computational prediction of molecular pathogen-host interactions based on dual transcriptome data. Front. Microbiol. *6*, 65. http://dx.doi.org/10.3389/fmicb.2015.00065

Shishkin, A.A., Giannoukos, G., Kucukural, A., Ciulla, D., Busby, M., Surka, C., Chen, J., Bhattacharyya, R.P., Rudy, R.F., Patel, M.M., *et al.* (2015). Simultaneous generation of many RNA-seq libraries in a single reaction. Nat. Methods *12*, 323–325. http://dx.doi.org/10.1038/nmeth.3313

Sims, D., Sudbery, I., Ilott, N.E., Heger, A., and Ponting, C.P. (2014). Sequencing depth and coverage: key considerations in genomic analyses. Nat. Rev. Genet. *15*, 121–132. http://dx.doi.org/10.1038/nrg3642

Soneson, C., and Delorenzi, M. (2013). A comparison of methods for differential expression analysis of RNA-seq data. BMC Bioinformatics *14*, 91. http://dx.doi.org/10.1186/1471-2105-14-91

Spatafora, J., Stajich, J., and Grigoriev, I. (2013). 1000 Fungal Genomes project. Phytopathology *103*, 137–137

Sundell, D., Mannapperuma, C., Netotea, S., Delhomme, N., Lin, Y.C., Sjödin, A., Van de Peer, Y., Jansson, S., Hvidsten, T.R., and Street, N.R. (2015). The Plant Genome Integrative Explorer Resource: PlantGenIE.org. New Phytol. *208*, 1149–1156. http://dx.doi.org/10.1111/nph.13557

Teixeira, P.J., Thomazella, D.P., Reis, O., do Prado, P.F., do Rio, M.C., Fiorin, G.L., Jose, J., Costa, G.G., Negri, V.A., Mondego, J.M., *et al.* (2014). High-resolution transcript profiling of the atypical biotrophic interaction between *Theobroma cacao* and the fungal pathogen *Moniliophthora perniciosa*. Plant Cell *26*, 4245–4269. http://dx.doi.org/10.1105/tpc.114.130807

Tello-Ruiz, M.K., Stein, J., Wei, S., Preece, J., Olson, A., Naithani, S., Amarasinghe, V., Dharmawardhana, P., Jiao, Y., Mulvaney, J., *et al.* (2016). Gramene 2016: comparative plant genomics and pathway resources. Nucleic Acids Res. *44*, D1133–D1140. http://dx.doi.org/10.1093/nar/gkv1179

Thimm, O., Bläsing, O., Gibon, Y., Nagel, A., Meyer, S., Krüger, P., Selbig, J., Müller, L.A., Rhee, S.Y., and Stitt, M. (2004). MAPMAN: a user-driven tool to display genomics data sets onto diagrams of metabolic pathways and other biological processes. Plant J. *37*, 914–939. http://dx.doi.org/10.1111/j.1365-313X.2004.02016.x

Trapnell, C., Hendrickson, D.G., Sauvageau, M., Goff, L., Rinn, J.L., and Pachter, L. (2013). Differential analysis of gene regulation at transcript resolution with RNA-seq. Nat. Biotechnol. *31*, 46–53. http://dx.doi.org/10.1038/nbt.2450

Trapnell, C., Roberts, A., Goff, L., Pertea, G., Kim, D., Kelley, D.R., Pimentel, H., Salzberg, S.L., Rinn, J.L., and Pachter, L. (2012). Differential gene and transcript expression analysis of RNA-seq experiments with TopHat and Cufflinks. Nat. Protoc. *7*, 562–578. http://dx.doi.org/10.1038/nprot.2012.016

Urban, M., Pant, R., Raghunath, A., Irvine, A.G., Pedro, H., and Hammond-Kosack, K.E. (2015). The Pathogen-Host Interactions database (PHI-base): additions and future developments. Nucleic Acids Res. *43*, D645–D655. http://dx.doi.org/10.1093/nar/gku1165

Wang, K., Singh, D., Zeng, Z., Coleman, S.J., Huang, Y., Savich, G.L., He, X., Mieczkowski, P., Grimm, S.A., Perou, C.M., *et al.* (2010). MapSplice: accurate mapping of RNA-seq reads for splice junction discovery. Nucleic Acids Res. *38*, e178. http://dx.doi.org/10.1093/nar/gkq622

Wang, Y.C., Lin, C., Chuang, M.T., Hsieh, W.P., Lan, C.Y., Chuang, Y.J., and Chen, B.S. (2013). Interspecies protein-protein interaction network construction for characterization of host-pathogen interactions: a *Candida albicans*-zebrafish interaction study. BMC Syst. Biol. 7. http://dx.doi.org/10.1186/1752-0509-7-79

Wegrzyn, J.L., Lee, J.M., Tearse, B.R., and Neale, D.B. (2008). TreeGenes: A forest tree genome database. Int J Plant Genomics *2008*, 412875. http://dx.doi.org/10.1155/2008/412875

Westermann, A.J., Gorski, S.A., and Vogel, J. (2012). Dual RNA-seq of pathogen and host. Nat. Rev. Microbiol. *10*, 618–630. http://dx.doi.org/10.1038/nrmicro2852

Wilhelm, B.T., and Landry, J.R. (2009). RNA-Seq-quantitative measurement of expression through massively parallel RNA-sequencing. Methods *48*, 249–257. http://dx.doi.org/10.1016/j.ymeth.2009.03.016

Wu, T.D., and Nacu, S. (2010). Fast and SNP-tolerant detection of complex variants and splicing in short reads. Bioinformatics *26*, 873–881. http://dx.doi.org/10.1093/bioinformatics/btq057

Yang, C., and Wei, H. (2015). Designing microarray and RNA-Seq experiments for greater systems biology discovery in modern plant genomics. Mol. Plant. *8*, 196–206. http://dx.doi.org/10.1016/j.molp.2014.11.012

Yazawa, T., Kawahigashi, H., Matsumoto, T., and Mizuno, H. (2013). Simultaneous transcriptome analysis of sorghum and *Bipolaris sorghicola* by using RNA-seq in combination with de novo transcriptome assembly. PLOS ONE *8*, e62460. http://dx.doi.org/10.1371/journal.pone.0062460

Yokotani, N., Tsuchida-Mayama, T., Ichikawa, H., Mitsuda, N., Ohme-Takagi, M., Kaku, H., Minami, E., and

Nishizawa, Y. (2014). OsNAC111, a blast disease-responsive transcription factor in rice, positively regulates the expression of defense-related genes. Mol. Plant-Microbe Interact. 27, 1027–1034. http://dx.doi.org/10.1094/mpmi-03-14-0065-r

Zhang, Z.H., Jhaveri, D.J., Marshall, V.M., Bauer, D.C., Edson, J., Narayanan, R.K., Robinson, G.J., Lundberg, A.E., Bartlett, P.F., Wray, N.R., *et al.* (2014). A comparative study of techniques for differential expression analysis on RNA-Seq data. PLOS ONE *9*, e103207. http://dx.doi.org/10.1371/journal.pone.0103207

Zuluaga, A.P., Sole, M., Lu, H., Gongora-Castillo, E., Vaillancourt, B., Coll, N., Buell, C.R., and Valls, M. (2015). Transcriptome responses to *Ralstonia solanacearum* infection in the roots of the wild potato *Solanum commersonii*. BMC Genomics *16*, 246. http://dx.doi.org/10.1186/s12864-015-1460-1

# Next-generation Sequencing Sheds New Light on Small RNAs in Plant Reproductive Development

9

Xiaobai Li

Zhejiang Academy of Agricultural Sciences, Hangzhou, China.

Correspondence: hufanfan1982815@gmail.com and lixiaobai@mail.zaas.ac.cn

https://doi.org/10.21775/9781910190654.10

## Abstract

Reproductive development is a key step of the plant life cycles and indicates the start of a new life cycle. The reproductive organs including flower, fruit and seed, have diverse and complex structures, which is a syndrome in the evolution of angiosperms. The development of plant reproductive organs depends on the correct spatial and temporal expression of numerous genes acting in concert to form regulatory networks. Small non-coding RNAs (sRNAs) play a key role in the reproductive development through different modes of sequence-specific interaction with their targets. The sRNAs guide transcriptional and post-transcriptional regulation to intensively integrate into the complex process. Next-generation sequencing (NGS) techniques have greatly extended the scientist's capabilities to identify and quantify sRNAs by supplying a massive number of sequences. In turn this has led to a greater understanding of the many complex roles and interactions that sRNAs are involved with during reproductive development. In this review, we provide an overview of the biogenesis and classification of plant sRNAs, and summarize the recent progress in the understanding of plant sRNA in the flower, and fruit/seed development. Also, we discuss NGS approaches to profile global sRNA expression, as well as the application of sRNA-seq/degradome-seq approaches to identify novel sRNAs and verify their targets related to the above development processes.

## Introduction

In plants, small, non-coding RNAs have been characterized as an important factor in post-transcriptional regulation. Regulatory small RNAs (sRNAs) generally range in size from 20 to 24 nucleotides (nt). They are products of their larger RNA precursors that have double-stranded duplexes and are sliced by endonuclease Dicer-like (DCL) proteins (Axtell, 2013). One strand of the duplex, the mature sRNA, is bounded to the Argonaute (AGO) protein to assemble a RNA-induced Silencing Complex (RISC), which allows hybridizing with the target RNAs. The target RNAs are silenced by repressing chromatin modifications, decreasing RNA stability, and lowering translational efficiency (Vaucheret, 2006). The AGO-bounded small-RNA assembly in plant works as a reservoir of sequence-specific negative regulators deployed for a wide variety of functions in many aspects of plant development (Axtell, 2013). The mutation or overexpression of sRNAs always leads to pleiotropic phenotypes including the developmental defects in leaves, flowers, and fruits (Buxdorf *et al.*, 2010; Wang *et al.*, 2011b). The underlying mechanisms of sRNA pleiotropism could be contributed to the spatial and temporal of action of sRNA and their downstream targets. The variability of sRNA expression patterns and targets may endow them with diverse functions in different organs or developmental stages. In this review, we overview the formation and action of plant sRNAs, and discuss next-generation sequencing (NGS) approaches (e.g. sRNA-seq/degradome-seq) to identify the sRNA and verify their targets. Mostly, we summarize the recent progress in understanding the roles of plant sRNA in reproductive regulation. We focus on the regulation of flowering time, male and female organ development, and fruit/seed development by sRNAs and their targets. Understanding how sRNAs function in plant reproductive development could provide new insights into improving the important agronomic traits during plant breeding

## Plant sRNA biogenesis and classification

Generally, plant sRNA can be categorized into two different groups, based on their biogenesis and function. One group, hairpin RNAs (hpRNAs), is derived from the single-stranded precursors that process an intramolecular and self-complementary 'hairpin structure'. The other group, small interfering RNAs (siRNAs), is derived from double-stranded precursors. Plants hpRNAs include microRNAs (miRNA) and some non-microRNAs (Fig. 9.1). In contrast to most animal miRNAs which bind their targets with imprecise complementarity to repress translation (Ambros *et al.*, 2003; Seggerson *et al.*, 2002), plant miRNAs commonly interact with their targets through perfect or near-perfect complementarity to directly degrade mRNA target rather than repress translation (Llave *et al.*, 2002; Rhoades *et al.*, 2002). Although chemically similar to miRNAs, siRNAs arise from double-stranded RNAs instead of specific fold-back structures (hairpin). During RNA interference (RNAi), siRNAs guide sequence-specific cleavage of RNAs (Denli and Hannon, 2003), as well nuclear events including histone and DNA methylation, resulting in transcriptional silencing (Reinhart and Bartel, 2002; Volpe *et al.*, 2002; Zilberman *et al.*, 2003). Despite the differences in origin, miRNAs and siRNAs are functionally interchangeable. Like miRNAs, most known plant siRNAs function in guiding target RNA degradation (Llave *et al.*, 2002; Reinhart and Bartel, 2002; Tang *et al.*, 2003).

Currently, most of known plant siRNAs fall into one of the three secondary groups: heterochromatic siRNAs (hc-siRNAs), secondary siRNAs (typically, trans-acting siRNA or ta-siRNA), or natural antisense transcript-derived siRNAs (natural-antisense siRNAs, or nat-siRNAs) (Fig. 9.1) (Axtell, 2013). Hc-siRNAs, the most abundant type of sRNAs, are derived from transposons or other repetitive sequences (Wei *et al.*, 2012). Their single-stranded RNA transcripts are transcribed by the DNA-dependent RNA polymerase IV (Pol IV) (Herr *et al.*, 2005; Onodera *et al.*, 2005), and converted into dsRNAs by RNA-dependent RNA polymerase 2 (RDR2) (Xie *et al.*, 2004). These dsRNAs are then processed by DCL3 to generate 23- to 24-nt-long hc-siRNAs (Wroblewski *et al.*, 2014). Secondary siRNAs production tends to be triggered by one or more upstream sRNAs, which could be either miRNAs or other siRNAs. The upstream sRNAs can also help recruit an RDR to synthesize complementary double-stranded RNA, which are then processed into secondary siRNA by DCL (Allen *et al.*, 2005; Axtell, 2013; Yoshikawa *et al.*, 2005). Ta-siRNAs are a representative class of the secondary siRNAs. Ta-siRNAs are transcribed from TAS transcripts by RNA polymerase II (Pol II). The TAS transcripts either go through AGO-mediated and miRNA-guided cleavage or are converted to double-stranded RNA by RDR6 and Suppressor of Gene Silencing 3 (SGS3) (Talmor-Neiman *et al.*, 2006). The resulting dsRNAs are further processed by DCL4 to produce 21-nt length siRNAs (Allen *et al.*, 2005). Nat-siRNAs precursors are formed by the hybridization of independently transcribed and complementary RNAs, which relies on the plant-specific RNA polymerase IV (PolIV) (Zhang *et al.*, 2012). An initial nat-siRNA from the complementary region of two NAT genes is produced through the action of DCL and RDR, and then the initial nat-siRNAs guide the cleavage of one NAT gene thus forms the sequential production of new nat-siRNAs. However, these new nat-siRNAs are not required for further cleavage, unlike tas-siRNA playing a role in targeting unlinked transcripts (Vaucheret, 2006). Nat-siRNAs are derived either from the opposite strands of the same locus (*cis*-Nat-siRNAs), or from genes with no overlap (*trans*-Nat-siRNAs). However, *trans*-NAT-siRNAs have not been found in plants until recently (Axtell, 2013). In previous *Arabidopsis* studies, the biogenesis of biotic- and abiotic-induced nat-siRNAs is dependent on DCL1 and/or DCL2, and RDR6 (Borsani *et al.*, 2005; Katiyar-Agarwal *et al.*, 2006). The different result in *Arabidopsis* and rice shows that the production of 23- to 28-nt nat-siRNAs are

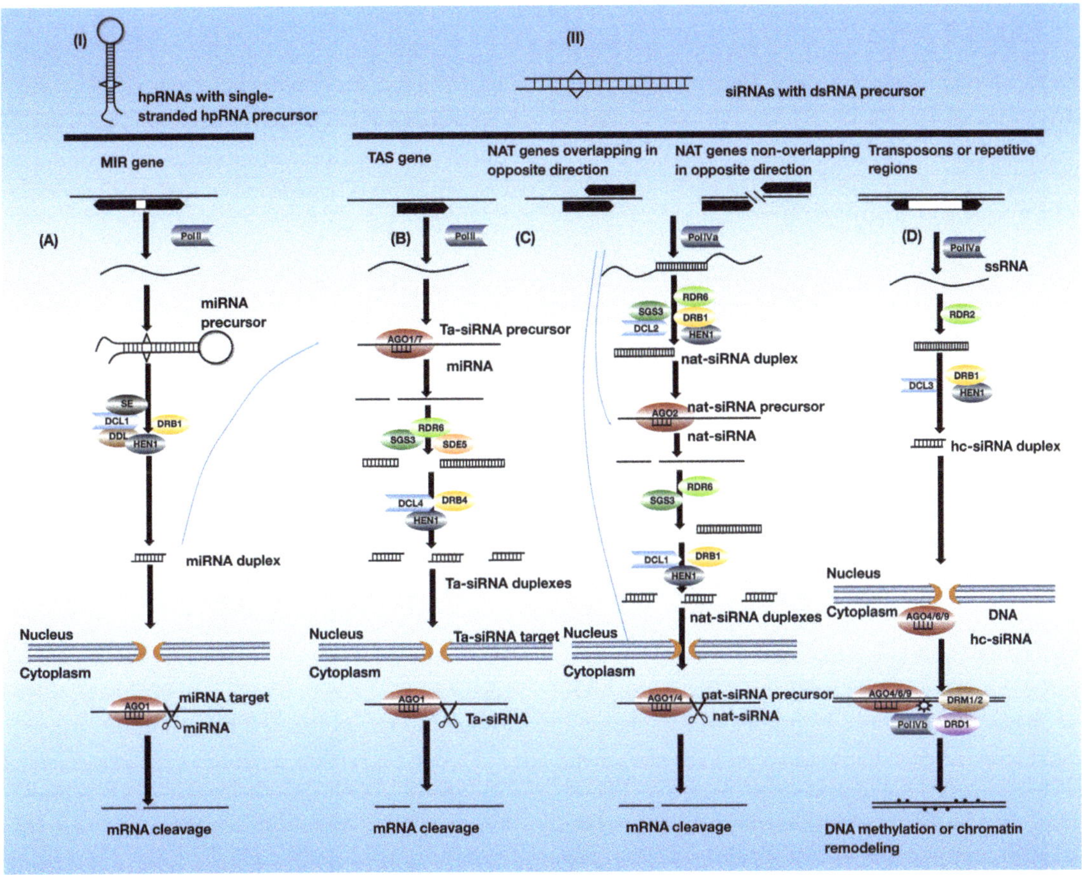


**Figure 9.1** miRNA, ta-siRNA, nat-siRNA, and hc-siRNA pathways in plants. The same colour is used to indicate members of the same gene family. Plant small RNAs have two groups. (I) One is derived from single-stranded precursors that process an intramolecular and self-complementary 'hairpin structure', called hairpin RNAs (hpRNAs), typically, miRNA, and (II) the other is derived from double-stranded precursors, and called small interfering RNAs (siRNAs), including ta-siRNA, nat-siRNA, and hc-siRNA. (A) The miRNA pathway: MIRNA genes are transcribed by Pol II into precursor RNAs that form hairpin stem–loops due to their self-complementary structure, a classical trait of miRNA precursors. Processing of miRNA precursors into a 21-to 22-nucleotide miRNA duplex with 2-nucleotide 39 overhangs requires Cap binding proteins CBP20 and CBP80, DAWDLE (DDL), SERRATE (SE)SE (a zinc finger protein), DRB1 (a dsRNA binding protein), and DCL1 (an RNaseIII). miRNA duplexes are methylated by HEN1 to protect against SDN exonuclease degradation. One strand of the miRNA duplex binds to AGO1. The miRNA-AGO1 are transported to the cytoplasm to complement mRNA targets, and AGO1 cleaves the target mRNAs. (B) The ta-siRNA pathway: ta-siRNAs are also transcribed by Pol II. miRNAs binds to AGO1 or AGO7 to guide cleavage of ta-siRNA precursors. After cleavage, fragments of ta-siRNA precursor are synthesized into dsRNA by RDR6, SGS3, and SILENCING DEFECTIVE5 (SDE5), and the dsRNA is cleaved into 21- to 22-nucleotide ta-siRNAs by the DRB4/DCL4/HEN1 complex. (C) nat-siRNA pathway: a pair of natural cis-antisense transcripts (overlapping or non-overlapping) are transcribed by Pol IV. Two transcripts generate a dsRNA molecular with their complementary end sequences. The dsRNA are cleaved by DCL2, assisted by NRPD1a, RDR6, and SGS3, and turned into a stable 24-nt nat-siRNA. This single 24-nucleotide nat-siRNA subsequently targets one of the cis-antisense gene pair transcripts for cleavage, and the cleaved RNA molecule is converted to dsRNA by RDR6 and SGS3. The RDR6/SGS3-synthesized dsRNA molecule is then processed into phased 21- nucleotide nat-siRNAs by the action of DCL1, assisted by DRB1 and HEN1. The phased 21-nucleotide nat-siRNAs, like the ta-siRNA class of endogenous sRNAs, are in turn used as guides to direct sequence-specific silencing of homologous mRNAs. However, the production of these 21-nt nat-siRNAs does not appear to be required for down-regulating the constitutive transcript. (D) Heterochromatin pathway: transposons and repeat DNA are transcribed by PolIV, which is then converted into dsRNA by RDR2-mediated synthesis of dsRNA. The dsRNA is processed by DCL3 into 24-nucleotide siRNAs. These siRNAs are methylated by HEN1 and loaded onto AGO4, AGO6, and AGO9, which induce DNA methylation through the DNA methyltransferase DRM1/2. DNA methylation of some loci also requires PolIV and the SWI2/SNF2-like chromatin remodelling factor DRD1.

dependent on DCL3 and RDR2, while 20- to 22-nt nat-siRNAs are processed by DCL1 and RDR6, (Borsani *et al.*, 2005; Ron *et al.*, 2010; Zhang *et al.*, 2012).

## NGS accelerating plant sRNAomics

Many efforts have been made to discover and identify sRNA from diverse plant species in recent years. To date, three main approaches, including forward genetics, bioinformatics prediction and direct cloning and sequencing, are used to identify sRNA and their targets, and define sRNA function in plants, (Liu *et al.*, 2014a). The methods of forward genetics primarily rely on mapping, isolating, cloning, and sequencing small cellular RNAs. Although these endeavours have drastically increased the amount of information available for sRNA and their target mRNAs, it is still impractical to count on these traditional genetic methods to empirically explore a large scale of complex and diverse sRNAs and their targets mRNAs (Liu *et al.*, 2014a). Thanks to the pairing between sRNAs and their targets, bioinformatic methods have become feasible and effective in identifying sRNAs. The precise or near-precise complementarity between miRNAs and mRNAs in plants, as well as their high conservation, makes systematic target identification easier in plants than in animals (Axtell and Bartel, 2005; Rhoades *et al.*, 2002). However, a number of non-conserved miRNAs are specifically expressed in certain species, tissue or developmental stages but at low levels, which makes it difficult to identify them (Mao *et al.*, 2012; Xu *et al.*, 2012).

In recent years, the next-generation sequencing (NGS) technologies have replaced many of the traditional methods for identifying sRNA and their targets. NGS has dramatically accelerated genomic studies by supplying a massive amount of sequences data, which allows comprehensive analysis of genomes, transcriptomes, and sRNAomics. The current NGS platforms include 454 Genome Sequencers (Roche Applied Science), Solexa Genome Analyser (Illumina), the SOLiD platform (Applied Biosystems), and the more recent Polonator (Dover/Harvard) and the Helicos sequencer (Helicos) and the Ion Torrent Personal Genome Machine (PGM) (Life Technologies). These NGS platforms have reduced the cost and time of required for traditional Sanger sequencing (Liu *et al.*, 2012; Loman *et al.*, 2012; Metzker, 2010; Shendure and Ji, 2008). The application of NGS has considerably advanced miRNA discovery in several non-model plants as it has provided a large amount of low-cost genomic and transcriptomic data for these plant species (Fang *et al.*, 2014). Furthermore, NGS also provides a new method for the validation of the splicing targets throughout the whole genome, i.e. degradome sequencing. Degradome sequencing can precisely identify sRNA target genes and explore their interactions which aids in clarifying the functions of these sRNAs in the diverse biological processes. NGS has revolutionized the traditional computational target prediction, and been successfully applied to screen for miRNA targets in many plants, for example *Arabidopsis* (Addo-Quaye *et al.*, 2008; German *et al.*, 2008), grapevine (Pantaleo *et al.*, 2010) maize (Zhao *et al.*, 2012), cucumber (Mao *et al.*, 2012), oilseed rape (Xu *et al.*, 2012), rice (Sun *et al.*, 2015) and soybean (Chen *et al.*, 2016).

Among the NGS platforms, Illumina-Solexa has been a popular platform, due to its relatively low cost, relatively short running time and relatively long reads. ABI SOLID yields the highest accuracy but requires a long time for sequencing runs. Compared with the other methods, Roche 454 generates the longest reads, which allows for more accurate read assembly and can span larger repetitive regions within the genome. However, Roche 454 is costly and low throughput and has a high error rate. Ion torrent is notable for its three different priced sequencing-chip reagents based on the throughout requirement, so that scientists have different choices according to their experiment design. However, like Roche 454, Ion Torrent PGM is apt to produce homopolymer-associated indel errors (Loman *et al.*, 2012). HeliScope overcomes the bias representation of templates and mutations in clonally amplified templates, but expensive and introduces higher error rates compared with other technologies (Metzker, 2010). Polonator is an open source platform with free software and protocols so that users are required to maintain and control reagents qualities. It may be the least expensive platform, but yields the shortest reads so far (Metzker, 2010). Interested readers can refer to several excellent reviews which introduce these NGS platforms in more detail (Liu *et al.*, 2012;

Loman *et al.*, 2012; Metzker, 2010; Shendure and Ji, 2008).

Since NGS' steps into the model species *Arabidopsis* (Fahlgren *et al.*, 2007; Lu *et al.*, 2005; Rajagopalan *et al.*, 2006; Sunkar and Zhu, 2004) and rice (Jones-Rhoades and Bartel, 2004; Sunkar *et al.*, 2005) for studying sRNAs, it has been also successfully applied in many other non-model plant species (Bi *et al.*, 2015; Khaldun *et al.*, 2015; Wu *et al.*, 2014; Xin *et al.*, 2015; Xu *et al.*, 2013; Yao *et al.*, 2015a; Zeng *et al.*, 2015). Plant miRNAs are one of the most widely studied groups of sRNAs. To date, a total of 8496 miRNAs from 73 Viridiplantae species have been registered in the latest miRBase (version 21, June 2014) (www.mirbase.org/) (Kozomara and Griffiths-Jones, 2014). For comparison, a total of 8,433 miRNAs from 121 plant species as well as the possible target genes and predicted interaction site have been deposited in the Plant Microrna Database (PMRD) (http://bioinformatics.cau.edu.cn/PMRD/) (Zhang *et al.*, 2010). Another resource for predicting and comparing plant microRNAs is MicroPC (μPC) (www3a.biotec.or.th/micropc/index.html), which provides the comprehensive information of plant miRNAs from two sources including the previously reported miRNAs from miRBase and the predicted miRNAs. It also collects the targets of these miRNA based on a large-scale expressed sequence tags (ESTs) analysis (Mhuantong and Wichadakul, 2009). In addition, a plant sRNA target analysis server (psRNATarget) (http://plantgrn.noble.org/psRNATarget/) has incorporated a series of recently discovered miRNA target recognition sites in plants (Dai and Zhao, 2011). To relieve the growing pressure of handling massive quantities of short-read sequences, numerous NGS-based sRNA transcriptome bioinformatics analysis tools have been developed (Li *et al.*, 2012; Zhou *et al.*, 2011). These tools include the standalone software used for deep sequencing miRNA data, such as MicroRazerS (Emde *et al.*, 2010), CASHX pipeline (Fahlgren *et al.*, 2009), mirDeep/mirDeep2 (Friedländer *et al.*, 2008, 2012), miRTRAP (Hendrix *et al.*, 2010), MIREAP (Song *et al.*, 2010), miRNAkey (Ronen *et al.*, 2010) MIReNA (Mathelier and Carbone, 2010), miRExpress (Wang *et al.*, 2009b), miREvo (Wen *et al.*, 2012), and miRVine (Belli Kullan *et al.*, 2015), as well as some web servers, such a mirTools (Zhu *et al.*, 2010), UEA sRNA toolkit (Moxon *et al.*, 2008b), and SeqBuster (Pantano *et al.*, 2010). Moreover, several databases offer the serves for miRNAs search and sRNAs annotation, such as deepBase (Yang *et al.*, 2010), miRBase (Griffiths-Jones *et al.*, 2008), and FANTOM4 EdgeExpressDB (Severin *et al.*, 2009). Among these tools, mirTools, UEA sRNA Toolkit, miRNAkey, miRExpress, miRExpress, FANTOM4 EdgeExpressDB, and miRVine can also be used to compare the expression of sRNAs based on NGS' data. Performance comparison and evaluation of some software tools are available in some previous reviews (Li *et al.*, 2012; Zhou *et al.*, 2011). Notably, several particular tools, such as psRNATarget, UEA sRNA tools (Stocks *et al.*, 2012), Target-Align (Xie and Zhang, 2010), mirplant (An *et al.*, 2014) and PsRobot (Wu *et al.*, 2012) are tailored to deal with plant sRNAs based on species specificity.

## SRAs profiling by NGS in plant reproductive development

Plant development goes through two distinct phases: vegetative and reproductive growth. Reproduction in angiosperms is referred to as a continuous process characterized by three distinct phases, i.e. flowering and pollination, fruits and seed dispersal, seed and seedling establishment. Flowers have their unique appearances, as a result of angiosperms' evolutionary success (Li *et al.*, 2016b). Their specialized and complex perianth morphologies have evolved in response to the selective pressure from pollination. The outmost whorl of flower (sepals) plays a protective role, while the inner whorl, the corolla, attracts insects with showy petals. The fruit develops from the flower, which is also the result of angiosperms' evolutionary success. The fruit primarily grows to accommodate the seeds, protects and nourishes them, and assists in their dispersal at maturity.

Recently, intensive studies on sRNA transcriptome have shed light on understanding the molecular mechanisms that control flower and fruit/seed development in both model (Chen *et al.*, 2011a; Lee *et al.*, 2010; Mohorianu *et al.*, 2011; Moxon *et al.*, 2008a; Zuo *et al.*, 2012, 2013) and non-model plants (Bi *et al.*, 2015; Khaldun *et al.*, 2015; Li *et al.*, 2015a; Liu *et al.*, 2014d; Xin *et al.*, 2015; Xu *et al.*, 2013; Yao *et al.*, 2015a; Zeng *et al.*, 2015). High-throughput parallel sequencing technologies have greatly improved the efficiency of finding novel sRNA functions under various conditions and from plant species. sRNA families have also been identified across various plant species, many of which are universal. These sRNAs are often involved in numerous crucial roles at each major stage of development and target the regulatory genes, such as transcription factors.

sRNAs related to flower development have been profiled in many crops, vegetable, fruit trees and ornamental flowers. In wild rice (*O. rufipogon*), 512 conserved rice miRNAs in 214 miRNA families and 290 new predicted miRNAs have been identified from three libraries generated at the vegetative and flowering stages (Chen *et al.*, 2013). Among these miRNAs, 187 miRNAs regulate the expression of flowering-related genes, including flowering-related miRNAs, such as oru-miR97, oru-miR117, oru-miR135, and orumiR137, etc., and early-flowering-related miRNAs, such as osa-miR160f, osa-miR164d, osa-miR167d, osa-miR169a, osa-miR172b, and oru-miR4, etc. In hickory (*Carya cathayensis Sarg.*), 39 known miRNAs in 23 families, as well as two novel and ten potential novel miRNAs in nine families, have been identified from two floral differentiation stages (Wang *et al.*, 2012b). A diverse set of miRNAs and their target genes may be related to hickory flower development. In radish, 94 known miRNAs in 21 conserved and 13 non-conserved miRNA families, as well as 44 potential novel miRNAs, have been identified from the two libraries at vegetative and reproductive stages (Nie *et al.*, 2015). Among them, 42 known and 17 novel miRNAs with different expression have been identified as bolting and flowering-related miRNAs, and 154 target transcripts for these miRNAs have been shown to be involved in plant development, signal transduction, and transcriptional regulation. In four varieties of Roses (*Rosa* sp.), 25 novel and 242 conserved miRNAs have been identified from floral tissues (Kim *et al.*, 2012a). Combinative analysis with transcriptomic data identified a serial of shared and species-specific miRNAs. They are involved in flower development. Degradome and sRNA analysis in the leaves, stalks and flower buds of *Phalaenopsis* exposed to low ambient temperature have identified 46 sRNA groups and their targets that are involved in flowering time and developmental processes under (An and Chan, 2012). In cymbidium, 48 conserved mature conserved miRNA in 20 families, and 45 novel miRNAs, as well as two ta-siRNAs, have been identified from flower tissue at different stages (Li *et al.*, 2015a). Being assisted with transcriptomic data, flower-development-related unigenes have been identified, including the MADS-box transcription factors targeted by miR156, miR172 and miR5179, as well as various hormone responding factors targeted by miR159.

Many sRNA have been discovered to be involved in a model plant-tomato fruit development. A large amount of the targets of the sRNAs are involved in ethylene (ET) or auxin (AUX) pathway, such as ethylene response factors (*AP2/ERF* and *ERF4*) or a key enzyme in ET biosynthesis (*1-aminocyclopropane-1-carboxylate oxidase 1, ACO1*), as well as two splice variants of constitutive *triple response 4* (*LeCTR4* and *LeCTR4*) regulated by miR1917 (Moxon *et al.*, 2008a). In another study, seven novel miRNA families, as well as 24 families including 103 conserved miRNAs and ten non-conserved miRNAs have been identified at three different fruit ripening stages and fruits by exogenous ET treatment (Zuo *et al.*, 2012). The targets of many these miRNAs are predicted to be transcription factors, such as *CNR* or *AP2a*, regulated by miR156/157 or miR172, respectively. Some targets are putatively involved in fruit ripening and softening, such as *pectate lyase* and *beta-galactosidase*, while a few are putatively involved in ET biosynthesis and signalling pathways, such as *ACS*, *EIN2* and *CTR1*. Two members of the TAS3 family (miRZ8 and miRZ9), 590 putative phased sRNAs and 125 cis-natural antisenses (nat-siRNAs) have also been identified in tomato fruit ripening process (Zuo *et al.*, 2013). Furthermore, the degradome has identified a total of 119 pairs of miRNA-mRNA e.g. AGO1/miR168, 106 of which are new, e.g. TAS3-mRNA/miR390 (Karlova *et al.*, 2013). AGO1/miR168 is involved in miRNA biogenesis, indicating a feedback loop regulation of fruit development. Moreover, the miRNAs/targets responding to ET also results in tomato pedicel abscission (Xu *et al.*, 2015). The correlation can vary between a miRNA and its target during tomato fruit ripening. Since miRNAs always negatively regulate their targets, the high number of positively correlated miRNA/target pairs suggests that mutual exclusion could be as widespread as temporal regulation. For example, miR164 and NAMCUC2 are positively related in flower opening, whereas, as the fruit develops, miR164 becomes negatively related to *NAM-CUC2* (Lopez-Gomollon *et al.*, 2012).

Next-generation sequencing has also been applied to identify sRNA and their expression pattern in the seed of many important crops, as well as miRNA target cleavage. In rice, 434 known miRNAs have been obtained from different rice grain filling stages. The predicted targets of the differentially expressed miRNAs may participate in signal transduction, carbohydrate, and nitrogen metabolism, the response to stimuli and epigenetic regulation (Yi *et al.*, 2013). Similarly, 457 known miRNAs and 13 novel miRNAs have been identified from superior and inferior spikelets during grain development stages. Genes targeted by those differentially expressed miRNAs, i.e. miR156, miR164, miR167, miR397, miR1861, and miR1867, have been considered to play roles in multiple developmental and signalling pathways related to plant hormone homeostasis and starch accumulation (Peng *et al.*, 2014). In maize, 40 known and 162 novel miRNA families have been identified from developing grain. Further analysis shows miR159, miR164, miR166, miR171, miR390, miR399, and miR529 families have putative roles in the embryogenesis of maize grain development by participating in transcriptional regulation and morphogenesis, while miR167 and miR528 families participate in metabolism process and stress response during nutrient storage (Li *et al.*, 2016a). In the developing fibres of cotton, sRNA sequencing has identified 47 conserved miRNA families and seven new miRNAs (Liu *et al.*, 2014d). Together with degradome sequencing, 140 targets of 30 conserved miRNAs and 38 targets of five new miRNAs have been confirmed to affect fibre development, including *SBP* and *MYB*, *LEUCINE-RICH RECEPTOR-LIKE PROTEIN KINASE* (*LRR-RLK*), *PECTATE LYASE* (*PLL*), *TUBULIN*, *UDP-GLUCURONIC ACID DECARBOXYLASE* (*UXS1*) and *CYTOCHROME C OXIDASE SUBUNIT 1* (*CO1*). miRNA156/157 in ovule and fibre development regulate mature fibre length.

Next-generation sequencing has considerably advanced fruit miRNA discovery in fruit species. In strawberry, 88 known and 1224 new miRNAs have been obtained, as well as 103 targets cleaved by 19 known and 55 new miRNAs families. Among them, 14 targets, including *NAC* transcription factor, *ARF*, and *MYB* transcription factors have been found to be involved in regulating fruit senescence (Xu *et al.*, 2013). In apple, 23 conserved, 10 less-conserved and 42 apple-specific miRNAs or families have been discovered from fruit and other vegetative organs (Xia *et al.*, 2012b). The miRNAs target 118 genes with a wide range of enzymatic and regulatory activities. In addition, over 70 diverse genes targeted by 100 miRNAs have been confirmed by degradome analysis. The confirmed genes are potentially involved in diverse aspects of plant growth and development. In pear, 2216 target genes of 188 known miRNAs and 1127 target genes of 184 novel miRNAs have been identified. The miRNAs are widely involved in the regulation of fruit development, including 11 miRNAs in the pathway of lignin biosynthesis, nine miRNAs in sugar and acid metabolism, and one regulating AUX response factor (Wu *et al.*, 2014). In date palm, 238 conserved miRNAs and defined 78 fruit-development-associated miRNAs have been identified. These miRNAs function mainly in regulating genes involved in starch/sucrose metabolisms and other carbon metabolic pathways (Xin *et al.*, 2015). In litchi, 296 miRNAs in 49 known miRNA families have been identified from pericarp under ambient storage and post-cold storage, as well as their 197 targets Of these, 14 miRNA–target pairs are actively involved in litchi fruit senescence-related processes, including energy regulation, anthocyanin metabolism, hormone signalling, and pathogen infection defence (Yao *et al.*, 2015a). In *Lycium chinense*, 60 conserved miRNAs in 31 families and 30 novel miRNAs have been identified from shoot tip and fruit. Five miRNAs are related to fruit maturation, lycopene biosynthesis and signalling pathways (Khaldun *et al.*, 2015). In another *Lycium* species, *L. barbarum*, 50 novel miRNAs, and 38 known miRNAs are from fruit at different developmental stages. The miRNA–target interactions for *L. barbarum* ripening regulators include miR156/157-*LbCNR* and -*LbWRKY8*, and miR171-*LbGRAS*. Regulatory interactions potentially control fruit quality and nutritional value via sugar and secondary metabolite accumulation, including miR156 targeting of *FRUCTOKINASE* and *1-deoxy-D-xylulose-5-phosphate synthase* (*DXS*) and miR164 targeting of *BETA-FRUCTOFURANOSIDASE* (Zeng *et al.*, 2015). In banana, a total of 125 known miRNAs and 26 novel miRNAs have been identified from fruit under ET or *1-methylcyclopropene* (1-MCP) treatment. The differentially expressed 82 miRNAs have been predicted to target 815 genes. Some of the target genes encode transcription factors and other functional proteins, including *SPL, AP2, EIN3, E3 ubiquitin ligase, β-galactosidase*, and *β-glucosidase* (Bi *et al.*, 2015).

In summary, comprehensive analyses of the transcriptome, sRNAs and degradome using NGS provides a useful platform for investigating miRNA-directed and non-miRNA-directed endonucleolytic cleavage in plants, and is a viable tool for exploring sRNA–target interactions involved in reproductive development.

## Roles of sRNAs in flowering time

Flowering is a crucial step of the reproduction in angiosperms. It starts from the floral transition and the floral patterning and go to the development of floral organs (Luo *et al.*, 2013). Floral transition is the switch from the vegetative stage to the reproductive growth. The process is influenced by diverse endogenous and exogenous cues such as age, hormones, photoperiod, and temperature (Andrés and Coupland, 2012; Bäurle and Dean, 2006; Srikanth and Schmid, 2011). Molecular and genetic studies of the model plant, *Arabidopsis*, have suggested there are five flowering time pathways, including age, autonomous, gibberellins (GAs), photoperiod and vernalization pathways (Amasino and Michaels, 2010). Further evidence indicates the extensive crosstalks, feedback or feed-forward loops between the components of these pathways. Many of these components are transcription factors, and a typical example is the MADS-box genes determining the identity of floral organ and patterned to the classic ABCDE model (Krizek and Fletcher, 2005; Li *et al.*, 2016b; Li *et al.*, 2015a; Li *et al.*, 2013b). The class A genes (e.g. *APETALA1, AP1*), B genes (e.g. *PISTILLATA, PI*, and *APETALA3, AP3*), C genes (e.g. *AGAMOUS, AG*), D genes (e.g. *SEEDSTICK, STK* and *SHATTERPROOF, SHP*), and E genes (e.g. *SEPALLATA, SEP*) are interacted and orchestrated to control the formation of the four whorls, i.e. sepal, petals, stamen, and ovule (Krizek and Fletcher, 2005; Theissen, 2001). In addition, other MADS-box genes, e.g. *FLOWERING LOCUS T (FT)* (Lee *et al.*, 2013) and *FLOWERING LOCUS C (FLC)* (Deng *et al.*, 2011), *SUPPRESSOR OF OVEREXPRESSION OF CO 1 (SOC1)* (Lee and Lee, 2010), *FRUITFULL (FUL)* and *LEAFY (LFY)* (Yamaguchi *et al.*, 2009) have been reported to regulate flowering time, while *AGAMOUS-LIKE 15 (AGL15), AGL18* (Adamczyk *et al.*, 2007) and *FOREVER YOUNG FLOWER (FYF)* (Chen *et al.*, 2011b) play some roles in flower senescence and abscission.

sRNAs work as the mediators of transcription factors for crosstalk to regulate the floral transition, and they are finally integrated into flowering pathways by feedback or feed-forward loops. In vascular plants, current studies have focused on miRNAs, such as the well-studied miR156 and miR172 (Fig. 9.2). The two miRNAs are a part of highly conserved regulatory module across all angiosperms (Huijser and Schmid, 2011). They arise independently but are negatively regulated by each other (Li and Zhang, 2016). miR156 targets a group of transcription factors, called *SQUAMOSA PROMOTER BINDING likes* (*SPB* or *SPLs*) family genes, such as *SPL3* and *SPL9* (Rhoades *et al.*, 2002; Xing *et al.*, 2010) (Table 9.1). The expression of miR156 enriches at the seedling stage and then drops progressively, which is observed in *Arabidopsis*, rice, maize, and poplar (Chuck *et al.*, 2007a; Wang *et al.*, 2011a; Wu and Poethig, 2006; Xie *et al.*, 2012). The decrease in miR156 expression results in accumulation of *SPLs*, which in turn promotes flowering (Kim *et al.*, 2012b). For example, *SPL9* promotes the transcription of *FUL, SOC1*, and *AGL42* (Wang *et al.*, 2009a), *SPL3* promotes the transcription of *FUL, LFY*, and *AP1* (Wang *et al.*, 2009a; Yamaguchi *et al.*, 2009), and *SPL4* and *SPL5* promote the transcription of *FUL, SOC1, LFY, AP1*, and *FT* (Yamaguchi *et al.*, 2009). Interestingly, these studies also indicate that *SPL3* and *SPL9* act in parallel to the floral inducer, *FT* (Fig. 9.2). In contrast to miR156, miR172 is scarcely expressed in plant juvenile phases of plants and accumulates over the developmental time. MiR172 acts downstream of miR156 and promotes flowering (Aukerman and Sakai, 2003; Chen, 2004; Jung *et al.*, 2007; Wu *et al.*, 2009). miR172 is regulated by miR156 through a bridge of *SPLs*. *SPLs* directly regulates the transcription of miR172 by binding its promoter region (Wu *et al.*, 2009). miR172 specifically targets *AP2, TOE1, TOE2, TOE3, SMZ*, and *SNZ* in *Arabidopsis* (Aukerman and Sakai, 2003; Mathieu *et al.*, 2009; Schmid *et al.*, 2003), and six *AP2-like* genes in maize, including *GL15* (Chuck *et al.*, 2007b) (Table 9.1). These genes act the

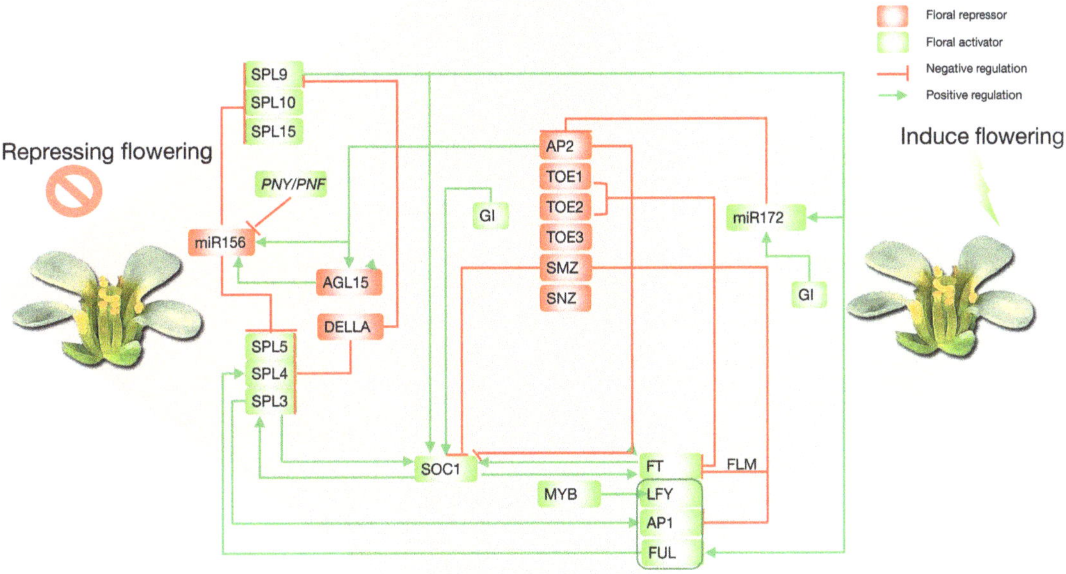


**Figure 9.2** The *miR156–SPL* and *miR172–AP2* modules participating in the regulatory networks of flowering. The miR156 pathway is shown on the left and the miR172 pathway on the right. miR156 down-regulates its target genes, SPL family TFs, while miR172 down-regulates the target genes of the *AP2*-like family. Flowering is induced by up-regulating expression of *SPL3*, *SPL4*, and *SPL5* genes. The *spl4* expression is activated by FUL. SPL3 regulates the expression of LFY, *AP1*, *FT*, and *FUL*, and *SPL9* directly regulates the expression of *AP1*, *SOC1*, and *FUL*. *The spl4* expression is activated by FUL. Flowering can be suppressed by expression of AP2-like genes, *AP2*, *SMZ*, *SNZ*, and *TOE1–3*. miR156 is positively regulated by *AP2* and *AGL15*. *TOE1* and *TOE3* repress flowering by down-regulating *FT*. *SMZ* negatively regulates the expression of *AP1*, *SOC1*, and *FT*, the latter through *FLM*. The *DELLA* protein represses *SPL9* and the expression of SPL3/4/5. SPL3/4/5, are highly induced by *PNY* and *PNF* through negative regulation of miR156. Green arrows and green boxes represent activation and floral activators, respectively. Red arrows and red boxes represent repression and flower repressors, respectively. *AGL15*, *AGAMOUS-LIKE15; AP2*, *APETALA2; PNY*, *PENNYWISE; PNF*, *POUND-FOOLISH; FT*, *FLOWERING LOCUS T; FLM*, *FLOWERING LOCUS M; FUL*, *FRUTIFUL; LFY*, *LEAFY; SOC1*, *SUPPRESSOR OF CONSTANS1; SMZ*, *SCHLAFMUTZE; SNZ*, *SCHNARCHZAPFEN; SPL*, *SQUAMOSA PROMOTER-BINDING PROTEIN-LIKE; TOE1–3*, *TARGET OF EAT1–3.*

downstream targets of miR156 (Wu *et al.*, 2009) and negatively regulate flowering time (Aukerman and Sakai, 2003; Jung *et al.*, 2007; Lauter *et al.*, 2005). Thus, miR156 and miR172 form a feedback loop for fine tuning the flowering response (Yant *et al.*, 2010).

In addition, miR156 and miR172 appear to mediate the interactions within other flowering pathways (Bergonzi *et al.*, 2013; Zhou *et al.*, 2013). In the GA pathway of *Arabidopsis*, GA promotes flowering through the degradation of several *DELLA* proteins, GA repressors, which can interact with *SPL9* (Harberd *et al.*, 2009; Yu *et al.*, 2012) (Fig. 9.2). The *SPL*-mediated activation of miR172 and floral identity genes eventually leads to floral transition (Yu *et al.*, 2012). The stability of *DELLA*, which directly affects *SPLs* and miR172 activity, is not only regulated by the GA signal but also by several other phytohormones, including AUX, ET, and abscisic acid (ABA) (Vanstraelen and Benkova, 2012). In the photoperiod pathway, miR172 is regulated by photoperiod through *GIGANTEA* (*GI*), a key regulator that mediates photoperiodic flowering (Jung *et al.*, 2007) (Fig. 9.2). *GI* can

**Table 9.1** SRNAs and their target genes involved in reproductive development, including flower and fruit/seed development

| miRNA | Target genes | Species | Function(s) of target genes in reproductive development | References |
|---|---|---|---|---|
| miR156 | *SPL2, SPL3, SPL4, SPL5, SPL9, SPL10, SPL15* | *Arabidopsis thaliana* | Floral organ identity and flowering time | Gandikota *et al*., 2007; Schwab *et al*., 2005; Schwarz *et al*., 2008; Wang *et al*., 2008; Wu *et al*., 2009; Wu and Poethig, 2006; Yamaguchi *et al*., 2009 |
| | *SPL9–2* | *Brassica rapa* ssp. *pekinensis* | Controls the heading time by shortening the seedling and rosette stages | Wang *et al*., 2014b |
| | *SBP1* | *Antirrhinum majus* | Controls flowering time | Preston and Hileman, 2010 |
| | *SBP3* | *Physcomitrella patens* | Negatively regulates the vegetative developmental transition from protonemata to leafy gametophores | Cho *et al*., 2012 |
| | Teosinte glume architecture1 (tga1) | *Zea mays* | Promotes flowering | Chuck *et al*., 2007a |
| | *SPL* | *Panicum virgatum* L. (switchgrass) | Promotes flowering | Chuck *et al*., 2011; Fu *et al*., 2012 |
| | *SBP* | *Solanum lycopersicum* | Promotes normal phase change | Salinas *et al*., 2012; Zhang *et al*., 2011 |
| | *SPL* | *Oryza sativa* | Promotes flowering | Xie *et al*., 2006, 2012 |
| | *SPL* | *Acacia confusa, Acacia colei, Eucalyptus globulus, Hedera helix, Quercus acutissima, Populus × canadensis* | Promotes juvenile-to-adult vegetative phase change | Wang *et al*., 2011a |
| | *AaSPL* | *Arabis alpina* | Promotes flowering | Bergonzi *et al*., 2013 |
| | *NtSPL* | *Nicotiana tabacum* | Promotes flowering | Zhang *et al*., 2015 |
| miR156/157 | *SlySBP2, SlySBP6b, SlySBP10, SlySBP13, SlySBP15* | *Solanum lycopersicum L.* | Flower and fruit morphology, fruit ripening | Ferreira e Silva *et al*., 2014; Karlova *et al*., 2013; Moxon *et al*., 2008a |
| | CNR | | Fruit development and ripening | Karlova *et al*., 2013; Moxon *et al*., 2008a |
| miR157 | *SPL* | *Torenia fournieri* | Promotes normal phase change and branching | Shikata *et al*., 2012 |
| miR159 | *AtMYB: AtMYB33, AtMYB65, AtMYB97, AtMYB101, AtMYB104, AtMYB120* | *Arabidopsis thaliana* | Promote flowering in response to GA and length of photoperiod | Achard *et al*., 2004; Alonso-Peral *et al*., 2010 |
| | *OsGAMYB OsGAMYBL1* | *Oryza sativa* | Promotes heading | Tsuji *et al*., 2006 |
| | *LtGAMYB* | *Lolium temulentum* | Promotes flowering | Achard *et al*., 2004; Woodger *et al*., 2003 |
| | *SsGAMYB* | *Sinningia speciosa* | Promotes flowering | Li *et al*., 2013a |
| miR160 | *ARF16, ARF17, ARF10* | *Solanum lycopersicum L.* | Fruit development and ripening | Karlova *et al*., 2013 |

**Table 9.1** Continued

| miRNA | Target genes | Species | Function(s) of target genes in reproductive development | References |
|---|---|---|---|---|
| miR164 | *CUC2* | *Solanum lycopersicum L.* | Fruit development and ripening | Moxon *et al.*, 2008a |
| | *CUC1*, *CUC2* | *Arabidopsis thaliana* | Flower development | Baker *et al.*, 2005; Laufs *et al.*, 2004 |
| | *GOBLET (GOB)* | *Arabidopsis thaliana* | Flower development | Baker *et al.*, 2005; Sieber *et al.*, 2007 |
| miR165/166 | *HD-ZIP III genes: ATHB15*, *ATHB8*, *REVOLUTA(REV)*, *PHABULOSA(PHB)*, *PHAVOLUTA(PHV)* | *Arabidopsis thaliana* | Flower development, embryo patterning, postembryonic and axial meristem initiation | Floyd and Bowman, 2004; Kim *et al.*, 2005; Reinhart *et al.*, 2002; Zhou *et al.*, 2007 |
| miR167 | *ARF6*, *ARF8* | *Arabidopsis thaliana* | Flower development | Ru *et al.*, 2006; Wu *et al.*, 2006 |
| | *GAMyb-like1*, *GAMyb-like2* | *Solanum lycopersicum* L. | Fruit development and ripening | Karlova *et al.*, 2013 |
| miR168 | *ARGONAUTE 1* | *Solanum lycopersicum* L. | Fruit development and ripening | Karlova *et al.*, 2013 |
| miR169 | *AtNF-Y family of TFs* | *Arabidopsis thaliana* | Repress and promote flowering in response to abiotic stress | Xu *et al.*, 2014 |
| | *NF-YA family of TFs* | Petunia and *Antirrhinum* | Flower development | Jones-Rhoades and Bartel, 2004; Zhao *et al.*, 2009 |
| miR171 | *LOST MERISTEM 1* | *Arabidopsis thaliana* | Delays flowering | Xue *et al.*, 2014 |
| | *SCARECROW-LIKE* | *Hordeum vulgare* | Promotes flowering | Curaba *et al.*, 2013 |
| miR170/ miR171 | *GRAS/SCR (Scarecrow-related)* | *Solanum lycopersicum* L. | Fruit development and ripening | Karlova *et al.*, 2013; Moxon *et al.*, 2008a |
| miR172 | *AP2-like: TOE1*, *TOE2*, *TOE3*, *SMZ*, *SNZ* | *Arabidopsis thaliana* | Negatively regulates induction of flowering | Aukerman and Sakai, 2003; Jung *et al.*, 2007; Mathieu *et al.*, 2009; Schmid *et al.*, 2003; Schwab *et al.*, 2005; Yant *et al.*, 2010 |
| | *CfTOE1* | *Cardamine flexuosa* | Repressor of flowering | Zhou *et al.*, 2013 |
| | *GLOSSY15* | *Zea mays* | Repressor of flowering | Lauter *et al.*, 2005 |
| | *ZmTOE1* | *Zea mays* | Repressor of flowering | Salvi *et al.*, 2007 |
| | *InAP2-like* | *Ipomoea nil* | Involved in photoperiodic flower induction | Glazińska *et al.*, 2009 |
| | *RELATED TO APETALA2 1(RAP1)* | *Solanum tuberosum* | Possible repressor of flowering and tuberization | Martin *et al.*, 2009 |
| | *SlAP2a*, *SlAP2b*, *SlAP2c*, *SlAP2e*, *SlAP2d* | *Solanum lycopersicum* L. | Fruit development and ripening | Itaya *et al.*, 2008; Karlova *et al.*, 2013; Moxon *et al.*, 2008a |
| | *Sly-TASI3a*, *Sly-TASI3b* | *Solanum lycopersicum* L. | Fruit development and ripening | Karlova *et al.*, 2013 |
| | *Cleistogamy1(Cly1)* | *Hordeum vulgare* | Promotes cleistogamy | Nair *et al.*, 2010 |
| | *AP2* | *Malus X domestica* | Fruit development and ripening | |
| miR319 | *TCP family of TFs* | *Arabidopsis thaliana* | Promotes flowering | Palatnik *et al.*, 2003 |

**Table 9.1** Continued

| miRNA | Target genes | Species | Function(s) of target genes in reproductive development | References |
|---|---|---|---|---|
| miR390 | *TAS3* | *Arabidopsis thaliana* | Promotes juvenility by negative regulation of*ARF3* | Fahlgren *et al.*, 2006 |
| miR393 | *AtTIR1, AFB 1–3* | *Arabidopsis thaliana* | Repressor of flowering | Chen *et al.*, 2011c |
| | *OsTIR1, OsAFB2* | *Oryza sativa* | Repressor of flowering | Xia *et al.*, 2012a |
| | *SlTIR1* | *Solanum lycopersicum* L. | Flower-to-fruit transition | Karlova *et al.*, 2013; Ren *et al.*, 2011 |
| miR396 | *GRF1, GRF2, GRF3, GRF4, GRF7, GRF8, GRF9* | *Arabidopsis thaliana* | Pistil development | Liang *et al.*, 2014 |
| | *GRF6, GRF10* | *Oryza sativa* | Inflorescence and spikelet development | Liu *et al.*, 2014b |
| miR399 | *PHOSPHATE 2* | *Arabidopsis thaliana* | Ambient temperature-responsive repressor of flowering | Kim *et al.*, 2011 |
| miR824 | *AGL16* | *Arabidopsis thaliana* | Represses flowering in certain genetic backgrounds and environmental conditions | Hu *et al.*, 2014 |
| miR858/ miR828 | *MYB2* | *Gossypium* | Fibre development | Guan *et al.*, 2014 |
| miR5200 | *FTL1, FTL2* | *Brachypodium distachyon* | Promotes flowering in response to photoperiod | Wu *et al.*, 2013 |
| miR4376 | *ACA10* | *Solanum lycopersicum* L. | Flower morphology and fruit yield | Wang *et al.*, 2011b |
| TAS3 | *ARF3, ARF4* | *Solanum lycopersicum* L. | Fruit development and ripening | Karlova *et al.*, 2013 |
| | *ARF3/ETTIN, ARF4* | *Arabidopsis thaliana* | Fruit development and ripening | Fahlgren *et al.*, 2006 |
| cis-nat-siRNA | *ARI14* | *Arabidopsis thaliana* | Facilitates sperm formation and double fertilization | Ron *et al.*, 2010 |
| Nat-siRNA | *DGD2* | *Phalaenopsis aphrodite* | Spike initiation and flowering | An and Chan, 2012 |

also positively regulate *CO* transcription, the core component responsible for measuring the distinction of day length (Fowler *et al.*, 1999; Park *et al.*, 1999). Notably, the miR172-dependent pathway in photoperiodic induction is independent to *CO* but requires the functional *FT*, the target of *CO* (Jung *et al.*, 2007). Moreover, many miR156-targeted *SPLs* are able to respond to the photoperiodic changes (Wang *et al.*, 2014a). For example, *SPL3, SPL4,* and *SPL5* are highly induced by photoperiod in a *PENNYWISE (PNY)*- and *POUND-FOOLISH (PNF)*- dependent manner (Lal *et al.*, 2011; Schmid *et al.*, 2003), and this process is carried out through negative regulation of miR156 (Lal *et al.*, 2011) (Fig. 9.2). Taken together, miR156/miR172 may integrate the signals from multiple phytohormones to coordinate floral development in response to environmental changes.

miR159 is another important component mediating the flowering pathway, which targets at least three *MYB* transcription factors in *Arabidopsis*, including *MYB33, MYB65* and *MYB101* (Achard *et al.*, 2004). *MYB33* responding to photoperiods or GA treatment promotes the floral transition through activation of *LFY* (Gocal *et al.*, 2001). Similar to miR172, miR159 can be promoted by GA through overcoming *DELLA*-mediated repression. In contrary to miR172, increased level of miR159 delays flowering only in short day plants (Achard *et al.*, 2004). However, another report is contradictory and states that overexpression of miR159 does not alter flowering time (Alonso-Simon *et al.*, 2010; Schwab *et al.*, 2005). Thus, *MYB33* and the related *MYB* genes' roles in flowering await further investigation.

miR393 is another regulator of flowering and targets two AUX receptors, *OsTIR1* and *OsAFB2* in rice. It promotes flowering by suppressing the sensitivity of *OsTIR1* and *OsAFB2* to AUX (Xia *et al.*, 2012a), suggesting a potential role of AUX signalling in flowering. Similarly to miR393, miR399 can induce early flowering by degrading its target transcription factors, *PHOSPHATE 2* (*PHO2*) in *Arabidopsis*. *PHO2* is known to maintain phosphate homeostasis and regulate flowering time (Kim *et al.*, 2011).

One particular gene, *DIGALACTOSYLDIACYLGLYCEROL SYNTHASE 2* (*DGD2*), has been identified to be targeted by the NATs unique to *Phalaenopsis* (An and Chan, 2012). *DGD2* affect the lipid content of chloroplast membranes under phosphate-starvation conditions (Ge *et al.*, 2011). In *Phalaenopsis*, it may play a unique role in regulating the lipid composition of the chloroplast membrane under low ambient temperature treatment. Low ambient temperature is required for spikelet initiation in *Phalaenopsis* (Blanchard and Runkle, 2006; Chen *et al.*, 2008), suggesting that the NATs play a key role in flowering time and/or developmental processes during the flowering phase transition.

## Roles of sRNAs in androecium development

The androecium is a male organ with an anther on the top. The anther contains haploid microspores derived from the meiosis of diploid sporogenous cells. The cytoplasm and cytoskeletons of microspores are reorganized and eventually develop into pollen grains. Pollen development is essential for male fertility, which relies heavily on somatic anther tissues, such as the tapetum. The formation of the tapetum ensures early pollen development, after which tapetum degeneration nourishes pollen into maturation (Twell, 2011). Genome-wide studies have revealed that many sRNA are involved in male organ development in plant species (Wei *et al.*, 2011; Yin and Shen, 2010; Zhang *et al.*, 2009). Some of these sRNAs participate in tapetum formation and degeneration, microspore formation, and pollen release (Table 9.1).

miR156-targeted *SPLs* (miR156 non-targeted SPL8) are necessary for maintaining anther fertility in *Arabidopsis*. Loss of miR156-targeted *SPLs* leads to a semi-sterile anther due to an abnormality of primary tapetum cells and primary sporogenous cells (Xing *et al.*, 2010).

miR159 post-transcriptionally regulates the conserved *GAMYB-like* genes which respond to GA signal (Gubler *et al.*, 2002; Murray *et al.*, 2003). Overexpression of miR159 or disruption of the GA biosynthesis pathway both delays flowering and reduces fertility (Achard *et al.*, 2004; Cheng *et al.*, 2004). In *Arabidopsis* and rice, the proper development of the pollen cells relies on the spatial restriction of *GAMYB-like* gene expression in the anthers (Millar and Gubler, 2005; Tsuji *et al.*, 2006). miR159 overexpression down-regulates anther *MYB33/MYB65* in *Arabidopsis* and anther *TaGAMYB1* in wheat, and suppresses tapetum degradation other than forms hypertrophic tapetum, which therefore affects microspore development and eventually leads to male sterility (Millar and Gubler, 2005; Tsuji *et al.*, 2006; Wang *et al.*, 2012a).

miR165 and miR166 differ by only one nucleotide. In *Arabidopsis*, it is involved in microspore development by regulating their target gene *REVOLUTA (REV)*, an *HD-ZIP III* family gene (Table 9.1). The expression of *REV* is negatively associated with that of *FILAMENTOUS FLOWER* (*FIL*). *REV* negatively regulates the establishment of anther polarity, while *FIL* regulates the development of microsporangia and microspore mother cells (Lian *et al.*, 2013). The biogenesis of plant miR165/166 requires HYPONASTIC LEAVES1 (HYL1), a double-stranded RNA-binding protein. HYL1 mediates the accurate processing of pre-miRNAs into mature miRNAs via DCL1. HYL1 deficiency-suppressing miR165/166 reduces the microsporangia by half (four in wild type) and decreases male fertility in *Arabidopsis* (Lian *et al.*, 2013). The balance between *REV* and *FIL* expression is regulated by HY1-dependent miR165/166 and required for the formation of inner microsporangia and anther connectives during stamen development.

MiR167 controls pollen development by directly targeting two *auxin response factor* genes (*ARFs*) *ARF6* and *ARF8* (Tabata *et al.*, 2010). In *Arabidopsis*, the dysfunction of miR167 indulges *ARF6* and *ARF8* ectopic expression in the connective cells of anther. The connective cells become too large to break for pollen release (Ru *et al.*, 2006; Wu *et al.*, 2006). Deep sequencing shows a high expression of tae-miR166/167 in the spikelet tissue of thermo-sensitive male-sterile wheat line suffered from cold-induced male sterility, where members of *ARF* families are the targets of tae-miR167 (Tang *et al.*, 2012). Similarly, nineteen members including the diverse isomiRs, have been identified for miR167 and most of them have shown a significant up-regulation in the anther tissue of male sterile mutants in citrus (Fang *et al.*, 2014).

MiR169 family members target the gene family *NF-YA* transcription factor. In Petunia and *Antirrhinum*, *NF-YA* transcription factors belong to the homeotic C-class genes during flower development (Jones-Rhoades and Bartel, 2004; Zhao *et al.*, 2009). These miR169-encoding genes are referred as *BLIND* (*BL*) in Petunia and *FISTULATA* (*FIS*) in *Antirrhinum*, and both of them repress the expression of *NF-YA* genes in inner two floral whorls during flower development. Loss of *FIS* and *BL*, or the lack of miR169, result in the conversion of petals to stamens in the second whorl (Cartolano *et al.*, 2007). Notably, such a role of miR169 in regulating of C-gene activity in Petunia and *Antirrhinum* does not appear in *Arabidopsis* (Cartolano *et al.*, 2007).

Cis-nat-siRNAs are also involved in pollen development (Ron *et al.*, 2010). This kind of sRNAs has only been reported in the previous studies on plants exposed to environmental stress (Borsani *et al.*, 2005; Katiyar-Agarwal *et al.*, 2006). In a recent study on *Arabidopsis*, cis-nat-siRNA-based regulation plays some key roles in reproductive function, as it facilitates sperm formation and double fertilization (Ron *et al.*, 2010). Two inversely transcribed genes, *KOKOPELLI* (*KPL*) and *ARIADNE14* (*ARI14*), generate a pair of sperm-specific nat-siRNAs. In the absence of *KPL*, the *ARI14* RNA level in sperm is increased and fertilization is impaired. Finally, the accumulation of *ARI14* transcripts leads to reduced seed set.

In addition, miR166 is highly accumulated in developing rice pollen, and eight other miRNAs (OsmiR528, OsmiR5793, OsmiR1432, OsmiR159, OsmiR812d, OsmiR2118c, OsmiR172d, and miR5498) are differentially expressed in rice anthers between a cytoplasmic male sterility (CMS) line and a maintainer line, suggesting these miRNAs play some roles in rice pollen development (Wei *et al.*, 2011). Eight miRNAs, Zma-miR601, Zma-miR602, Zma-miR603, Zma-miR604, Zma-miR605, Zma-miR606, Zma-miR607, and Zma-miR397 families have also been identified in a maize cytoplasmic male sterile line. Their 18 potential targets of these miRNAs are considered to be involved in a number of biological processes during pollen development (Shen *et al.*, 2011). Four other families, miR397, miR399, miR408, and miR535 have exhibited a significantly different expression between male sterile mutant and wild type in citrus (Fang *et al.*, 2014). Three members of miR397 (crt-miR397.1, crtmiR397.2, crt-miR397.3) have been shown to be up-regulated in the anther tissue of male sterile mutant in citrus and they targeted *BRANCHED-CHAIN-AMINOACID AMINOTRANSFERASE 2* (*BCAT2*) (Fang *et al.*, 2014). miR397 also targets some oxidative genes like *LACCASES*, *PEROXIDASE*, and *TUBULIN*, which play an important role in pollen viability and cell wall formation during pollen development in *Arabidopsis* (Khraiwesh *et al.*, 2012).

## Roles of sRNAs in gynoecium development

The gynoecium is the female reproductive organ and is composed of one or more fused carpels. Carpels have three parts: a stigma at the top, a style, and an ovary. The ovary accommodates ovules, which are differentiated for fertilization and subsequent embryogenesis. The gametocyte development and morphological maturity for pollination determine female fertility.

Many of the miRNAs in androecium also have roles in the gynoecium. The miR156-targeted *SPLs* not only negatively regulate anther fertility but also redundantly control gynoecium patterning (Table 9.1). In *Arabidopsis*, down-regulation of miR156-targeted *SPLs* clearly results in a shortened style and an apically swollen ovary narrowing into an elongated gynophore (Xing *et al.*, 2013). The up-regulation of miR156 and down-regulation of *SPL8* (miR156 non-targeting *SPL8*) both enhance the impairment of pollen tube in *ETTIN* mutant. In addition, *ETTIN* is an *ARF3*, possibly facilitates AUX production or accumulation, and determines apical-basal patterning during ovary formation (Xing *et al.*, 2013). Thus, miR156-targeted *SPLS* and *SPL8* redundantly control gynoecium patterning through interfering with AUX homeostasis and signalling. Additionally, miR167-targeted *ARF6* and *ARF8* both regulate gynoecium and stamen development in immature flowers (Ru *et al.*, 2006). In tomato, the down-regulation of miR167-targeted *ARF6* and *ARF8* leads to the absence of trichomes on the styles, the failure of pollen tube formation and sterility (Liu *et al.*, 2014c). The roles of miR167 and *ARFs* are conserved across different species (Li *et al.*, 2015b). In *Arabidopsis*, miR165 and miR166 also both regulate gynoecium development. Both of them target the same *HD-ZIP III* genes *ATHB15*, *ATHB8*, *REV*, *PHABULOSA* (*PHB*) and *PHAVOLUTA* (*PHV*) (Table 9.1) (Floyd and Bowman, 2004; Reinhart *et al.*, 2002; Zhou *et al.*, 2007), and regulate these target genes in different ways (Jung *et al.*, 2007). Overexpression of miR165 in *Arabidopsis MERISTEM ENLARGEMENT 1* (*MEN1*) mutant results in a decrease in the expression of all of these *HD-ZIP III* target genes and causes developmental defects in the SAM such as an enlarged apical meristem and short and sterile carpels (Kim *et al.*, 2005). MiR166 has also been shown to control embryonic SAM development in rice and maize (Nagasaki *et al.*, 2007; Nogueira *et al.*, 2007).

A set of miRNAs in vegetative growth also function in gynoecium formation and female fertility. In *Arabidopsis*, miR396 has an essential roles in leaf growth by regulating *GROWTH REGULATING FACTOR* (*GRF*), which interacts with *GRF-interacting factor 1* (*GIF1*) to participate in cell proliferation (Liu *et al.*, 2009). Besides, miR396 controls carpel number by down-regulating its targets *GRF6/10* (Liang *et al.*, 2014). MiR164 is another sRNA regulating vegetative and gynoecium development. miR164 targets *Arabidopsis CUP-SHAPED COTYLEDON* (*CUC*) genes *CUC1* and *CUC2* to control the boundary size of meristems (Laufs *et al.*, 2004), shoot apical meristem formation, and the initiation of leaf margin development (Hibara *et al.*, 2003; Kamiuchi *et al.*, 2014; Nikovics *et al.*, 2006). Moreover, miR164 are involved in tomato boundary formation in different organs by regulating two *NAM* genes, *GOBLET* (*GOB*) and *SlNAM2*. In leaf, *GOB* down-regulation causes the leaflets to have a smoother margin. Similarly, in *Arabidopsis* flowers, *SlNAM2* down-regulation leads to an abnormal fusion of the sepal and whorl (Berger *et al.*, 2009; Hendelman *et al.*, 2013) and a defect in carpel closure (Baker *et al.*, 2005; Sieber *et al.*, 2007). In maize, the *TASSELSEED4* (*TS4*) gene, encodes *zma*-miR172, controls tassel formation and induces carpel abortion by degrading *INDETERMINATE SPIKELET1* (*IDS1*), an *APETALA2* (*AP2*)-*LIKE* transcription factor. However, in ears, *IDS1* expression is abundant at the base of the initiating primordial carpel. Thus, the regulation of *IDS1* by zma-miR172 results in the differences in primordial initiation, and sex determination (Chuck *et al.*, 2007b). In addition, the down-regulation of *SBPs* by *zma*-miR156 results in feminized male tassels (Hultquist and Dorweiler, 2008). The regulatory networks of sRNAs are pervasive and appear to cross regulate among many vegetative growth and reproductive growth pathways.

## Roles of sRNAs in fruit/seed development

Fruits/seeds are derived from the maturation of one or more flower(s), and the gynoecium of the flower(s) forms the whole or part of the fruit. The gynoecium expands to appropriately accommodate the seeds, while developing into a fruit, and the walls of ovary become pericarps. The fruit set and growth is determined in part by the seeds. In such a way, seed and fruit development occur as a coordinated processes.

To date, the gene regulation networks underlying fruit development have been largely unravelled for model plants, such as *Arabidopsis* and tomato. *Arabidopsis* has dehiscent fruits (standard silique), composed of the three main elements along the fruit mediolateral axis: the valve, the valve margin, and the replum. The fruit elements are determined by the genes during the early stages of gynoecium development and the genes in subsequent specific domains. Many of these genes encode transcription factors. For example, an MADS-Box transcription factor, *RIPENING-INHIBITOR* (*RIN*), is involved in regulating most of the tomato ripening phenomena, e.g. pro-vitamin A and carotenoid accumulation, softening, and the production of flavour compounds, etc. The *RIN* also acts upstream of both ET- and non-ethylene-mediated ripening control (Vrebalov *et al.*, 2002). An *SBP*-box transcription factor, *COLORLESS NONRIPENING* (*CNR*), is a positive regulator of tomato fruit ripening, and plays roles in tomato fruit colour formation and pericarp texture (Manning *et al.*, 2006). Another transcription factor, *AP2*, is a negative regulator of ripening in tomato in response to ET (Karlova *et al.*, 2011). In addition to these fruit-ripening genes, some others control tomato fruit size and shape, for instance, *Fw2.2* encodes a plant-specific protein and participates in cell division in fruit specifically to regulate fruit size (Cong and Tanksley, 2006). An orthologue *Arabidopsis YABBY2*, *FASCIATED* (*FAS*) is associated with a flat tomato fruit shape and the number of locules (Cong *et al.*, 2008; Rodríguez *et al.*, 2011). *LOCULE NUMBER* (*LC*), is near to the tomato orthologue of *Arabidopsis WUS*, is a homeodomain transcription factor gene also regulating the number of locules (Mayer *et al.*, 1998). Another two genes, *SUN* and *OVATE*, control fruit elongation shape (Liu *et al.*, 2002; Xiao *et al.*, 2008). More information on tomato fruit development can be obtained from a previous review (Karlova *et al.*, 2014).

Many transcription factor mRNAs are the targets of sRNA. Through bridging these transcription factors, sRNAs are extensively integrated into genetic networks to regulate fruit development (Table 9.1). In *Arabidopsis*, miR172 plays a critical role in fruit morphogenesis, allowing proper fruit growth after fertilization by targeting the transcriptional repressor *AP2* (and *TOE3*) (José Ripoll *et al.*, 2015). *AP2* (and *TOE3*) represses ovary and silique growth by modulating cell division and expansion through inhibiting the expression of *AG* and *FUL* (Fig. 9.3) (José Ripoll *et al.*, 2015; Yant *et al.*, 2010). In the absence of AP2, FUL acts together with ARF proteins (such as ARF8) to directly activate the

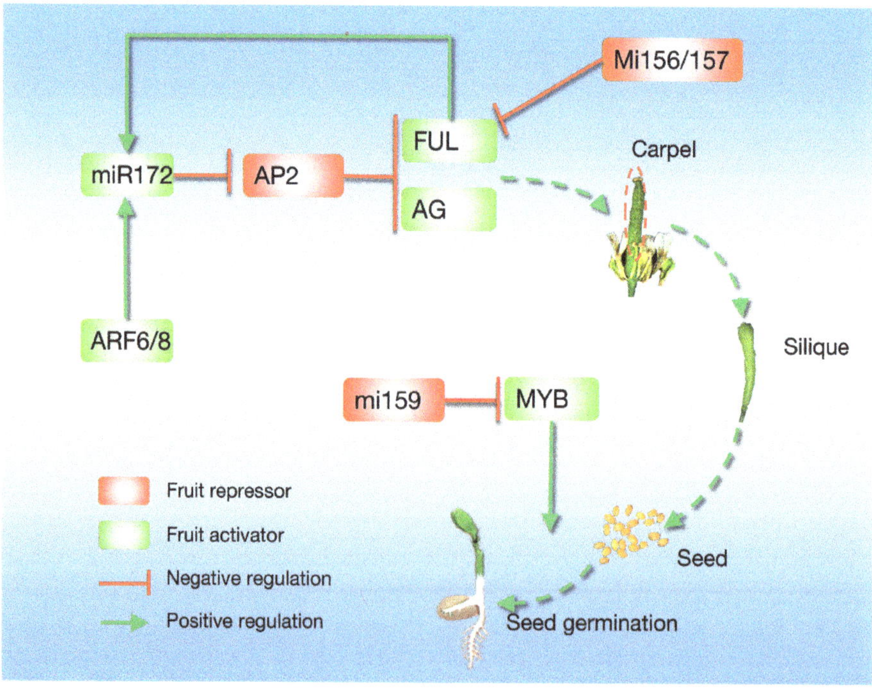


**Figure 9.3** Model for the regulation of fruit growth in *Arabidopsis* by small RNA. MiRNA172 negatively regulates AP2 by inhibiting its translation. AP2 directly binds to the promoter and represses the expression of both FUL and AG which are important for carpel and *Arabidopsis* silique development. An elevated level of miRNA172 inhibits *Arabidopsis* silique growth. In seeds, the regulation of MYB by miR159 and GA for the aleurone cells development and death is required for germination. FUL is a well-characterized regulator of cell differentiation by miR156- and miR157- during early stages of *Arabidopsis* fruit development.

miR172 valve-specific miR172-encoding gene so as to promote valve growth (three main domains outside of silique: the valves, the valve margin, and the replum). Interestingly, the opposite impact of miRNA172 occurs in apple, and is that miRNA172 inhibits apple fruit growth through negative regulation of AP2, which is required for sepal and fruit flesh cortex development (Yao *et al.*, 2015b). In apple, overexpression of miRNA172 results in a decreased number of hypanthium cells before pollination and small size of it after pollination. In conclusion, the impact of miRNA172 on fruit growth may be achieved initially by influencing the size of floral organs, which contribute tissues to fruit development and then modulating cell division and expansion in these tissues in response to fertilization-induced signals and signal transduction. However, its roles are species specific.

MiR159's regulations on *GAMYB-like* genes act not only in leaf and flower (including anthers), but also in seeds through GA signalling pathway (Gubler *et al.*, 2002; Murray *et al.*, 2003). In *Arabidopsis* seeds, the regulation of three *GAMYB-LIKE* genes (*AtMYB33/65/101*) by miR159 and GA controlling the aleurone cells development and death, is required for germination (Fig. 9.3) (Alonso-Simon *et al.*, 2010). The seed dormancy/germination process is well known to be antagonistically regulated by both GA and ABA. Accordingly, miR159 acts as a negative regulator, while *AtMYB33* and *AtMYB101* serve as positive regulators of germination responding to ABA (Reyes and Chua, 2007). Some *MYBs* are synchronously targeted by other miRNAs, such as miR858, and miR828 (Guan *et al.*, 2014; Pang *et al.*, 2009). miR828 and miR858 have been proven to coordinate *Arabidopsis* trichome and cotton fibre development via *MYBs* (Guan *et al.*, 2014). Cotton fibre is a kind of special single-celled trichomes initiated from the epidermal layer of cotton ovule and its development shares some similar regulatory mechanisms with leaf trichome development (Wan *et al.*, 2014). *MYB* family transcription factors are the critical regulator in fibre development, such as the *Arabidopsis R2R3 MYB* domain transcription factor *GLABROUS1* (*GL1*) (Larkin *et al.*, 1993), and a cotton *GL1-LIKE MYB* transcription factor (referred as *GaMYB2*) (Wang *et al.*, 2004). Interestingly, in cotton, the reduced expression of *GhMYB25-like* gene leads to a fibreless phenotype but no effect on the formation of other trichomes (Walford *et al.*, 2011). In apple, miR159-*MYBs* may regulate both male organ and embryo development, while MiR828- and miR858-*MYBs* are linked with diverse biological processes and metabolism pathways underlying cell wall formation, lignification, anthocyanin biosynthesis, cell fate and identity, plant development and the response to biotic and abiotic stresses (Xia *et al.*, 2012b).

Many miRNA-target regulations are active from flowering to fruiting. For example, miR156-SPL/SBP box node, works not only in flower development, but also in diverse aspects of fruit development, including phase transition, plant architecture, trichome distribution, embryonic patterning and anthocyanin biosynthesis (Chuck *et al.*, 2007a; Gou *et al.*, 2011; Mohorianu *et al.*, 2011; Nodine and Bartel, 2010; Schwab *et al.*, 2005; Wang *et al.*, 2009a; Yu *et al.*, 2010). In tomato, *S. lycopersicum SBP* box genes (*SlySBP*) have been identified to be the target of miR156 and miR157 (Moxon *et al.*, 2008b; Salinas *et al.*, 2012). As discussed in flower development, miR156-targeted *SPL3* positively and directly regulates the MADS box genes *APETALA1* (*AP1*), *FUL* and *LFY* (the central regulator of flowering) (Fig. 9.3) (Yamaguchi *et al.*, 2009). Interestingly, *FUL* is a well-characterized regulator of cell differentiation by miR156- and miR157- during the early stages of *Arabidopsis* fruit development (Fig. 9.3) (Gu *et al.*, 1998). Besides, *FUL*-like genes play important roles in the development of the fruit wall in two basal eudicot Papaveraceae species during fruit maturation (Pabón-Mora *et al.*, 2012). Tomato MADS box gene *FUL1/TDR4*, an orthologue of *AtFUL*, is induced during ripening and probably the target of *CNR*, a *SlySBP* family member (Bemer *et al.*, 2012; Seymour *et al.*, 2013). Another target of tomato *SBPs* target, the MADS-box gene *MACROCALYX* (*MC*), not only controls sepal and inflorescence, but also regulates the tomato fruit abscission zone by working with another MADS box gene, *JOINTLESS* (Nakano *et al.*, 2012; Vrebalov *et al.*, 2002). Moreover, overexpressing the *AtMIR156b* precursor indirectly induces the expression of *LeT6/TKN2* (a *KNOX*-like class I gene) and *GOB* (a *NAM/CUC*-like gene) in developing the ovaries of transgenic plants and the ovaries of the natural mutant *Mouse-ear* (*Me*), which exhibits abnormal flower and fruit morphologies (Ferreira e Silva *et al.*, 2014). The abnormal morphologies include extra carpels and ectopic structures in fruits. In another example, miR396 targets *GRF6/10*, which combines with *GIF1* to form a transcription co-activator. In rice, overexpressing OsmiR396 and knocking down *GRF6* result in phenotypes with open husks and sterile lemma in seed (Liu *et al.*, 2014b). *GRF6* and *GRF10* transactivate the rice *JMJD2* family *JMJC* gene *706* (*OsJMJ706*) and *CRINKLY4 RECEPTOR-LIKE KINASE* (*OsCR4*) responding to GA, which is required for floral development into normal seed (Liu *et al.*, 2014b). miR397 regulates rice *LACCASE* (*LAC*), a regulator involved in brassinosteroids signalling, and also plays a crucial role in the productivity of other crop plants (Zhang *et al.*, 2013). Through genome-wide identification and screening, OsmiR397 is highly expressed in rice seeds, and implicated in the regulation of several key yield-related factors, such as vascular bundle formation, panicle branches numbers, effective grains and tiller numbers, grain hull, and endosperm size (Chen *et al.*, 2011a; Zhang *et al.*, 2013). The yield-related factors, in turn, increased grain yield in miR397 up-regulating plants. Additionally, the regulatory interaction between miR397 and *LAC* has been predicted to be conserved in many species, including tobacco, populus, and *Arabidopsis* (Jones-Rhoades and Bartel, 2004). miR4376 regulates the expression of an auto-inhibited $Ca^{2+}$-ATPase in tomato, *ACA10* which plays a critical role in tomato flower morphology and fruit yield (Wang *et al.*, 2011b). $Ca^{2+}$ signatures, and oscillations, in the cytoplasm or organelles are critical for signal transduction and are regulated by $Ca^{2+}$ influx through the activities of $Ca^{2+}$ channels and $Ca^{2+}$ efflux through the activities of high affinity $Ca^{2+}$-ATPases (pumps) or low-affinity $Ca^{2+}$ exchangers (Dodd *et al.*, 2010; Kudla *et al.*, 2010; McAinsh and Pittman, 2009). There are 14 $Ca^{2+}$-ATPases in *Arabidopsis* and rice. Four of them are endoplasmic reticulum-type $Ca^{2+}$-ATPases and ten are ACAs (Baxter *et al.*, 2003). Based on sequence alignments and intron positions, *Arabidopsis ACA10*, *ACA8*, and *ACA9* are clustered together and their encoded proteins are located in the plasma membrane (Baxter *et al.*, 2003). In *Arabidopsis*, ACA9 is necessary for normal pollen tube growth and fertilization (Schiøtt *et al.*, 2004), whereas ACA10 is critical for inflorescence growth and juvenile-to-adult phase transition under long-day growth conditions (George *et al.*, 2008). miR4376 is highly accumulated in leaves and young flower buds but later diminished in young green fruits and almost absent in mature fruits. The most significant impact of miR4376 or *SlACA10* overexpression is the drastically yield-loss of mature fruits (Wang *et al.*, 2011b).

## Conclusions and future perspectives

The discoveries of sRNAs in plants and the growing evidence of their involvement in a variety of functional roles have offered a new insight into plant reproductive biology. Notably, different sRNA families and their different members can make very specific contributions to the spatial and temporal control of their targets. The confirmed sRNAs targets are highly interconnected. The spatial and temporal expression patterns of sRNA and the cross-talks between various sRNA-regulated modules may enables sRNAs to function in different organs or developmental stages. For example, miR156 and miR172 have essential functions in the whole reproductive processes, including androecium, gynoecium, and fruit development. Thus, altering the expression levels of specific plant sRNAs makes it possible to regulate the key transcription factors and entire downstream gene regulatory networks. In practice, some important processes such as plant growth and stature, flowering, seed set and yield are obvious targets for genetic engineering. sRNA networks play a central role in these important processes, which makes them appeal biotechnologies for developing the varieties with improved key agronomic traits. For example, male fertility-related miRNAs have the potential to be utilized in hybrid breeding. In numerous studies, both sRNA targets have been quantitatively and spatially manipulated, e.g. introducing mutations into target genes or specific sRNA loci that fine-tune sRNAs control. In contrast, the mutants or transgenic varieties provide the evidence that sRNA regulate nodulation in regulating many plant reproductive developmental processes.

The sRNA-mediated regulation networks underlying plant reproductive development are much more complicated than initially expected. The high-throughput data, such as transcriptome and degradome, have been widely used to identify miRNA sequences and to validate their target genes in plants. Being assisted by the appropriate bioinformatics tools, people can predict the sRNAs of interests and their targets. However, many sRNA sequences remain uncharacterized in sequencing projects, indicating far more sRNA loci than previously expected. These uncharacterized sRNAs are likely to have a broader range of functions in plant reproductive development than currently known. The large gap between uncharacterized sRNA and the confirmation of their functions awaits more experimental and computational studies to fill in future.

## References


Achard, P., Herr, A., Baulcombe, D.C., and Harberd, N.P. (2004). Modulation of floral development by a gibberellin-regulated microRNA. Development *131*, 3357–3365. http://dx.doi.org/10.1242/dev.01206

Adamczyk, B.J., Lehti-Shiu, M.D., and Fernandez, D.E. (2007). The MADS domain factors AGL15 and AGL18 act redundantly as repressors of the floral transition in *Arabidopsis*. Plant J. *50*, 1007–1019.

Addo-Quaye, C., Eshoo, T.W., Bartel, D.P., and Axtell, M.J. (2008). Endogenous siRNA and miRNA targets identified by sequencing of the *Arabidopsis* degradome. Curr. Biol. *18*, 758–762. http://dx.doi.org/10.1016/j.cub.2008.04.042

Allen, E., Xie, Z., Gustafson, A.M., and Carrington, J.C. (2005). microRNA-directed phasing during trans-acting siRNA biogenesis in plants. Cell *121*, 207–221.

Alonso-Peral, M.M., Li, J., Li, Y., Allen, R.S., Schnippenkoetter, W., Ohms, S., White, R.G., and Millar, A.A. (2010). The microRNA159-regulated GAMYB-like genes inhibit growth and promote programmed cell death in *Arabidopsis*. Plant Physiol. *154*, 757–771. http://dx.doi.org/10.1104/pp.110.160630

Alonso-Simón, A., Neumetzler, L., García-Angulo, P., Encina, A.E., Acebes, J.L., Álvarez, J.M., and Hayashi, T. (2010). Plasticity of xyloglucan composition in bean (*Phaseolus vulgaris*)-cultured cells during habituation and dehabituation to lethal concentrations of dichlobenil. Mol. Plant 3, 603–609.

Amasino, R.M., and Michaels, S.D. (2010). The timing of flowering. Plant Physiol. *154*, 516–520. http://dx.doi.org/10.1104/pp.110.161653

Ambros, V., Bartel, B., Bartel, D.P., Burge, C.B., Carrington, J.C., Chen, X., Dreyfuss, G., Eddy, S.R., Griffiths-Jones, S., Marshall, M., *et al.* (2003). A uniform system for microRNA annotation. RNA 9, 277–279.

An, F.M., and Chan, M.T. (2012). Transcriptome-wide characterization of miRNA-directed and non-miRNA-directed endonucleolytic cleavage using Degradome analysis under low ambient temperature in Phalaenopsis aphrodite subsp. formosana. Plant Cell Physiol. 53, 1737–1750.

An, J., Lai, J., Sajjanhar, A., Lehman, M.L., and Nelson, C.C. (2014). miRPlant: an integrated tool for identification of plant miRNA from RNA sequencing data. BMC Bioinf. *15*, 275. http://dx.doi.org/10.1186/1471-2105-15-275

Andrés, F., and Coupland, G. (2012). The genetic basis of flowering responses to seasonal cues. Nat. Rev. Genet. *13*, 627–639. http://dx.doi.org/10.1038/nrg3291

Aukerman, M.J., and Sakai, H. (2003). Regulation of flowering time and floral organ identity by a MicroRNA and its APETALA2-like target genes. Plant Cell *15*, 2730–2741. http://dx.doi.org/10.1105/tpc.016238

Axtell, M.J. (2013). Classification and comparison of small RNAs from plants. Annu. Rev. Plant Biol. *64*, 137–159. http://dx.doi.org/10.1146/annurev-arplant-050312-120043

Axtell, M.J., and Bartel, D.P. (2005). Antiquity of microRNAs and their targets in land plants. Plant Cell *17*, 1658–1673.

Baker, C.C., Sieber, P., Wellmer, F., and Meyerowitz, E.M. (2005). The early extra petals1 mutant uncovers a role for microRNA miR164c in regulating petal number in *Arabidopsis*. Curr. Biol. *15*, 303–315.

Bäurle, I., and Dean, C. (2006). The timing of developmental transitions in plants. Cell *125*, 655–664.

Baxter, I., Tchieu, J., Sussman, M.R., Boutry, M., Palmgren, M.G., Gribskov, M., Harper, J.F., and Axelsen, K.B. (2003). Genomic comparison of P-type ATPase ion pumps in *Arabidopsis* and rice. Plant Physiol. *132*, 618–628. http://dx.doi.org/10.1104/pp.103.021923

Belli Kullan, J., Lopes Paim Pinto, D., Bertolini, E., Fasoli, M., Zenoni, S., Tornielli, G.B., Pezzotti, M., Meyers, B.C., Farina, L., Pè, M.E., *et al.* (2015). miRVine: a microRNA expression atlas of grapevine based on small RNA sequencing. BMC Genomics *16*, 393. http://dx.doi.org/10.1186/s12864-015-1610-5

Bemer, M., Karlova, R., Ballester, A.R., Tikunov, Y.M., Bovy, A.G., Wolters-Arts, M., Rossetto, P.d.e.B., Angenent, G.C., and de Maagd, R.A. (2012). The tomato FRUITFULL homologs TDR4/FUL1 and MBP7/FUL2 regulate ethylene-independent aspects of fruit ripening. Plant Cell *24*, 4437–4451. http://dx.doi.org/10.1105/tpc.112.103283

Berger, Y., Harpaz-Saad, S., Brand, A., Melnik, H., Sirding, N., Alvarez, J.P., Zinder, M., Samach, A., Eshed, Y., and Ori, N. (2009). The NAC-domain transcription factor GOBLET specifies leaflet boundaries in compound tomato leaves. Development *136*, 823–832.

Bergonzi, S., Albani, M.C., Ver Loren van Themaat, E., Nordström, K.J., Wang, R., Schneeberger, K., Moerland, P.D., and Coupland, G. (2013). Mechanisms of age-dependent response to winter temperature in perennial flowering of Arabis alpina. Science *340*, 1094–1097. http://dx.doi.org/10.1126/science.1234116

Bi, F., Meng, X., Ma, C., and Yi, G. (2015). Identification of miRNAs involved in fruit ripening in Cavendish bananas by deep sequencing. BMC Genomics *16*, 776. http://dx.doi.org/10.1186/s12864-015-1995-1

Blanchard, M.G., and Runkle, E.S. (2006). Temperature during the day, but not during the night, controls flowering of Phalaenopsis orchids. J. Exp. Bot. *57*, 4043–4049.

Borsani, O., Zhu, J., Verslues, P.E., Sunkar, R., and Zhu, J.K. (2005). Endogenous siRNAs derived from a pair of natural cis-antisense transcripts regulate salt tolerance in *Arabidopsis*. Cell *123*, 1279–1291.

Buxdorf, K., Hendelman, A., Stav, R., Lapidot, M., Ori, N., and Arazi, T. (2010). Identification and characterization of a novel miR159 target not related to MYB in tomato. Planta *232*, 1009–1022. http://dx.doi.org/10.1007/s00425-010-1231-9

Cartolano, M., Castillo, R., Efremova, N., Kuckenberg, M., Zethof, J., Gerats, T., Schwarz-Sommer, Z., and Vandenbussche, M. (2007). A conserved microRNA module exerts homeotic control over Petunia hybrida and Antirrhinum majus floral organ identity. Nat. Genet. *39*, 901–905.

Chen, C.J., liu, Q., Zhang, Y.C., Qu, L.H., Chen, Y.Q., and Gautheret, D. (2011a). Genome-wide discovery and analysis of microRNAs and other small RNAs from rice embryogenic callus. RNA. Biol. *8*, 538–547.

Chen, H., Arsovski, A.A., Yu, K., and Wang, A. (2016). Genome-Wide Investigation Using sRNA-Seq, Degradome-Seq and Transcriptome-Seq Reveals Regulatory Networks of microRNAs and Their Target Genes in Soybean during Soybean mosaic virus Infection. PLOS ONE *11*, e0150582.

Chen, M.K., Hsu, W.H., Lee, P.F., Thiruvengadam, M., Chen, H.I., and Yang, C.H. (2011b). The MADS box gene, *FOREVER YOUNG FLOWER*, acts as a repressor controlling floral organ senescence and abscission in *Arabidopsis*. Plant J. *68*, 168–185. http://dx.doi.org/10.1111/j.1365-313X.2011.04677.x

Chen, W.H., Tseng, Y.C., Liu, Y.C., Chuo, C.M., Chen, P.T., Tseng, K.M., Yeh, Y.C., Ger, M.J., and Wang, H.L. (2008). Cool-night temperature induces spike emergence and affects photosynthetic efficiency and metabolizable carbohydrate and organic acid pools in Phalaenopsis aphrodite. Plant Cell Rep. *27*, 1667–1675.

Chen, X. (2004). A microRNA as a translational repressor of APETALA2 in *Arabidopsis* flower development. Science *303*, 2022–2025. http://dx.doi.org/10.1126/science.1088060

Chen, Z., Li, F., Yang, S., Dong, Y., Yuan, Q., Wang, F., Li, W., Jiang, Y., Jia, S., and Pei, X. (2013). Identification and functional analysis of flowering related microRNAs in common wild rice (*Oryza rufipogon* Griff.). PLOS ONE *8*, e82844. http://dx.doi.org/10.1371/journal.pone.0082844

Chen, Z.H., Bao, M.L., Sun, Y.Z., Yang, Y.J., Xu, X.H., Wang, J.H., Han, N., Bian, H.W., and Zhu, M.Y. (2011c). Regulation of auxin response by miR393-targeted transport inhibitor response protein 1 is involved in normal development in *Arabidopsis*. Plant Mol. Biol. *77*, 619–629. http://dx.doi.org/10.1007/s11103-011-9838-1

Cheng, H., Qin, L., Lee, S., Fu, X., Richards, D.E., Cao, D., Luo, D., Harberd, N.P., and Peng, J. (2004). Gibberellin regulates *Arabidopsis* floral development via suppression of DELLA protein function. Development *131*, 1055–1064.

Cho, S.H., Coruh, C., and Axtell, M.J. (2012). miR156 and miR390 regulate tasiRNA accumulation and developmental timing in Physcomitrella patens. Plant Cell *24*, 4837–4849. http://dx.doi.org/10.1105/tpc.112.103176

Chuck, G., Cigan, A.M., Saeteurn, K., and Hake, S. (2007a). The heterochronic maize mutant Corngrass1 results from overexpression of a tandem microRNA. Nat. Genet. 39, 544–549.

Chuck, G., Meeley, R., Irish, E., Sakai, H., and Hake, S. (2007b). The maize tasselseed4 microRNA controls sex determination and meristem cell fate by targeting Tasselseed6/indeterminate spikelet1. Nat. Genet. *39*, 1517–1521.

Chuck, G.S., Tobias, C., Sun, L., Kraemer, F., Li, C., Dibble, D., Arora, R., Bragg, J.N., Vogel, J.P., Singh, S., *et al.* (2011). Overexpression of the maize Corngrass1 microRNA prevents flowering, improves digestibility, and increases starch content of switchgrass. Proc. Natl. Acad. Sci. U. S. A. *108*, 17550–17555.

Cong, B., Barrero, L.S., and Tanksley, S.D. (2008). Regulatory change in YABBY-like transcription factor led to evolution of extreme fruit size during tomato domestication. Nat. Genet. *40*, 800–804. http://dx.doi.org/10.1038/ng.144

Cong, B., and Tanksley, S.D. (2006). FW2.2 and cell cycle control in developing tomato fruit: a possible example of gene co-option in the evolution of a novel organ. Plant Mol. Biol. *62*, 867–880. http://dx.doi.org/10.1007/s11103-006-9062-6

Curaba, J., Talbot, M., Li, Z., and Helliwell, C. (2013). Overexpression of microRNA171 affects phase transitions and floral meristem determinancy in barley. BMC Plant Biol. *13*, 6. http://dx.doi.org/10.1186/1471-2229-13-6

Dai, X., and Zhao, P.X. (2011). psRNATarget: a plant small RNA target analysis server. Nucleic Acids Res. *39*, W155–9. http://dx.doi.org/10.1093/nar/gkr319

Deng, W., Ying, H., Helliwell, C.A., Taylor, J.M., Peacock, W.J., and Dennis, E.S. (2011). FLOWERING LOCUS C (FLC) regulates development pathways throughout the life cycle of *Arabidopsis*. Proc. Natl. Acad. Sci. U.S.A. *108*, 6680–6685. http://dx.doi.org/10.1073/pnas.1103175108

Denli, A.M., and Hannon, G.J. (2003). RNAi: an ever-growing puzzle. Trends Biochem. Sci *28*, 196–201.

Dodd, A.N., Kudla, J., and Sanders, D. (2010). The language of calcium signaling. Annu. Rev. Plant Biol. *61*, 593–620. http://dx.doi.org/10.1146/annurev-arplant-070109-104628

Emde, A.K., Grunert, M., Weese, D., Reinert, K., and Sperling, S.R. (2010). MicroRazerS: rapid alignment of small RNA reads. Bioinformatics *26*, 123–124. http://dx.doi.org/10.1093/bioinformatics/btp601

Fahlgren, N., Howell, M.D., Kasschau, K.D., Chapman, E.J., Sullivan, C.M., Cumbie, J.S., Givan, S.A., Law, T.F., Grant, S.R., Dangl, J.L., *et al.* (2007). High-throughput sequencing of *Arabidopsis* microRNAs: evidence for frequent birth and death of MIRNA genes. PLOS ONE *2*, e219. http://dx.doi.org/10.1371/journal.pone.0000219

Fahlgren, N., Montgomery, T.A., Howell, M.D., Allen, E., Dvorak, S.K., Alexander, A.L., and Carrington, J.C. (2006). Regulation of AUXIN RESPONSE FACTOR3 by TAS3 ta-siRNA affects developmental timing and patterning in *Arabidopsis*. Curr. Biol. *16*, 939–944.

Fahlgren, N., Sullivan, C.M., Kasschau, K.D., Chapman, E.J., Cumbie, J.S., Montgomery, T.A., Gilbert, S.D., Dasenko, M., Backman, T.W., Givan, S.A., *et al.* (2009). Computational and analytical framework for small RNA profiling by high-throughput sequencing. RNA *15*, 992–1002. http://dx.doi.org/10.1261/rna.1473809

Fang, Y.-N., Qiu, W.-M., Wang, Y., Wu, X.-M., Xu, Q., and Guo, W.-W. (2014). Identification of differentially expressed microRNAs from a male sterile Ponkan mandarin (*Citrus reticulata* Blanco) and its fertile wild type by small RNA and degradome sequencing. Tree Genet. Genomes *10*, 1567–1581.

Ferreira e Silva, G.F., Silva, E.M., Azevedo, M.d.a.S., Guivin, M.A., Ramiro, D.A., Figueiredo, C.R., Carrer, H., Peres, L.E., and Nogueira, F.T. (2014). microRNA156-targeted SPL/SBP box transcription factors regulate tomato ovary and fruit development. Plant J. *78*, 604–618. http://dx.doi.org/10.1111/tpj.12493

Floyd, S.K., and Bowman, J.L. (2004). Gene regulation: ancient microRNA target sequences in plants. Nature *428*, 485–486. http://dx.doi.org/10.1038/428485a

Fowler, S., Lee, K., Onouchi, H., Samach, A., Richardson, K., Morris, B., Coupland, G., and Putterill, J. (1999). GIGANTEA: a circadian clock-controlled gene that regulates photoperiodic flowering in *Arabidopsis* and encodes a protein with several possible membrane-spanning domains. EMBO J. *18*, 4679–4688.

Friedländer, M.R., Chen, W., Adamidi, C., Maaskola, J., Einspanier, R., Knespel, S., and Rajewsky, N. (2008). Discovering microRNAs from deep sequencing data using miRDeep. Nat. Biotechnol. *26*, 407–415. http://dx.doi.org/10.1038/nbt1394

Friedländer, M.R., Mackowiak, S.D., Li, N., Chen, W., and Rajewsky, N. (2012). miRDeep2 accurately identifies known and hundreds of novel microRNA genes in seven animal clades. Nucleic Acids Res. *40*, 37–52. http://dx.doi.org/10.1093/nar/gkr688

Fu, C., Sunkar, R., Zhou, C., Shen, H., Zhang, J.Y., Matts, J., Wolf, J., Mann, D.G., Stewart, C.N., Jr., Tang, Y., *et al.* (2012). Overexpression of miR156 in switchgrass (Panicum virgatum L.) results in various morphological alterations and leads to improved biomass production. Plant Biotechnol. J. *10*, 443–452.

Gandikota, M., Birkenbihl, R.P., Höhmann, S., Cardon, G.H., Saedler, H., and Huijser, P. (2007). The miRNA156/157 recognition element in the 3' UTR of the *Arabidopsis* SBP box gene SPL3 prevents early flowering by translational inhibition in seedlings. Plant J. *49*, 683–693.

Ge, C., Georgiev, A., Ohman, A., Wieslander, A., and Kelly, A.A. (2011). Tryptophan residues promote membrane association for a plant lipid glycosyltransferase involved in phosphate stress. J. Biol. Chem. *286*, 6669–6684.

George, L., Romanowsky, S.M., Harper, J.F., and Sharrock, R.A. (2008). The ACA10 Ca2+-ATPase regulates adult vegetative development and inflorescence architecture in *Arabidopsis*. Plant Physiol. *146*, 716–728.

German, M.A., Pillay, M., Jeong, D.H., Hetawal, A., Luo, S., Janardhanan, P., Kannan, V., Rymarquis, L.A., Nobuta, K., German, R., *et al.* (2008). Global identification of microRNA-target RNA pairs by parallel analysis of RNA ends. Nat. Biotechnol. *26*, 941–946. http://dx.doi.org/10.1038/nbt1417

Glazińska, P., Zienkiewicz, A., Wojciechowski, W., and Kopcewicz, J. (2009). The putative miR172 target gene InAPETALA2-like is involved in the photoperiodic flower induction of Ipomoea nil. J. Plant Physiol. *166*, 1801–1813. http://dx.doi.org/10.1016/j.jplph.2009.05.011

Gocal, G.F., Sheldon, C.C., Gubler, F., Moritz, T., Bagnall, D.J., MacMillan, C.P., Li, S.F., Parish, R.W., Dennis, E.S., Weigel, D., *et al.* (2001). GAMYB-like genes, flowering, and gibberellin signaling in *Arabidopsis*. Plant Physiol. *127*, 1682–1693.

Gou, J.Y., Felippes, F.F., Liu, C.J., Weigel, D., and Wang, J.W. (2011). Negative regulation of anthocyanin biosynthesis in *Arabidopsis* by a miR156-targeted SPL transcription factor. Plant Cell *23*, 1512–1522. http://dx.doi.org/10.1105/tpc.111.084525

Griffiths-Jones, S., Saini, H.K., van Dongen, S., and Enright, A.J. (2008). miRBase: tools for microRNA genomics. Nucleic Acids Res. *36*, D154–8.

Gu, Q., Ferrándiz, C., Yanofsky, M.F., and Martienssen, R. (1998). The FRUITFULL MADS-box gene mediates cell differentiation during *Arabidopsis* fruit development. Development *125*, 1509–1517.

Guan, X., Pang, M., Nah, G., Shi, X., Ye, W., Stelly, D.M., and Chen, Z.J. (2014). miR828 and miR858 regulate homoeologous MYB2 gene functions in *Arabidopsis* trichome and cotton fibre development. Nat. Commun. *5*, 3050. http://dx.doi.org/10.1038/ncomms4050

Gubler, F., Chandler, P.M., White, R.G., Llewellyn, D.J., and Jacobsen, J.V. (2002). Gibberellin signaling in barley aleurone cells. Control of SLN1 and GAMYB expression. Plant Physiol. *129*, 191–200. http://dx.doi.org/10.1104/pp.010918

Harberd, N.P., Belfield, E., and Yasumura, Y. (2009). The angiosperm gibberellin-GID1-DELLA growth regulatory mechanism: how an "inhibitor of an inhibitor" enables flexible response to fluctuating environments.. Plant Cell *21*, 1328–1339.

Hendelman, A., Stav, R., Zemach, H., and Arazi, T. (2013). The tomato NAC transcription factor SlNAM2 is involved in flower-boundary morphogenesis. J. Exp. Bot. *64*, 5497–5507. http://dx.doi.org/10.1093/jxb/ert324

Hendrix, D., Levine, M., and Shi, W. (2010). miRTRAP, a computational method for the systematic identification of miRNAs from high throughput sequencing data. Genome Biol. *11*, R39. http://dx.doi.org/10.1186/gb-2010-11-4-r39

Hibara, K., Takada, S., and Tasaka, M. (2003). CUC1 gene activates the expression of SAM-related genes to induce adventitious shoot formation. Plant J. *36*, 687–696.

Hu, J.Y., Zhou, Y., He, F., Dong, X., Liu, L.Y., Coupland, G., Turck, F., and de Meaux, J. (2014). miR824-Regulated AGAMOUS-LIKE16 Contributes to Flowering Time Repression in *Arabidopsis*. Plant Cell *26*, 2024–2037.

Huijser, P., and Schmid, M. (2011). The control of developmental phase transitions in plants. Development *138*, 4117–4129. http://dx.doi.org/10.1242/dev.063511

Hultquist, J.F., and Dorweiler, J.E. (2008). Feminized tassels of maize mop1 and ts1 mutants exhibit altered levels of miR156 and specific SBP-box genes. Planta *229*, 99–113. http://dx.doi.org/10.1007/s00425-008-0813-2

Itaya, A., Bundschuh, R., Archual, A.J., Joung, J.G., Fei, Z., Dai, X., Zhao, P.X., Tang, Y., Nelson, R.S., and Ding, B. (2008). Small RNAs in tomato fruit and leaf development. Biochim. Biophys. Acta *1779*, 99–107.

Jones-Rhoades, M.W., and Bartel, D.P. (2004). Computational identification of plant microRNAs and their targets, including a stress-induced miRNA.

Mol. Cell *14*, 787–799. http://dx.doi.org/10.1016/j.molcel.2004.05.027

José Ripoll, J., Bailey, L.J., Mai, Q.A., Wu, S.L., Hon, C.T., Chapman, E.J., Ditta, G.S., Estelle, M., and Yanofsky, M.F. (2015). microRNA regulation of fruit growth. Nat. Plants *1*, 15036. http://dx.doi.org/10.1038/nplants.2015.36

Jung, J.H., Seo, Y.H., Seo, P.J., Reyes, J.L., Yun, J., Chua, N.H., and Park, C.M. (2007). The GIGANTEA-regulated microRNA172 mediates photoperiodic flowering independent of CONSTANS in *Arabidopsis*. Plant Cell *19*, 2736–2748.

Kamiuchi, Y., Yamamoto, K., Furutani, M., Tasaka, M., and Aida, M. (2014). The CUC1 and CUC2 genes promote carpel margin meristem formation during *Arabidopsis* gynoecium development. Front. Plant Sci. *5*, 165.

Karlova, R., Chapman, N., David, K., Angenent, G.C., Seymour, G.B., and de Maagd, R.A. (2014). Transcriptional control of fleshy fruit development and ripening. J. Exp. Bot. *65*, 4527–4541. http://dx.doi.org/10.1093/jxb/eru316

Karlova, R., Rosin, F.M., Busscher-Lange, J., Parapunova, V., Do, P.T., Fernie, A.R., Fraser, P.D., Baxter, C., Angenent, G.C., and de Maagd, R.A. (2011). Transcriptome and metabolite profiling show that APETALA2a is a major regulator of tomato fruit ripening. Plant Cell *23*, 923–941. http://dx.doi.org/10.1105/tpc.110.081273

Karlova, R., van Haarst, J.C., Maliepaard, C., van de Geest, H., Bovy, A.G., Lammers, M., Angenent, G.C., and de Maagd, R.A. (2013). Identification of microRNA targets in tomato fruit development using high-throughput sequencing and degradome analysis. J. Exp. Bot. *64*, 1863–1878. http://dx.doi.org/10.1093/jxb/ert049

Katiyar-Agarwal, S., Morgan, R., Dahlbeck, D., Borsani, O., Villegas, A., Zhu, J.K., Staskawicz, B.J., and Jin, H. (2006). A pathogen-inducible endogenous siRNA in plant immunity. Proc. Natl. Acad. Sci. U.S.A. *103*, 18002–18007.

Khaldun, A.B., Huang, W., Liao, S., Lv, H., and Wang, Y. (2015). Identification of microRNAs and target genes in the fruit and shoot tip of Lycium chinense: a traditional Chinese medicinal plant. PLOS ONE *10*, e0116334. http://dx.doi.org/10.1371/journal.pone.0116334

Khraiwesh, B., Zhu, J.K., and Zhu, J. (2012). Role of miRNAs and siRNAs in biotic and abiotic stress responses of plants. Biochim. Biophys. Acta *1819*, 137–148. http://dx.doi.org/10.1016/j.bbagrm.2011.05.001

Kim, J., Jung, J.H., Reyes, J.L., Kim, Y.S., Kim, S.Y., Chung, K.S., Kim, J.A., Lee, M., Lee, Y., Narry Kim, V., *et al.* (2005). microRNA-directed cleavage of ATHB15 mRNA regulates vascular development in *Arabidopsis* inflorescence stems. Plant J. *42*, 84–94.

Kim, J., Park, J.H., Lim, C.J., Lim, J.Y., Ryu, J.Y., Lee, B.W., Choi, J.P., Kim, W.B., Lee, H.Y., Choi, Y., *et al.* (2012a). Small RNA and transcriptome deep sequencing proffers insight into floral gene regulation in Rosa cultivars. BMC Genomics *13*, 657. http://dx.doi.org/10.1186/1471-2164-13-657

Kim, J.J., Lee, J.H., Kim, W., Jung, H.S., Huijser, P., and Ahn, J.H. (2012b). The microRNA156-SQUAMOSA PROMOTER BINDING PROTEIN-LIKE3 module regulates ambient temperature-responsive flowering via FLOWERING LOCUS T in *Arabidopsis*. Plant Physiol. *159*, 461–478. http://dx.doi.org/10.1104/pp.111.192369.

Kim, W., Ahn, H.J., Chiou, T.J., and Ahn, J.H. (2011). The role of the miR399-PHO2 module in the regulation of flowering time in response to different ambient temperatures in *Arabidopsis* thaliana. Mol. Cells. *32*, 83–88. http://dx.doi.org/10.1007/s10059-011-1043-1

Kozomara, A., and Griffiths-Jones, S. (2014). miRBase: annotating high confidence microRNAs using deep sequencing data. Nucleic Acids Res. *42*, D68–73. http://dx.doi.org/10.1093/nar/gkt1181

Krizek, B.A., and Fletcher, J.C. (2005). Molecular mechanisms of flower development: an armchair guide. Nat. Rev. Genet. *6*, 688–698.

Kudla, J., Batistic, O., and Hashimoto, K. (2010). Calcium signals: the lead currency of plant information processing. Plant Cell *22*, 541–563. http://dx.doi.org/10.1105/tpc.109.072686

Lal, S., Pacis, L.B., and Smith, H.M. (2011). Regulation of the SQUAMOSA PROMOTER-BINDING PROTEIN-LIKE genes/microRNA156 module by the homeodomain proteins PENNYWISE and POUND-FOOLISH in *Arabidopsis*. Mol. Plant *4*, 1123–1132.

Larkin, J.C., Oppenheimer, D.G., Pollock, S., and Marks, M.D. (1993). *Arabidopsis* GLABROUS1 Gene Requires Downstream Sequences for Function. Plant Cell *5*, 1739–1748. http://dx.doi.org/10.1105/tpc.5.12.1739

Laufs, P., Peaucelle, A., Morin, H., and Traas, J. (2004). MicroRNA regulation of the CUC genes is required for boundary size control in *Arabidopsis* meristems. Development *131*, 4311–4322. http://dx.doi.org/10.1242/dev.01320

Lauter, N., Kampani, A., Carlson, S., Goebel, M., and Moose, S.P. (2005). microRNA172 down-regulates glossy15 to promote vegetative phase change in maize. Proc. Natl. Acad. Sci. U.S.A. *102*, 9412–9417.

Lee, H., Yoo, S.J., Lee, J.H., Kim, W., Yoo, S.K., Fitzgerald, H., Carrington, J.C., and Ahn, J.H. (2010). Genetic framework for flowering-time regulation by ambient temperature-responsive miRNAs in *Arabidopsis*. Nucleic. Acids Res. *38*, 3081–3093. http://dx.doi.org/10.1093/nar/gkp1240

Lee, J., and Lee, I. (2010). Regulation and function of SOC1, a flowering pathway integrator. J. Exp. Bot. *61*, 2247–2254. http://dx.doi.org/10.1093/jxb/erq098

Lee, R., Baldwin, S., Kenel, F., McCallum, J., and Macknight, R. (2013). FLOWERING LOCUS T genes control onion bulb formation and flowering. Nat. Commun. *4*, 2884. http://dx.doi.org/10.1038/ncomms3884

Li, C., and Zhang, B. (2016). MicroRNAs in Control of Plant Development. J. Cell. Physiol. *231*, 303–313. http://dx.doi.org/10.1002/jcp.25125

Li, D., Liu, Z., Gao, L., Wang, L., Gao, M., Jiao, Z., Qiao, H., Yang, J., Chen, M., Yao, L., *et al.* (2016a). Genome-Wide Identification and Characterization of microRNAs in Developing Grains of Zea mays L. PLOS ONE *11*, e0153168. http://dx.doi.org/10.1371/journal.pone.0153168

Li, X., Bian, H., Song, D., Ma, S., Han, N., Wang, J., and Zhu, M. (2013a). Flowering time control in ornamental gloxinia (Sinningia speciosa) by manipulation of miR159 expression. Ann. Bot. *111*, 791–799. http://dx.doi.org/10.1093/aob/mct034

Li, X., Jackson, A., Xie, M., Wu, D., Tsai, W.C., and Zhang, S. (2016b). Proteomic insights into floral biology. Biochim. Biophys. Acta *1864*, 1050–1060. http://dx.doi.org/10.1016/j.bbapap.2016.02.023

Li, X., Jin, F., Jin, L., Jackson, A., Ma, X., Shu, X., Wu, D., and Jin, G. (2015a). Characterization and comparative profiling of the small RNA transcriptomes in two phases of flowering in Cymbidium ensifolium. BMC Genomics *16*, 622. http://dx.doi.org/10.1186/s12864-015-1764-1

Li, X., Luo, J., Yan, T., Xiang, L., Jin, F., Qin, D., Sun, C., and Xie, M. (2013b). Deep sequencing-based analysis of the *Cymbidium ensifolium* floral transcriptome. PLOS ONE *8*, e85480. http://dx.doi.org/10.1371/journal.pone.0085480

Li, Y., Zhang, Z., Liu, F., Vongsangnak, W., Jing, Q., and Shen, B. (2012). Performance comparison and evaluation of software tools for microRNA deep-sequencing data analysis. Nucleic. Acids Res. *40*, 4298–4305. http://dx.doi.org/10.1093/nar/gks043

Li, Z.F., Zhang, Y.C., and Chen, Y.Q. (2015b). miRNAs and lncRNAs in reproductive development. Plant Sci. *238*, 46–52. http://dx.doi.org/10.1016/j.plantsci.2015.05.017

Lian, H., Li, X., Liu, Z., and He, Y. (2013). HYL1 is required for establishment of stamen architecture with four microsporangia in *Arabidopsis*. J. Exp. Bot. *64*, 3397–3410. http://dx.doi.org/10.1093/jxb/ert178

Liang, G., He, H., Li, Y., Wang, F., and Yu, D. (2014). Molecular mechanism of microRNA396 mediating pistil development in *Arabidopsis*. Plant Physiol. *164*, 249–258. http://dx.doi.org/10.1104/pp.113.225144

Liu, B., Li, J., and Cairns, M.J. (2014a). Identifying miRNAs, targets and functions. Briefings Bioinf. *15*, 1–19. http://dx.doi.org/10.1093/bib/bbs075

Liu, D., Song, Y., Chen, Z., and Yu, D. (2009). Ectopic expression of miR396 suppresses GRF target gene expression and alters leaf growth in *Arabidopsis*. Physiol. Plant *136*, 223–236. http://dx.doi.org/10.1111/j.1399-3054.2009.01229.x

Liu, H., Guo, S., Xu, Y., Li, C., Zhang, Z., Zhang, D., Xu, S., Zhang, C., and Chong, K. (2014b). OsmiR396d-regulated OsGRFs function in floral organogenesis in rice through binding to their targets OsJMJ706 and OsCR4. Plant Physiol. *165*, 160–174. http://dx.doi.org/10.1104/pp.114.235564

Liu, J., Van Eck, J., Cong, B., and Tanksley, S.D. (2002). A new class of regulatory genes underlying the cause of pear-shaped tomato fruit. Proc. Natl. Acad. Sci. U.S.A. *99*, 13302–13306. http://dx.doi.org/10.1073/pnas.162485999

Liu, L., Li, Y., Li, S., Hu, N., He, Y., Pong, R., Lin, D., Lu, L., and Law, M. (2012). Comparison of next-generation sequencing systems. J. Biomed. Biotechnol. *2012*, 251364. http://dx.doi.org/10.1155/2012/251364

Liu, N., Wu, S., Van Houten, J., Wang, Y., Ding, B., Fei, Z., Clarke, T.H., Reed, J.W., and van der Knaap, E. (2014c). Down-regulation of AUXIN RESPONSE FACTORS 6 and 8 by microRNA 167 leads to floral development defects and female sterility in tomato. J. Exp. Bot. *65*, 2507–2520. http://dx.doi.org/10.1093/jxb/eru141

Liu, N., Tu, L., Tang, W., Gao, W., Lindsey, K., and Zhang, X. (2014d). Small RNA and degradome profiling reveals a role for miRNAs and their targets in the developing fibers of Gossypium barbadense. Plant J. *80*, 331–344. http://dx.doi.org/10.1111/tpj.12636

Llave, C., Kasschau, K.D., Rector, M.A., and Carrington, J.C. (2002). Endogenous and silencing-associated small RNAs in plants. Plant Cell *14*, 1605–1619.

Loman, N.J., Misra, R.V., Dallman, T.J., Constantinidou, C., Gharbia, S.E., Wain, J., and Pallen, M.J. (2012). Performance comparison of benchtop high-throughput sequencing platforms. Nat. Biotechnol. *30*, 434–439. http://dx.doi.org/10.1038/nbt.2198

Lopez-Gomollon, S., Mohorianu, I., Szittya, G., Moulton, V., and Dalmay, T. (2012). Diverse correlation patterns between microRNAs and their targets during tomato fruit development indicates different modes of microRNA actions. Planta *236*, 1875–1887.

Lu, C., Tej, S.S., Luo, S., Haudenschild, C.D., Meyers, B.C., and Green, P.J. (2005). Elucidation of the small RNA component of the transcriptome. Science *309*, 1567–1569.

Luo, Y., Guo, Z., and Li, L. (2013). Evolutionary conservation of microRNA regulatory programs in plant flower development. Dev. Biol. *380*, 133–144. http://dx.doi.org/10.1016/j.ydbio.2013.05.009

Manning, K., Tör, M., Poole, M., Hong, Y., Thompson, A.J., King, G.J., Giovannoni, J.J., and Seymour, G.B. (2006). A naturally occurring epigenetic mutation in a gene encoding an SBP-box transcription factor inhibits tomato fruit ripening. Nat. Genet. *38*, 948–952.

Mao, W., Li, Z., Xia, X., Li, Y., and Yu, J. (2012). A combined approach of high-throughput sequencing and degradome analysis reveals tissue specific expression of microRNAs and their targets in cucumber. PLOS ONE *7*, e33040. http://dx.doi.org/10.1371/journal.pone.0033040

Martínez-Laborda, A., and Vera, A. (2009). *Arabidopsis* fruit development. In Fruit Development and Seed Dispersal, L. Ostergaard, ed. (West Sussex, United Kingdom: Blackwell Publishing Ltd.), pp. 172–203.

Martin, A., Adam, H., Díaz-Mendoza, M., Zurczak, M., González-Schain, N.D., and Suárez-López, P. (2009). Graft-transmissible induction of potato tuberization by the microRNA miR172. Development *136*, 2873–2881. http://dx.doi.org/10.1242/dev.031658

Mathelier, A., and Carbone, A. (2010). MIReNA: finding microRNAs with high accuracy and no learning at genome scale and from deep sequencing data. Bioinformatics *26*, 2226–2234. http://dx.doi.org/10.1093/bioinformatics/btq329

Mathieu, J., Yant, L.J., Mürdter, F., Küttner, F., and Schmid, M. (2009). Repression of flowering by the miR172 target SMZ. PLOS Biol. *7*, e1000148. http://dx.doi.org/10.1371/journal.pbio.1000148

Mayer, K.F., Schoof, H., Haecker, A., Lenhard, M., Jürgens, G., and Laux, T. (1998). Role of WUSCHEL in regulating stem cell fate in the *Arabidopsis* shoot meristem. Cell *95*, 805–815.

McAinsh, M.R., and Pittman, J.K. (2009). Shaping the calcium signature. New. Phytol. *181*, 275–294. http://dx.doi.org/10.1111/j.1469-8137.2008.02682.x

Metzker, M.L. (2010). Sequencing technologies - the next generation. Nat. Rev. Genet. *11*, 31–46. http://dx.doi.org/10.1038/nrg2626

Mhuantong, W., and Wichadakul, D. (2009). MicroPC (microPC): A comprehensive resource for predicting and comparing plant microRNAs. BMC Genomics *10*, 366. http://dx.doi.org/10.1186/1471-2164-10-366

Millar, A.A., and Gubler, F. (2005). The *Arabidopsis* GAMYB-like genes, MYB33 and MYB65, are microRNA-regulated genes that redundantly facilitate anther development. Plant Cell *17*, 705–721.

Mohorianu, I., Schwach, F., Jing, R., Lopez-Gomollon, S., Moxon, S., Szittya, G., Sorefan, K., Moulton, V., and Dalmay, T. (2011). Profiling of short RNAs during fleshy fruit development reveals stage-specific sRNAome expression patterns. Plant J. *67*, 232–246.

Moxon, S., Jing, R., Szittya, G., Schwach, F., Rusholme Pilcher, R.L., Moulton, V., and Dalmay, T. (2008a). Deep sequencing of tomato short RNAs identifies microRNAs targeting genes involved in fruit ripening. Genome Res. *18*, 1602–1609. http://dx.doi.org/10.1101/gr.080127.108

Moxon, S., Schwach, F., Dalmay, T., Maclean, D., Studholme, D.J., and Moulton, V. (2008b). A toolkit for analysing large-scale plant small RNA datasets. Bioinformatics *24*, 2252–2253. http://dx.doi.org/10.1093/bioinformatics/btn428

Murray, F., Kalla, R., Jacobsen, J., and Gubler, F. (2003). A role for HvGAMYB in anther development. Plant J. 33, 481–491.

Nagasaki, H., Itoh, J., Hayashi, K., Hibara, K., Satoh-Nagasawa, N., Nosaka, M., Mukouhata, M., Ashikari, M., Kitano, H., Matsuoka, M., *et al.* (2007). The small interfering RNA production pathway is required for shoot meristem initiation in rice. Proc. Natl. Acad. Sci. U.S.A. *104*, 14867–14871.

Nair, S.K., Wang, N., Turuspekov, Y., Pourkheirandish, M., Sinsuwongwat, S., Chen, G., Sameri, M., Tagiri, A., Honda, I., Watanabe, Y., *et al.* (2010). Cleistogamous flowering in barley arises from the suppression of microRNA-guided HvAP2 mRNA cleavage. Proc. Natl. Acad. Sci. U.S.A. *107*, 490–495. http://dx.doi.org/10.1073/pnas.0909097107

Nakano, T., Kimbara, J., Fujisawa, M., Kitagawa, M., Ihashi, N., Maeda, H., Kasumi, T., and Ito, Y. (2012). MACROCALYX and JOINTLESS interact in the transcriptional regulation of tomato fruit abscission zone development. Plant Physiol. *158*, 439–450. http://dx.doi.org/10.1104/pp.111.183731

Nie, S.S., Xu, L., Wang, Y., Huang, D.Q., Muleke, E.M., Sun, X.C., Wang, R.H., Xie, Y., Gong, Y.Q., and Liu, L.W. (2015). Identification of bolting-related microRNAs and their targets reveals complex miRNA-mediated flowering-time regulatory networks in radish (Raphanus sativus L.). Sci. Rep. 5, 14034.

Nikovics, K., Blein, T., Peaucelle, A., Ishida, T., Morin, H., Aida, M., and Laufs, P. (2006). The balance between the MIR164A and CUC2 genes controls leaf margin serration in *Arabidopsis*. Plant Cell *18*, 2929–2945.

Nodine, M.D., and Bartel, D.P. (2010). MicroRNAs prevent precocious gene expression and enable pattern formation during plant embryogenesis. Genes Dev. *24*, 2678–2692.

Nogueira, F.T., Madi, S., Chitwood, D.H., Juarez, M.T., and Timmermans, M.C. (2007). Two small regulatory RNAs establish opposing fates of a developmental axis. Genes. Dev. *21*, 750–755.

Pabón-Mora, N., Ambrose, B.A., and Litt, A. (2012). Poppy APETALA1/FRUITFULL orthologs control flowering time, branching, perianth identity, and fruit development. Plant Physiol. *158*, 1685–1704. http://dx.doi.org/10.1104/pp.111.192104

Palatnik, J.F., Allen, E., Wu, X., Schommer, C., Schwab, R., Carrington, J.C., and Weigel, D. (2003). Control of leaf morphogenesis by microRNAs. Nature *425*, 257–263. http://dx.doi.org/10.1038/nature01958

Pang, M., Woodward, A.W., Agarwal, V., Guan, X., Ha, M., Ramachandran, V., Chen, X., Triplett, B.A., Stelly, D.M., and Chen, Z.J. (2009). Genome-wide analysis reveals rapid and dynamic changes in miRNA and siRNA sequence and expression during ovule and fiber development in allotetraploid cotton (Gossypium hirsutum L.). Genome Biol. *10*, R122.

Pantaleo, V., Szittya, G., Moxon, S., Miozzi, L., Moulton, V., Dalmay, T., and Burgyan, J. (2010). Identification of grapevine microRNAs and their targets using high-throughput sequencing and degradome analysis. Plant J. *62*, 960–976. http://dx.doi.org/10.1111/j.0960-7412.2010.04208.x

Pantano, L., Estivill, X., and Marti, E. (2010). SeqBuster, a bioinformatic tool for the processing and analysis of small RNAs datasets, reveals ubiquitous miRNA modifications in human embryonic cells. Nucleic. Acids Res. 38, e34.

Park, D.H., Somers, D.E., Kim, Y.S., Choy, Y.H., Lim, H.K., Soh, M.S., Kim, H.J., Kay, S.A., and Nam, H.G. (1999). Control of circadian rhythms and photoperiodic flowering by the *Arabidopsis* GIGANTEA gene. Science *285*, 1579–1582.

Peng, T., Sun, H., Qiao, M., Zhao, Y., Du, Y., Zhang, J., Li, J., Tang, G., and Zhao, Q. (2014). Differentially expressed microRNA cohorts in seed development may contribute to poor grain filling of inferior spikelets in rice. BMC Plant Biol. *14*, 196. http://dx.doi.org/10.1186/s12870-014-0196-4

Preston, J.C., and Hileman, L.C. (2010). SQUAMOSA-PROMOTER BINDING PROTEIN 1 initiates flowering in Antirrhinum majus through the activation of meristem identity genes. Plant J. *62*, 704–712.

Rajagopalan, R., Vaucheret, H., Trejo, J., and Bartel, D.P. (2006). A diverse and evolutionarily fluid set of microRNAs in *Arabidopsis* thaliana. Genes. Dev. *20*, 3407–3425.

Reinhart, B.J., and Bartel, D.P. (2002). Small RNAs correspond to centromere heterochromatic repeats. Science *297*, 1831. http://dx.doi.org/10.1126/science.1077183

Reinhart, B.J., Weinstein, E.G., Rhoades, M.W., Bartel, B., and Bartel, D.P. (2002). MicroRNAs in plants. Genes. Dev. *16*, 1616–1626. http://dx.doi.org/10.1101/gad.1004402

Ren, Z., Li, Z., Miao, Q., Yang, Y., Deng, W., and Hao, Y. (2011). The auxin receptor homologue in Solanum lycopersicum stimulates tomato fruit set and leaf morphogenesis. J. Exp. Bot. *62*, 2815–2826. http://dx.doi.org/10.1093/jxb/erq455

Reyes, J.L., and Chua, N.H. (2007). ABA induction of miR159 controls transcript levels of two MYB factors during *Arabidopsis* seed germination. Plant J. *49*, 592–606.

Rhoades, M.W., Reinhart, B.J., Lim, L.P., Burge, C.B., Bartel, B., and Bartel, D.P. (2002). Prediction of plant microRNA targets. Cell *110*, 513–520.

Rodríguez, G.R., Muños, S., Anderson, C., Sim, S.C., Michel, A., Causse, M., Gardener, B.B., Francis, D., and van der Knaap, E. (2011). Distribution of SUN, OVATE, LC, and FAS in the tomato germplasm and the relationship to fruit shape diversity. Plant Physiol. *156*, 275–285. http://dx.doi.org/10.1104/pp.110.167577

Ron, M., Alandete Saez, M., Eshed Williams, L., Fletcher, J.C., and McCormick, S. (2010). Proper regulation of a sperm-specific cis-nat-siRNA is essential for double fertilization in *Arabidopsis*. Genes. Dev. *24*, 1010–1021. http://dx.doi.org/10.1101/gad.1882810

Ronen, R., Gan, I., Modai, S., Sukacheov, A., Dror, G., Halperin, E., and Shomron, N. (2010). miRNAkey: a software for microRNA deep sequencing analysis. Bioinformatics *26*, 2615–2616. http://dx.doi.org/10.1093/bioinformatics/btq493

Ru, P., Xu, L., Ma, H., and Huang, H. (2006). Plant fertility defects induced by the enhanced expression of microRNA167. Cell Res. *16*, 457–465.

Salinas, M., Xing, S., Hohmann, S., Berndtgen, R., and Huijser, P. (2012). Genomic organization, phylogenetic comparison and differential expression of the SBP-box family of transcription factors in tomato. Planta *235*, 1171–1184.

Salvi, S., Sponza, G., Morgante, M., Tomes, D., Niu, X., Fengler, K.A., Meeley, R., Ananiev, E.V., Svitashev, S., Bruggemann, E., *et al.* (2007). Conserved noncoding genomic sequences associated with a flowering-time quantitative trait locus in maize. Proc. Natl. Acad. Sci. U.S.A. *104*, 11376–11381.

Schiøtt, M., Romanowsky, S.M., Baekgaard, L., Jakobsen, M.K., Palmgren, M.G., and Harper, J.F. (2004). A plant plasma membrane Ca2+ pump is required for normal pollen tube growth and fertilization. Proc. Natl. Acad. Sci. U.S.A. *101*, 9502–9507. http://dx.doi.org/10.1073/pnas.0401542101

Schmid, M., Uhlenhaut, N.H., Godard, F., Demar, M., Bressan, R., Weigel, D., and Lohmann, J.U. (2003). Dissection of floral induction pathways using global expression analysis. Development *130*, 6001–6012. http://dx.doi.org/10.1242/dev.00842

Schwab, R., Palatnik, J.F., Riester, M., Schommer, C., Schmid, M., and Weigel, D. (2005). Specific effects of microRNAs on the plant transcriptome. Dev. Cell *8*, 517–527.

Schwarz, S., Grande, A.V., Bujdoso, N., Saedler, H., and Huijser, P. (2008). The microRNA regulated SBP-box genes SPL9 and SPL15 control shoot maturation in *Arabidopsis*. Plant Mol. Biol. 67, 183–195. http://dx.doi.org/10.1007/s11103-008-9310-z

Seggerson, K., Tang, L., and Moss, E.G. (2002). Two genetic circuits repress the Caenorhabditis elegans heterochronic gene lin-28 after translation initiation. Dev. Biol. *243*, 215–225. http://dx.doi.org/10.1006/dbio.2001.0563

Severin, J., Waterhouse, A.M., Kawaji, H., Lassmann, T., van Nimwegen, E., Balwierz, P.J., de Hoon, M.J., Hume, D.A., Carninci, P., Hayashizaki, Y., *et al.* (2009). FANTOM4 EdgeExpressDB: an integrated database of promoters, genes, microRNAs, expression dynamics and regulatory interactions. Genome Biol. *10*, R39.

Seymour, G.B., Østergaard, L., Chapman, N.H., Knapp, S., and Martin, C. (2013). Fruit development and ripening. Annu. Rev. Plant Biol. *64*, 219–241. http://dx.doi.org/10.1146/annurev-arplant-050312-120057

Shen, Y., Zhang, Z., Lin, H., Liu, H., Chen, J., Peng, H., Cao, M., Rong, T., and Pan, G. (2011). Cytoplasmic male sterility-regulated novel microRNAs from maize. Funct. Integr. Genomics. *11*, 179–191. http://dx.doi.org/10.1007/s10142-010-0202-3

Shendure, J., and Ji, H. (2008). Next-generation DNA sequencing. Nat. Biotechnol. *26*, 1135–1145. http://dx.doi.org/10.1038/nbt1486

Shikata, M., Yamaguchi, H., Sasaki, K., and Ohtsubo, N. (2012). Overexpression of *Arabidopsis* miR157b induces bushy architecture and delayed phase transition in Torenia fournieri. Planta *236*, 1027–1035. http://dx.doi.org/10.1007/s00425-012-1649-3

Sieber, P., Wellmer, F., Gheyselinck, J., Riechmann, J.L., and Meyerowitz, E.M. (2007). Redundancy and specialization among plant microRNAs: role of the MIR164 family in developmental robustness. Development *134*, 1051–1060.

Song, C., Fang, J., Wang, C., Guo, L., Nicholas, K.K., and Ma, Z. (2010). MiR-RACE, a new efficient approach to determine the precise sequences of computationally identified trifoliate orange (Poncirus trifoliata) microRNAs. PLOS ONE *5*, e10861. Srikanth, A., and Schmid, M. (2011). Regulation of flowering time: all roads lead to Rome. Cell. Mol. Life Sci. *68*, 2013–2037. http://dx.doi.org/10.1007/s00018-011-0673-y

Stocks, M.B., Moxon, S., Mapleson, D., Woolfenden, H.C., Mohorianu, I., Folkes, L., Schwach, F., Dalmay, T., and Moulton, V. (2012). The UEA sRNA workbench: a suite of tools for analysing and visualizing next generation sequencing microRNA and small RNA datasets. Bioinformatics *28*, 2059–2061. http://dx.doi.org/10.1093/bioinformatics/bts311

Sun, Z., He, Y., Li, J., Wang, X., and Chen, J. (2015). Genome-wide characterization of rice black streaked dwarf virus-responsive microRNAs in rice leaves and roots by small RNA and degradome sequencing. Plant Cell Physiol. *56*, 688–699. http://dx.doi.org/10.1093/pcp/pcu213

Sunkar, R., Girke, T., Jain, P.K., and Zhu, J.K. (2005). Cloning and characterization of microRNAs from rice. Plant Cell *17*, 1397–1411.

Sunkar, R., and Zhu, J.K. (2004). Novel and stress-regulated microRNAs and other small RNAs from *Arabidopsis*. Plant Cell *16*, 2001–2019. http://dx.doi.org/10.1105/tpc.104.022830

Tabata, R., Ikezaki, M., Fujibe, T., Aida, M., Tian, C.E., Ueno, Y., Yamamoto, K.T., Machida, Y., Nakamura, K., and Ishiguro, S. (2010). *Arabidopsis* auxin response factor6 and 8 regulate jasmonic acid biosynthesis and floral organ development via repression of class 1 KNOX genes. Plant Cell Physiol. *51*, 164–175. http://dx.doi.org/10.1093/pcp/pcp176

Talmor-Neiman, M., Stav, R., Klipcan, L., Buxdorf, K., Baulcombe, D.C., and Arazi, T. (2006). Identification of trans-acting siRNAs in moss and an RNA-dependent

RNA polymerase required for their biogenesis. Plant J. *48*, 511–521.
Tang, G., Reinhart, B.J., Bartel, D.P., and Zamore, P.D. (2003). A biochemical framework for RNA silencing in plants. Genes. Dev. *17*, 49–63. http://dx.doi.org/10.1101/gad.1048103
Tang, Z., Zhang, L., Xu, C., Yuan, S., Zhang, F., Zheng, Y., and Zhao, C. (2012). Uncovering small RNA-mediated responses to cold stress in a wheat thermosensitive genic male-sterile line by deep sequencing. Plant Physiol. *159*, 721–738. http://dx.doi.org/10.1104/pp.112.196048
Theissen, G. (2001). Development of floral organ identity: stories from the MADS house. Curr. Opin. Plant Biol. *4*, 75–85.
Tsuji, H., Aya, K., Ueguchi-Tanaka, M., Shimada, Y., Nakazono, M., Watanabe, R., Nishizawa, N.K., Gomi, K., Shimada, A., Kitano, H., *et al.* (2006). GAMYB controls different sets of genes and is differentially regulated by microRNA in aleurone cells and anthers. Plant J. *47*, 427–444.
Twell, D. (2011). Male gametogenesis and germline specification in flowering plants. Sex. Plant Reprod. *24*, 149–160. http://dx.doi.org/10.1007/s00497-010-0157-5
Vanstraelen, M., and Benková, E. (2012). Hormonal interactions in the regulation of plant development. Annu. Rev. Cell Dev. Biol. *28*, 463–487. http://dx.doi.org/10.1146/annurev-cellbio-101011-155741
Vaucheret, H. (2006). Post-transcriptional small RNA pathways in plants: mechanisms and regulations. Genes. Dev. *20*, 759–771.
Volpe, T.A., Kidner, C., Hall, I.M., Teng, G., Grewal, S.I., and Martienssen, R.A. (2002). Regulation of heterochromatic silencing and histone H3 lysine-9 methylation by RNAi. Science *297*, 1833–1837. http://dx.doi.org/10.1126/science.1074973
Vrebalov, J., Ruezinsky, D., Padmanabhan, V., White, R., Medrano, D., Drake, R., Schuch, W., and Giovannoni, J. (2002). A MADS-box gene necessary for fruit ripening at the tomato ripening-inhibitor (rin) locus. Science *296*, 343–346. http://dx.doi.org/10.1126/science.1068181
Walford, S.A., Wu, Y., Llewellyn, D.J., and Dennis, E.S. (2011). GhMYB25-like: a key factor in early cotton fibre development. Plant J. *65*, 785–797. http://dx.doi.org/10.1111/j.1365-313X.2010.04464.x
Wan, Q., Zhang, H., Ye, W.X., Wu, H.T., and Zhang, T.Z. (2014). Genome-wide transcriptome profiling revealed cotton fuzz fiber development having a similar molecular model as *Arabidopsis* trichome. PLOS ONE 9, e97313.
Wang, J.W., Czech, B., and Weigel, D. (2009a). miR156-regulated SPL transcription factors define an endogenous flowering pathway in *Arabidopsis thaliana*. Cell *138*, 738–749. http://dx.doi.org/10.1016/j.cell.2009.06.014
Wang, J.W., Park, M.Y., Wang, L.J., Koo, Y., Chen, X.Y., Weigel, D., and Poethig, R.S. (2011a). miRNA control of vegetative phase change in trees. PLOS Genet. *7*, e1002012. http://dx.doi.org/10.1371/journal.pgen.1002012
Wang, J.W., Schwab, R., Czech, B., Mica, E., and Weigel, D. (2008). Dual effects of miR156-targeted SPL genes and CYP78A5/KLUH on plastochron length and organ size in *Arabidopsis* thaliana. Plant Cell *20*, 1231–1243. http://dx.doi.org/10.1105/tpc.108.058180
Wang, S., Wang, J.W., Yu, N., Li, C.H., Luo, B., Gou, J.Y., Wang, L.J., and Chen, X.Y. (2004). Control of plant trichome development by a cotton fiber MYB gene. Plant Cell *16*, 2323–2334. http://dx.doi.org/10.1105/tpc.104.024844
Wang, W., Li, G., Zhao, J., Chu, H., Lin, W., Zhang, D., Wang, Z., and Liang, W. (2014a). Dwarf Tiller1, a Wuschel-related homeobox transcription factor, is required for tiller growth in rice. PLOS Genet. *10*, e1004154. http://dx.doi.org/10.1371/journal.pgen.1004154
Wang, W.C., Lin, F.M., Chang, W.C., Lin, K.Y., Huang, H.D., and Lin, N.S. (2009b). miRExpress: analyzing high-throughput sequencing data for profiling microRNA expression. BMC Bioinf. *10*, 328. http://dx.doi.org/10.1186/1471-2105-10-328
Wang, Y., Itaya, A., Zhong, X., Wu, Y., Zhang, J., van der Knaap, E., Olmstead, R., Qi, Y., and Ding, B. (2011b). Function and evolution of a MicroRNA that regulates a Ca2+-ATPase and triggers the formation of phased small interfering RNAs in tomato reproductive growth. Plant Cell *23*, 3185–3203. http://dx.doi.org/10.1105/tpc.111.088013
Wang, Y., Sun, F., Cao, H., Peng, H., Ni, Z., Sun, Q., and Yao, Y. (2012a). TamiR159 directed wheat TaGAMYB cleavage and its involvement in anther development and heat response. PLOS ONE *7*, e48445. http://dx.doi.org/10.1371/journal.pone.0048445
Wang, Y., Wu, F., Bai, J., and He, Y. (2014b). BrpSPL9 (Brassica rapa ssp. pekinensis SPL9) controls the earliness of heading time in Chinese cabbage. Plant Biotechnol. J. *12*, 312–321. http://dx.doi.org/10.1111/pbi.12138
Wang, Z.J., Huang, J.Q., Huang, Y.J., Li, Z., and Zheng, B.S. (2012b). Discovery and profiling of novel and conserved microRNAs during flower development in *Carya cathayensis* via deep sequencing. Planta *236*, 613–621. http://dx.doi.org/10.1007/s00425-012-1634-x
Wei, L.Q., Yan, L.F., and Wang, T. (2011). Deep sequencing on genome-wide scale reveals the unique composition and expression patterns of microRNAs in developing pollen of *Oryza sativa*. Genome Biol. *12*, R53. http://dx.doi.org/10.1186/gb-2011-12-6-r53
Wei, W., Ba, Z., Gao, M., Wu, Y., Ma, Y., Amiard, S., White, C.I., Rendtlew Danielsen, J.M., Yang, Y.G., and Qi, Y. (2012). A role for small RNAs in DNA double-strand break repair. Cell *149*, 101–112. http://dx.doi.org/10.1016/j.cell.2012.03.002
Wen, M., Shen, Y., Shi, S., and Tang, T. (2012). miREvo: an integrative microRNA evolutionary analysis platform for next-generation sequencing experiments. BMC Bioinf. *13*, 140.
Woodger, F.J., Millar, A., Murray, F., Jacobsen, J.V., and Frank, G. (2003). The role of GAMYB transcription factors in GA-regulated gene expression. J. Plant Growth Regul. *22*, 176–184.
Wroblewski, T., Matvienko, M., Piskurewicz, U., Xu, H., Martineau, B., Wong, J., Govindarajulu, M., Kozik, A., and Michelmore, R.W. (2014). Distinctive profiles of small RNA couple inverted repeat-induced post-transcriptional gene silencing with endogenous RNA silencing pathways in *Arabidopsis*. RNA *20*, 1987–1999. http://dx.doi.org/10.1261/rna.046532.114

Wu, G., Park, M.Y., Conway, S.R., Wang, J.W., Weigel, D., and Poethig, R.S. (2009). The sequential action of miR156 and miR172 regulates developmental timing in *Arabidopsis*. Cell *138*, 750–759. http://dx.doi.org/10.1016/j.cell.2009.06.031

Wu, G., and Poethig, R.S. (2006). Temporal regulation of shoot development in *Arabidopsis* thaliana by miR156 and its target SPL3. Development *133*, 3539–3547.

Wu, H.J., Ma, Y.K., Chen, T., Wang, M., and Wang, X.J. (2012). PsRobot: a web-based plant small RNA meta-analysis toolbox. Nucleic Acids Res. *40*, W22–8. http://dx.doi.org/10.1093/nar/gks554

Wu, J., Wang, D., Liu, Y., Wang, L., Qiao, X., and Zhang, S. (2014). Identification of miRNAs involved in pear fruit development and quality. BMC Genomics *15*, 953. http://dx.doi.org/10.1186/1471-2164-15-953

Wu, L., Liu, D., Wu, J., Zhang, R., Qin, Z., Liu, D., Li, A., Fu, D., Zhai, W., and Mao, L. (2013). Regulation of FLOWERING LOCUS T by a microRNA in Brachypodium distachyon. Plant Cell *25*, 4363–4377. http://dx.doi.org/10.1105/tpc.113.118620

Wu, M.F., Tian, Q., and Reed, J.W. (2006). *Arabidopsis* microRNA167 controls patterns of ARF6 and ARF8 expression, and regulates both female and male reproduction. Development *133*, 4211–4218.

Xia, K., Wang, R., Ou, X., Fang, Z., Tian, C., Duan, J., Wang, Y., and Zhang, M. (2012a). OsTIR1 and OsAFB2 downregulation via OsmiR393 overexpression leads to more tillers, early flowering and less tolerance to salt and drought in rice. PLOS ONE *7*, e30039. http://dx.doi.org/10.1371/journal.pone.0030039

Xia, R., Zhu, H., An, Y.Q., Beers, E.P., and Liu, Z. (2012b). Apple miRNAs and tasiRNAs with novel regulatory networks. Genome Biol. *13*, R47. http://dx.doi.org/10.1186/gb-2012-13-6-r47

Xiao, H., Jiang, N., Schaffner, E., Stockinger, E.J., and van der Knaap, E. (2008). A retrotransposon-mediated gene duplication underlies morphological variation of tomato fruit. Science *319*, 1527–1530.

Xie, F., and Zhang, B. (2010). Target-align: a tool for plant microRNA target identification. Bioinformatics *26*, 3002–3003. http://dx.doi.org/10.1093/bioinformatics/btq568

Xie, K., Shen, J., Hou, X., Yao, J., Li, X., Xiao, J., and Xiong, L. (2012). Gradual increase of miR156 regulates temporal expression changes of numerous genes during leaf development in rice. Plant Physiol. *158*, 1382–1394. http://dx.doi.org/10.1104/pp.111.190488

Xie, K., Wu, C., and Xiong, L. (2006). Genomic organization, differential expression, and interaction of SQUAMOSA promoter-binding-like transcription factors and microRNA156 in rice. Plant Physiol. *142*, 280–293.

Xin, C., Liu, W., Lin, Q., Zhang, X., Cui, P., Li, F., Zhang, G., Pan, L., Al-Amer, A., Mei, H., *et al.* (2015). Profiling microRNA expression during multi-staged date palm (Phoenix dactylifera L.) fruit development. Genomics *105*, 242–251.

Xing, S., Salinas, M., Garcia-Molina, A., Hohmann, S., Berndtgen, R., and Huijser, P. (2013). SPL8 and miR156-targeted SPL genes redundantly regulate *Arabidopsis* gynoecium differential patterning. Plant J. *75*, 566–577.

Xing, S., Salinas, M., Höhmann, S., Berndtgen, R., and Huijser, P. (2010). miR156-targeted and nontargeted SBP-box transcription factors act in concert to secure male fertility in *Arabidopsis*. Plant Cell *22*, 3935–3950. http://dx.doi.org/10.1105/tpc.110.079343

Xu, M.Y., Dong, Y., Zhang, Q.X., Zhang, L., Luo, Y.Z., Sun, J., Fan, Y.L., and Wang, L. (2012). Identification of miRNAs and their targets from Brassica napus by high-throughput sequencing and degradome analysis. BMC Genomics *13*, 421. http://dx.doi.org/10.1186/1471-2164-13-421

Xu, M.Y., Zhang, L., Li, W.W., Hu, X.L., Wang, M.B., Fan, Y.L., Zhang, C.Y., and Wang, L. (2014). Stress-induced early flowering is mediated by miR169 in *Arabidopsis* thaliana. J. Exp. Bot. *65*, 89–101. http://dx.doi.org/10.1093/jxb/ert353

Xu, T., Wang, Y., Liu, X., Lv, S., Feng, C., Qi, M., and Li, T. (2015). Small RNA and degradome sequencing reveals microRNAs and their targets involved in tomato pedicel abscission. Planta *242*, 963–984. http://dx.doi.org/10.1007/s00425-015-2318-0

Xu, X., Yin, L., Ying, Q., Song, H., Xue, D., Lai, T., Xu, M., Shen, B., Wang, H., and Shi, X. (2013). High-throughput sequencing and degradome analysis identify miRNAs and their targets involved in fruit senescence of Fragaria ananassa. PLOS ONE *8*, e70959. http://dx.doi.org/10.1371/journal.pone.0070959

Xue, X.Y., Zhao, B., Chao, L.M., Chen, D.Y., Cui, W.R., Mao, Y.B., Wang, L.J., and Chen, X.Y. (2014). Interaction between two timing microRNAs controls trichome distribution in *Arabidopsis*. PLOS Genet. *10*, e1004266. http://dx.doi.org/10.1371/journal.pgen.1004266

Yamaguchi, A., Wu, M.F., Yang, L., Wu, G., Poethig, R.S., and Wagner, D. (2009). The microRNA-regulated SBP-Box transcription factor SPL3 is a direct upstream activator of LEAFY, FRUITFULL, and APETALA1. Dev. Cell *17*, 268–278. http://dx.doi.org/10.1016/j.devcel.2009.06.007

Yang, J.H., Shao, P., Zhou, H., Chen, Y.Q., and Qu, L.H. (2010). deepBase: a database for deeply annotating and mining deep sequencing data. Nucleic Acids Res. *38*, D123–30. http://dx.doi.org/10.1093/nar/gkp943

Yant, L., Mathieu, J., Dinh, T.T., Ott, F., Lanz, C., Wollmann, H., Chen, X., and Schmid, M. (2010). Orchestration of the floral transition and floral development in *Arabidopsis* by the bifunctional transcription factor APETALA2. Plant Cell *22*, 2156–2170. http://dx.doi.org/10.1105/tpc.110.075606

Yao, F., Zhu, H., Yi, C., Qu, H., and Jiang, Y. (2015a). MicroRNAs and targets in senescent litchi fruit during ambient storage and post-cold storage shelf life. BMC Plant Biol. *15*, 181. http://dx.doi.org/10.1186/s12870-015-0509-2

Yao, J.L., Xu, J., Cornille, A., Tomes, S., Karunairetnam, S., Luo, Z., Bassett, H., Whitworth, C., Rees-George, J., Ranatunga, C., *et al.* (2015b). A microRNA allele that emerged prior to apple domestication may underlie fruit size evolution. Plant J. *84*, 417–427. http://dx.doi.org/10.1111/tpj.13021

Yi, R., Zhu, Z., Hu, J., Qian, Q., Dai, J., and Ding, Y. (2013). Identification and expression analysis of microRNAs at the grain filling stage in rice(Oryza sativa L.)via deep sequencing. PLOS ONE *8*, e57863. http://dx.doi.org/10.1371/journal.pone.0057863

Yin, Z.J., and Shen, F.F. (2010). Identification and characterization of conserved microRNAs and their target genes in wheat (Triticum aestivum). Genet. Mol. Res. *9*, 1186–1196. http://dx.doi.org/10.4238/vol9-2gmr805

Yoshikawa, M., Peragine, A., Park, M.Y., and Poethig, R.S. (2005). A pathway for the biogenesis of trans-acting siRNAs in *Arabidopsis*. Genes. Dev. *19*, 2164–2175.

Yu, N., Cai, W.J., Wang, S., Shan, C.M., Wang, L.J., and Chen, X.Y. (2010). Temporal control of trichome distribution by microRNA156-targeted SPL genes in *Arabidopsis* thaliana. Plant Cell *22*, 2322–2335. http://dx.doi.org/10.1105/tpc.109.072579

Yu, S., Galvão, V.C., Zhang, Y.C., Horrer, D., Zhang, T.Q., Hao, Y.H., Feng, Y.Q., Wang, S., Schmid, M., and Wang, J.W. (2012). Gibberellin regulates the *Arabidopsis* floral transition through miR156-targeted SQUAMOSA promoter binding-like transcription factors. Plant Cell *24*, 3320–3332.

Zeng, S., Liu, Y., Pan, L., Hayward, A., and Wang, Y. (2015). Identification and characterization of miRNAs in ripening fruit of Lycium barbarum L. using high-throughput sequencing. Front. Plant Sci. *6*, 778. http://dx.doi.org/10.3389/fpls.2015.00778

Zhang, L., Chia, J.M., Kumari, S., Stein, J.C., Liu, Z., Narechania, A., Maher, C.A., Guill, K., McMullen, M.D., and Ware, D. (2009). A genome-wide characterization of microRNA genes in maize. PLOS Genet. 5, e1000716. http://dx.doi.org/10.1371/journal.pgen.1000716

Zhang, T., Wang, J., and Zhou, C. (2015). The role of miR156 in developmental transitions in Nicotiana tabacum. Sci. China Life Sci. *58*, 253–260. http://dx.doi.org/10.1007/s11427-015-4808-5

Zhang, X., Xia, J., Lii, Y.E., Barrera-Figueroa, B.E., Zhou, X., Gao, S., Lu, L., Niu, D., Chen, Z., Leung, C., *et al.* (2012). Genome-wide analysis of plant nat-siRNAs reveals insights into their distribution, biogenesis and function. Genome Biol. *13*, R20. http://dx.doi.org/10.1186/gb-2012-13-3-r20

Zhang, X., Zou, Z., Zhang, J., Zhang, Y., Han, Q., Hu, T., Xu, X., Liu, H., Li, H., and Ye, Z. (2011). Over-expression of sly-miR156a in tomato results in multiple vegetative and reproductive trait alterations and partial phenocopy of the sft mutant. FEBS Lett. *585*, 435–439. http://dx.doi.org/10.1016/j.febslet.2010.12.036

Zhang, Y.C., Yu, Y., Wang, C.Y., Li, Z.Y., Liu, Q., Xu, J., Liao, J.Y., Wang, X.J., Qu, L.H., Chen, F., *et al.* (2013). Overexpression of microRNA OsmiR397 improves rice yield by increasing grain size and promoting panicle branching. Nat. Biotechnol. *31*, 848–852. http://dx.doi.org/10.1038/nbt.2646

Zhang, Z., Yu, J., Li, D., Zhang, Z., Liu, F., Zhou, X., Wang, T., Ling, Y., and Su, Z. (2010). PMRD: plant microRNA database. Nucleic Acids Res. *38*, D806–13. http://dx.doi.org/10.1093/nar/gkp818

Zhao, B., Ge, L., Liang, R., Li, W., Ruan, K., Lin, H., and Jin, Y. (2009). Members of miR-169 family are induced by high salinity and transiently inhibit the NF-YA transcription factor. BMC Mol. Biol. *10*, 29. http://dx.doi.org/10.1186/1471-2199-10-29

Zhao, M., Tai, H., Sun, S., Zhang, F., Xu, Y., and Li, W.X. (2012). Cloning and characterization of maize miRNAs involved in responses to nitrogen deficiency. PLOS ONE *7*, e29669. http://dx.doi.org/10.1371/journal.pone.0029669

Zhou, C.M., Zhang, T.Q., Wang, X., Yu, S., Lian, H., Tang, H., Feng, Z.Y., Zozomova-Lihová, J., and Wang, J.W. (2013). Molecular basis of age-dependent vernalization in Cardamine flexuosa. Science *340*, 1097–1100. http://dx.doi.org/10.1126/science.1234340

Zhou, G.K., Kubo, M., Zhong, R., Demura, T., and Ye, Z.H. (2007). Overexpression of miR165 affects apical meristem formation, organ polarity establishment and vascular development in *Arabidopsis*. Plant Cell Physiol. *48*, 391–404.

Zhou, L., Li, X., Liu, Q., Zhao, F., and Wu, J. (2011). Small RNA transcriptome investigation based on next-generation sequencing technology. J. Genet. Genomics. *38*, 505–513. http://dx.doi.org/10.1016/j.jgg.2011.08.006

Zhu, E., Zhao, F., Xu, G., Hou, H., Zhou, L., Li, X., Sun, Z., and Wu, J. (2010). mirTools: microRNA profiling and discovery based on high-throughput sequencing. Nucleic Acids Res. *38*, W392–7. http://dx.doi.org/10.1093/nar/gkq393

Zilberman, D., Cao, X., and Jacobsen, S.E. (2003). ARGONAUTE4 control of locus-specific siRNA accumulation and DNA and histone methylation. Science *299*, 716–719. http://dx.doi.org/10.1126/science.1079695

Zuo, J., Zhu, B., Fu, D., Zhu, Y., Ma, Y., Chi, L., Ju, Z., Wang, Y., Zhai, B., and Luo, Y. (2012). Sculpting the maturation, softening and ethylene pathway: the influences of microRNAs on tomato fruits. BMC Genomics *13*, 7. http://dx.doi.org/10.1186/1471-2164-13-7

Zuo, J., Fu, D., Zhu, Y., Qu, G., Tian, H., Zhai, B., Ju, Z., Gao, C., Wang, Y., Luo, Y., *et al.* (2013). SRNAome parsing yields insights into tomato fruit ripening control. Physiol. Plant *149*, 540–553. http://dx.doi.org/10.1111/ppl.12055

# ChIP-seq: A Powerful Tool for Studying Protein–DNA Interactions in Plants

10

Xifeng Chen[1†], Vijai Bhadauria[2†] and Bojun Ma[1*]

[1]College of Chemistry and Life Sciences, Zhejiang Normal University, Jinhua, China.
[2]Crop Development Centre and Department of Plant Sciences, University of Saskatchewan, Saskatoon, SK, Canada.

*Correspondence: mbj@zjnu.cn
†These authors contributed equally

https://doi.org/10.21775/9781910190654.11

## Abstract

DNA-binding proteins, including transcription factors, epigenetic and chromatin modifiers, control gene expression in plants. To map the binding sites of DNA-binding proteins in the genome is crucial for decoding gene regulatory networks. Chromatin immunoprecipitation (ChIP) followed by high-throughput DNA sequencing (ChIP-seq) is a widely used approach to identify the DNA regions bound by a specific protein *in vivo*. The information generated from ChIP-seq has tremendously advanced our understanding of the mechanisms of transcription factors, cofactors and histone modifications in regulating gene expression. In this review, we reviewed the recent research advances of ChIP-seq in plants, including a description of the ChIP-seq workflow and its various applications in plants, and in addition, provided a perspective for the potential advances of ChIP-seq.

## Introduction

Genome-wide mapping of protein–DNA interactions and epigenetic markers is essential for full understanding of transcriptional regulation. Chromatin immunoprecipitation (ChIP) is an effective method for isolating the DNA fragments binding by a specific protein *in vivo* (Solomon *et al.*, 1988). This technology can co-precipitate proteins of interest with DNAs fragments, which are physically associated with the proteins. Previously, ChIP assay was often followed by a hybridization of the DNA fragments isolated to microarray (called ChIP-chip), in order to identify the sequences of DNA fragments in genome wide. However, along with the rapid development of DNA-sequencing technology, the DNA fragments obtained from ChIP have been sequenced directly by high-throughput sequencing (called ChIP-seq). Due to the greater resolution, the coverage and specificity of ChIP-seq are significantly improved compared with ChIP-chip (Luo and Lam, 2014). Thus, ChIP-seq has become a popular method for profiling chromatin modifications and transcription factor (TF)-binding sites in eukaryote genomes. In this review, we describe the workflow of ChIP-seq optimized for plants, summarize the applications of ChIP-seq in plants, and provide perspective regarding its potential advances.

## ChIP-seq workflow for plants

### ChIP assay

The protocol of ChIP assay for model plants (rice and *Arabidopsis*) was developed by Zhu *et al.* (2012). First, the DNA-binding protein is crosslinked to DNA *in vivo* by treating cells with formaldehyde. Next, the cells are lysed and chromatin is sheared by sonication into small fragments followed by immunoprecipitation of protein–DNA complexes using the specific protein antibodies. To ensure high-quality data, the choices of antibody and negative control are critical. Many antibodies are commercially available against the epitope tags, such as GFP, MYC or FLAG, which are usually fused to interest proteins in transgenic plants (Table 10.1). The implementation of epitope tags removes the muddled process in the synthesis of a certain antibody, and non-transgenic plants or plants only expressing the tags are the perfect negative controls. Certainly, ChIP can also be performed using a specific antibody against the interest protein itself in wild-type plants, which avoids the need to generate transgenic plants with complex genetic backgrounds. However, it is difficult to design a perfect negative control and thus increases the chance of false positives. The last step is reversing the crosslinks to release the DNA fragments from proteins followed by purification of the released DNA fragments. The quality of purified DNA can be tested by quantitative real-time PCR (ChIP-qPCR) to determine the enrichment of genomic regions of interest.

### Sequencing and data analysis

For high-throughput sequencing, a library of the purified DNA fragments are created and sequenced by multiple platforms (Chaitankar *et al.*, 2016). Presently, Illumina platforms (Genome Analyzer II or HiSeq X) are mostly used for ChIP-seq (Table 10.1). After sequencing, converting the raw ChIP-Seq data into biological insight requires sequential data analysis (Fig. 10.1). Generally, the raw reads from sequencing are firstly run quality control by program Fastqc (www.bioinformatics.bbsrc.ac.uk/projects/fastqc/) or Fastx-toolkit (http://hannonlab.cshl.edu/fastx_toolkit/) to remove the adapt and other contaminative sequences, and then mapped to the reference genome using a short-read mapping program, such as Bowtie (http://bowtie-bio.sourceforge.net/index.shtml). Following successful mapping reads to genome, the next task is genome-wide identification of the putative binding sites of the interest protein, or called peaks calling. This process can be performed by various algorithms (Chaitankar *et al.*, 2016), and the program MACS (http://liulab.dfci.harvard.edu/MACS/), QuEST (http://mendel.stanford.edu/sidowlab/downloads/quest/) and CisGenome (www.biostat.jhsph.edu/~hji/cisgenome/) are mostly used for plants (Table 10.1). Lastly, the discovery of binding sequence motifs can be performed by programs MEME, DREME or MEME-ChIP (http://meme-suite.org/).

## Applications of ChIP-seq in plants

### Genome-wide identification of TF's binding sites and target genes

Regulation of gene expression is very complex in plant cells. The binding of TFs to specific DNA target sequences (or *cis*-regulatory elements) is the fundamental basis of gene regulatory networks. ChIP-seq has enabled precise determination of genome-wide distribution of TF binding sites *in vivo*. This has been illustrated by recent studies of many plant TFs involved in regulation of development, circadian clock, flowering time, hormone signalling, abiotic or biotic stress in different tissues or at different developmental stages of both monocotyledonous and dicotyledonous plants (Table 10.1).

The antibodies mostly used to immunoprecipitate the DNA-binding protein are those commercially available ones raised against the epitomic tags such as GFP, HA, FlAG or MYC (Table 10.1). Accordingly, the transgenic plants that ectopically express tagged protein of interest are used for ChIP-seq. This is predominantly used for the ChIP-seq analysis in *Arabidopsis* because of its high efficient transformation system. For the plant species that are recalcitrant to transformation, the specific monoclonal or polyclonal antibodies are often used for ChIP-seq (Table 10.1).

Illumina GA II or HiSeq X is the mostly used deep sequencing platform for ChIP-seq (Table 10.1). After mapping reads to the reference genomes, the algorithm to identify the peaks enriched with reads in genomes were mostly performed by the program MACS, or by program QuEST, Cisgenome, ERANGE in few cases (Table 10.1). The number of peaks identified varies from hundreds to tens of thousands according to different interest proteins (Table 10.1). According to the genomic features, location of the peaks can be categorized into upstream sequence from transcription start sites (TSS), 5′ UTR, 3′ UTR, exon or intron. The regions distributed near TSS are promoter's site of genes, which are consider to be the targets for the TFs. Moreover, the expression of these putative target genes can be further validated by RNA-seq, microarray or qPCR (Table 10.1). Searching the common motifs in binding sites from ChIP-seq data likely leads to identify the cis regulatory elements of TFs, such as G-box (CACGTG) for PRRs and PIFs, EE (AAATATCT) for CCA1 (Table 10.1).

## Genome-wide discovery of histone marks

Histone marks (e.g. methylation and acetylation) play a substantial role in regulating gene expression throughout plant growth and development as well as in responses to different environmental stresses. ChIP-seq can perform genome-wide evaluation of the distribution of modified histones *in vivo*. Du *et al.* (2013) employed ChIP-seq to explore the di- and tri-methylation of H3K4 (H3K4me2 and H3K4me3) and acylation of H3K9 and H3K27 (H3K9ac and H3K27ac) in rice seedlings. They discovered that the histone marks were enriched around the TSSs. Using ChIP-seq, Brusslan *et al.* (2012) discovered that the enrichment degree of H3K4me3 mark were corresponded to the up- or down-regulation of genes throughout senescence in *Arabidopsis*. Hussey *et al.* (2015) used ChIP-seq to determine the role of the histone modification H3K4me3 in the epigenomic regulation of xylogenesis in *Eucalyptus grandis* tree. They discovered that H3K4me3 enrichment can activate the transcription of genes involved in developing xylem.

Hu *et al.* (2012) characterized the genome-wide distribution and alteration of H3K9ac in maize throughout cold stress with ChIP-seq. Their outcomes suggested that the genic regions showed a substantial complete decrease of H3K9ac due to cold stimulus. In *Arabidopsis*, the H3K4me3 abundance changed robustly located on the genes responding to dehydration stress, and these genes showed broader H3K4me3 distribution profiles from ChIP-seq (van Dijk *et al.*, 2010). Similarly in rice, the alterations in H3K4me3, determined with ChIP-seq, were positively correlated with transcript alterations in response to drought stress, and H3K4me3 modification levels were increased mainly for genes with low expression levels and decreased for genes with high expression levels (Zong *et al.*, 2013). Using ChIP-seq, Ayyappan *et al.* (2015) found H3K9me2 and H4K12ac marks impacted a great number of stress-responsive genes in common bean under rust (*Uromyces appendiculatus*) stress.

## Genome-wide analysis of the distribution of RNA polymerase

All eukaryotes have three DNA-dependent RNA polymerases (Pol I, II and III) that are necessary for the transcription. Liu *et al* (2014) identified 6499 ChIP-seq peaks for RNA pol II in poplar, and discovered the distribution of peaks was highly coincident with genic areas and greatly corresponded with transcript abundance. Moreover, plants have two additional polymerases (Pol IV and Pol V) that function in the RNA-directed DNA methylation pathway. Recently, the ChIP-seq protocols for studying of the genome-wide distribution of Pol II (de Lorenzo, 2015) and Pol V (Rowley *et al.*, 2013) were described in *Arabidopsis*. Zhong *et al.* (2012) pinpointed the targets of Pol V in *Arabidopsis* on a genome-wide scale with ChIP-seq, and discovered that Pol V is enriched at promoters and evolutionarily recent transposons. This localization pattern is greatly corresponded with Pol V-dependent DNA methylation and small RNA accumulation.

## Improvement for sequencing of repetitive genomic sequences

The heterochromatin or microsatellites with greatly replicated sequences are difficult to sequence by traditional techniques. Centromeric sequences are greatly replicated with tandem repeats, such as satellite DNA and retrotransposons, and greatly changeable in various species, even among tightly related organisms (Smith *et al.*, 2011). Nevertheless, a highly conserved function of the centromere is correlated with its epigenetic characteristics, including the histone H3 variant CENH3 in plants (Birchler *et al.*, 2011). Guo *et al.* (2016) performed ChIP-seq with CENH3 specific antibodies, and identified the ectopic genomic sequences present at the new centromeres of the 4DS chromosome in wheat (*Triticum aestivum*). Similarly, Nagaki *et al.* (2015) isolated the centromeric DNA sequences of sunflower (*Helianthus annuus*) by CENH3-ChIP-seq, and identified two different types of repetitive DNA sequences: the tandem repeat sequences and the long-interspersed-nuclear-element-like sequences. Additionally, Kowar *et al.* (2016) implemented CENH3-ChIP-seq analysis revealed that the centromeres of sugar beet (*Beta

**Table 10.1** ChIP-seq analysis of transcription factors in plants

| TF | Species | Biological process | Tissue | Anti-body | Sequencing platform | Algorithm for peaks | Binding regions | Expression analysis | Target genes | Common cis-element | Reference |
|---|---|---|---|---|---|---|---|---|---|---|---|
| PRR1 | *Arabidopsis* | Circadian clock | Seedling | Anti-GFP | Illumina GA II | ERANGE | 867 | Microarray | 301 | CACGTG or AATATCT | Huang *et al.*, 2012 |
| PRR5 | *Arabidopsis* | Circadian clock | Plant | Anti-GFP | Illumina GA II | MACS | 542 | Microarray | 64 | CACGTG | Nakamichi *et al.*, 2012 |
| PRR7 | *Arabidopsis* | Circadian clock | Seedling | Anti-HA | Illumina | QuEST | 73 | Microarray | 83 | CACGTG | Liu *et al.*, 2013 |
| PRR9 | *Arabidopsis* | Circadian clock | Plant | Anti-HA | Illumina HiSeq 2500 | MACS | – | Microarray | 150 | CACGTG | Liu *et al.*, 2016 |
| CCA1 | *Arabidopsis* | Circadian clock | Plant | Anti-FLAG | Illumina GA II | MACS | 449 | Microarray | 113 | AAAATATCT or AAAAATCT | Kamioka *et al.*, 2016 |
| FHY3 | *Arabidopsis* | Circadian clock. | Seedling | Anti-FLAG | Illumina GA II | MACS | 785 | Microarray | 71 | CACGCGC | Ouyang *et al.*, 2011 |
| PCFS4 | *Arabidopsis* | Flowering time | Seedling | Anti-POD | Illumina GA II | CisGenome | 892 | cDNA/EST | 7 | – | Xing *et al.*, 2013 |
| SWP73B | *Arabidopsis* | Development | Seedling | Anti-GFP | Illumina HiSeq 2500 | MACS | – | Microarray | 377 | – | Sacharowski *et al.*, 2015 |
| PIF1 | *Arabidopsis* | Development | Seedling | Anti-MYC | Illumina HiSeq | MACS | 3027 | RNA-seq | 304 | CACGTG or CACATG | Pfeiffer *et al.*, 2014 |
| PIF3 | *Arabidopsis* | Development | Seedling | Anti-MYC | Illumina GA II | MACS | 1064 | RNA-seq | 136 | CACGTG or CACATG | Pfeiffer *et al.*, 2014 |
| PIF4 | *Arabidopsis* | Development | Seedling | Anti-MYC | Illumina HiSeq | MACS | 2710 | RNA-seq | 198 | CACGTG or CACATG | Pfeiffer *et al.*, 2014 |
| PIF5 | *Arabidopsis* | Development | Seedling | Anti-HA | Illumina | MACS | 2058 | RNA-seq | 236 | CACGTG or CACATG | Pfeiffer *et al.*, 2014 |
| FHY3 | *Arabidopsis* | Development | Inflorescence | Anti-FLAG | Illumina | MACS | 1885 | RNA-seq | 238 | CACGCGC | Li *et al.*, 2016 |
| WRKY33 | *Arabidopsis* | Hormone | Plant | Anti-HA | Illumina HiSeq 2500 | QuEST | 1680 | RNA-seq | 318 | TTGACT/C | Liu *et al.*, 2015c |
| AL5 | *Arabidopsis* | Abiotic stress | Seedling | Anti-AL5 | Illumina GA II | MACS | 193 | qPCR | 8 | – | Wei *et al.*, 2015 |

| | | | | | | | | | | | |
|---|---|---|---|---|---|---|---|---|---|---|---|
| ASR5 | Rice | Metal stress | Root | Anti-ASR5 | Illumina HiSeq 2000 | MACS | 649 | RNA-seq | 104 | GGCCCA(T/A) | Arenhart *et al.*, 2014 |
| MADS58 | Rice | Development | Panicle | Anti-MADS58 | Illumina | MACS | 520 | qPCR | 13 | – | Chen *et al.*, 2015a |
| CCA1 | Maize | Circadian clock | Aerial tissue | Anti-CCA1 | Illumina HiSeq 2500 | MACS | 10136 | qPCR | 3 | AAAATA or AAGAAA | Ko *et al.*, 2016 |
| O2 | Maize | Development | Endosperm | Anti-O2 | Illumina GA II | MACS | 1686 | RNA-seq | 35 | TGACGTGG | Li *et al.*, 2015 |
| NAC | Soybean | Development | Cotyledon | Anti-NAC | Illumina HiSeq 2000 | MACS | 8246 | RNA-seq | 72 | G[AT]G[AG]G[AG]GA, C[AC]C[GA][TC][GA]CC or TGGGCC | Shamimuzzaman *et al.*, 2013 |
| YABBY | Soybean | Development | Cotyledon | Anti-YABBY | Illumina HiSeq 2000 | MACS | 18064 | RNA-seq | 96 | CC[CA][TC]C[TA][CT]C, GA[AG]AGAAA or CCCCAC | Shamimuzzaman *et al.*, 2013 |
| ASR1 | Tomato | Drought stress | Leaf | Anti-ASR1 | Illumina HiSeq 2000 | MACS | 225 | qPCR | 3 | (A/T)(A/G)AGCCCA | Ricardi *et al.*, 2014 |
| ERF68 | Tomato | Defence | Leaf | Anti-HA | Illumina | MACS | 2997 | qPCR | 8 | – | Liu and Cheng, 2016 |
| VRN1 | Barley | Vernalization | Seedling | Anti-HA | Illumina HiSeq 2000 | MACS | 514 | RNA-seq | 37 | CC[A/T]$_6$GG or CATGCATG | Deng *et al.*, 2015 |
| ARK1 | Populus | Woody growth, | Vascular | Anti-ARK1 | Illumina GA II | MACS | 14463 | RNA-seq | 866 | AAACCCT/A | Liu *et al.*, 2015b |
| MaASR | Banana | Drought stress | Plant | Anti-MaASR | Illumina | MACS | 410 | Microarray | 10 | – | Zhang *et al.*, 2015 |

Algorithm for peaks, the algorithm method used in genome-wide peaks calling; Binding regions, the number of TF's binding regions from peaks-calling analysis; Biological process, the biological process that the TF involved; Common *cis*-element, the Common *cis*-elements which may be bound by the TF; Expression analysis, the methods used for analysing the expression of TF's target genes; Sequencing Platform, the Platform used for high-throughput DNA sequencing; Species, the species that the TF was from; Target genes, the number of TF's target genes validated by expression; TF, transcription factor; Tissue, the tissue that used in ChIP-seq assay.
–, information was not described in the published article.

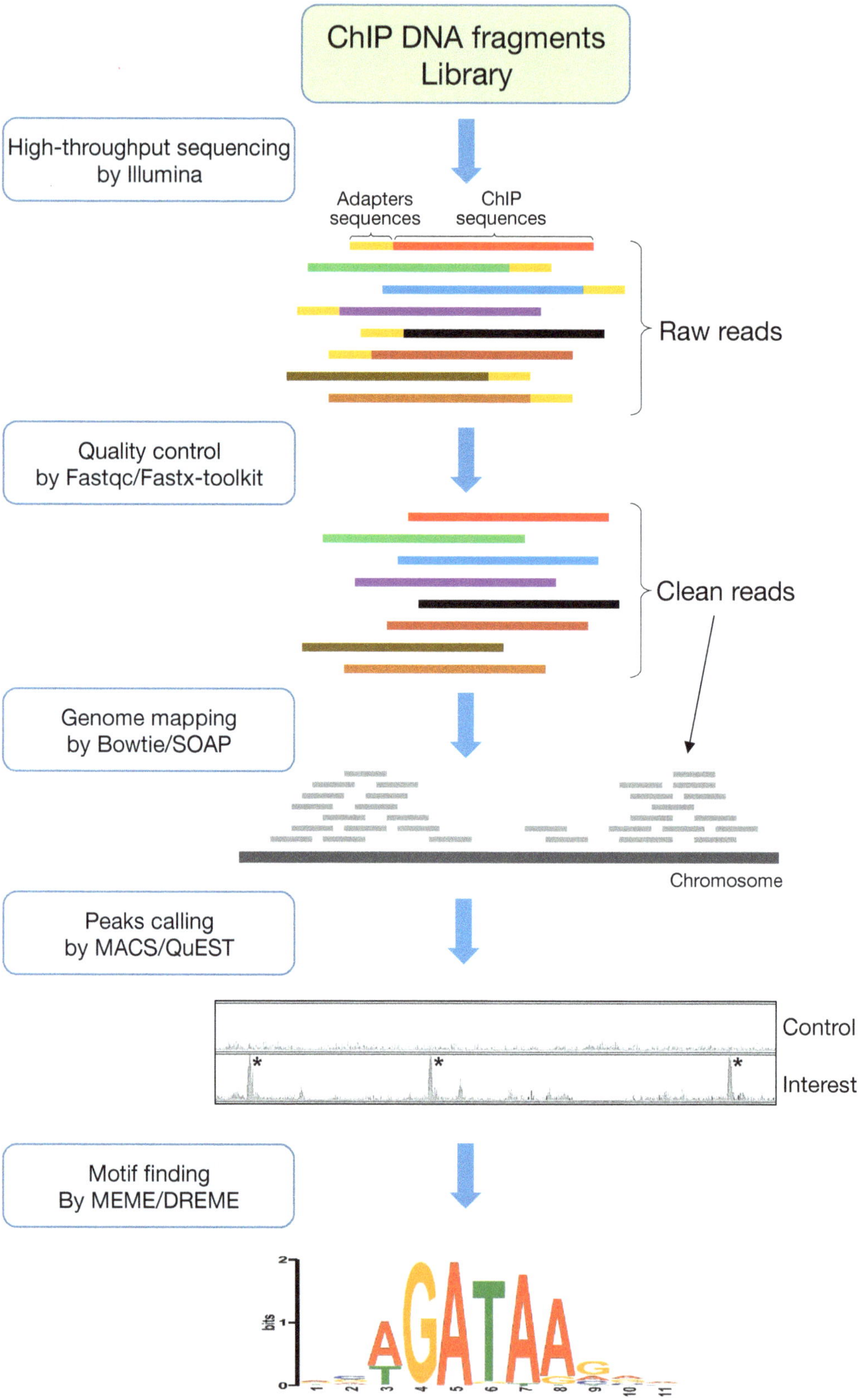


**Figure 10.1** Overview of workflow for ChIP-seq data analysis.

*vulgaris*) containing the satellite pBV and the Ty3-gypsy retrotransposon Beetle7. Consequently, ChIP-seq could be more efficient in sequencing of repeat regions, and the distinctive sequences flanking repeats are helpful in aligning the reads to the genome.

### Annotation of putative missing genes in genome

Since the histone modifications have a high concurrence possibility with active transcript loci, ChIP-seq may be implemented in predicting unannotated genes. For example, comparing with the rice genome annotation resources MSU/TIGR (http://rice.plantbiology.msu.edu/) and RAP–DB (http://rapdb.dna.affrc.go.jp/), Du *et al.* (2013) discovered 6,629 RAP–DB genes which were lost annotation in MSU/TIGR. Of these unannotated genes, they found that 471 genes had both H3K4me3 and H3K9ac enrichment inside their TSS regions, and proved the existence of these genes by expression analysis. Though the mRNA-seq could trace missing genes systematically, ChIP-seq is one of the best approaches for characterizing lost genes in organisms whose genomes have yet to be annotated to a large degree.

## Perspective

Genome-wide evaluations of epigenomic and transcriptomic profiles provide extensive resources for determining epigenetic regulatory mechanisms. ChIP-seq has been established as a main tool for comprehending these mechanisms. Even with the remarkable advancement in high-throughput sequencing platforms in the recent years, the DNA sequencing capacity increased rapidly with the lower cost. However, the greater challenges for biologists are very complicated bioinformatic processing of immense amount of ChIP-seq data using different programs (Fig. 10.1). This will require development of user-friendly and durable software to evaluate data and closer interaction between experimentalists and bioinformaticians. For example, ENCODE (Encyclopedia of DNA Elements; Landt *et al.*, 2012), offered a group of working procedures and guidelines for ChIP-seq, and address antibody validation, experimental replication, sequencing depth, and data quality assessment experiments. Unfortunately, ENCODE is designed only for human and animals so far. ChIPseek (Chen *et al.*, 2014) is a web-based analysis tool of ChIP data for animals and plants (*Arabidopsis* and rice), and provides a statistical summary of the dataset including a histogram of peak length distribution and of distances to the nearest TSS, and offers charts of genomic areas to get a complete perspective on the dataset. This is appropriate for biologists to analyse ChIP data without needing any programming expertise. Moreover, there are additional powerful ChIP-seq tools for specific plant spices, such as PRI-CAT (Muiño *et al.*, 2011) and *CressInt* (Chen *et al.*, 2015b) for *Arabidopsis*, and PopGenIE (Liu *et al.*, 2015a) for *Populus*. ChIP-seq knowledge is being rapidly added to public repositories that will culminate in the inception of databases to analyse gene expression and protein–protein networks. Such comprehensive analyses facilitate the systematic elucidation of diverse biological activities intricately cooperating in the genome.

## Acknowledgements

We are grateful to Dr Jianzhong Liu for critical reading of this manuscript and his valuable comments.

## Funding

This work was supported by the National Key Research and Development Program of China (2016YFD0100405); the Ministry of Agriculture of China (2016ZX08009003-001); the National Natural Science Foundation of China (31371593, 31671650); and the Natural Science Foundation of Zhejiang Province (LZ14C130001).

## References

Arenhart, R.A., Bai, Y., de Oliveira, L.F., Neto, L.B., Schunemann, M., Maraschin Fdos, S., Mariath, J., Silverio, A., Sachetto-Martins, G., Margis, R., *et al.* (2014). New insights into aluminum tolerance in rice: the ASR5 protein binds the STAR1 promoter and other aluminum-responsive genes. Mol. Plant 7, 709–721. http://dx.doi.org/10.1093/mp/sst160

Ayyappan, V., Kalavacharla, V., Thimmapuram, J., Bhide, K.P., Sripathi, V.R., Smolinski, T.G., Manoharan, M., Thurston, Y., Todd, A., and Kingham, B. (2015). Genome-wide profiling of histone modifications (H3K9me2 and H4K12ac) and gene expression in Rust (*Uromyces appendiculatus*) inoculated common bean (*Phaseolus vulgaris* L.). PLOS ONE *10*, e0132176. http://dx.doi.org/10.1371/journal.pone.0132176

Birchler, J.A., Gao, Z., Sharma, A., Presting, G.G., and Han, F. (2011). Epigenetic aspects of centromere function in plants. Curr. Opin. Plant Biol. *14*, 217–222. http://dx.doi.org/10.1016/j.pbi.2011.02.004

Brusslan, J.A., Rus Alvarez-Canterbury, A.M., Nair, N.U., Rice, J.C., Hitchler, M.J., and Pellegrini, M. (2012). Genome-wide evaluation of histone methylation changes associated with leaf senescence in *Arabidopsis*. PLOS ONE *7*, e33151. http://dx.doi.org/10.1371/journal.pone.0033151

Chaitankar, V., Karakülah, G., Ratnapriya, R., Giuste, F.O., Brooks, M.J., and Swaroop, A. (2016). Next generation sequencing technology and genomewide data analysis: Perspectives for retinal research. Prog. Retin. Eye Res. *55*, 1–31.

Chen, R., Shen, L.P., Wang, D.H., Wang, F.G., Zeng, H.Y., Chen, Z.S., Peng, Y.B., Lin, Y.N., Tang, X., Deng, M.H., *et al.* (2015a). A gene expression profiling of early rice stamen development that reveals inhibition of photosynthetic genes by OsMADS58. Mol. Plant *8*, 1069–1089. http://dx.doi.org/10.1016/j.molp.2015.02.004

Chen, T.W., Li, H.P., Lee, C.C., Gan, R.C., Huang, P.J., Wu, T.H., Lee, C.Y., Chang, Y.F., and Tang, P. (2014). ChIPseek, a web-based analysis tool for ChIP data. BMC Genomics *15*, 539. http://dx.doi.org/10.1186/1471-2164-15-539

Chen, X., Ernst, K., Soman, F., Borowczak, M., and Weirauch, M.T. (2015b). *CressInt*: a user-friendly web resource for genome-scale exploration of gene regulation in *Arabidopsis thaliana*. Curr. Plant Biol. 3-4, 48–55. http://dx.doi.org/10.1016/j.cpb.2015.09.001

de Lorenzo, L. (2015). Genome-wide analysis of distribution of RNA polymerase II isoforms using ChIP-seq. Methods Mol. Biol. *1255*, 209–221. http://dx.doi.org/10.1007/978-1-4939-2175-1_18

Deng, W., Casao, M.C., Wang, P., Sato, K., Hayes, P.M., Finnegan, E.J., and Trevaskis, B. (2015). Direct links between the vernalization response and other key traits of cereal crops. Nat. Commun. *6*, 5882. http://dx.doi.org/10.1038/ncomms6882

Du, Z., Li, H., Wei, Q., Zhao, X., Wang, C., Zhu, Q., Yi, X., Xu, W., Liu, X.S., Jin, W., *et al.* (2013). Genome-wide analysis of histone modifications: H3K4me2, H3K4me3, H3K9ac, and H3K27ac in *Oryza sativa* L. *Japonica*. Mol. Plant *6*, 1463–1472. http://dx.doi.org/10.1093/mp/sst018

Guo, X., Su, H., Shi, Q., Fu, S., Wang, J., Zhang, X., Hu, Z., and Han, F. (2016). *De novo* centromere formation and centromeric sequence expansion in wheat and its wide hybrids. PLOS Genet. *12*, e1005997. http://dx.doi.org/10.1371/journal.pgen.1005997

Hu, Y., Zhang, L., He, S., Huang, M., Tan, J., Zhao, L., Yan, S., Li, H., Zhou, K., Liang, Y., *et al.* (2012). Cold stress selectively unsilences tandem repeats in heterochromatin associated with accumulation of H3K9ac. Plant Cell Environ. *35*, 2130–2142. http://dx.doi.org/10.1111/j.1365-3040.2012.02541.x

Huang, W., Pérez-García, P., Pokhilko, A., Millar, A.J., Antoshechkin, I., Riechmann, J.L., and Mas, P. (2012). Mapping the core of the *Arabidopsis* circadian clock defines the network structure of the oscillator. Science *336*, 75–79. http://dx.doi.org/10.1126/science.1219075

Hussey, S.G., Mizrachi, E., Groover, A., Berger, D.K., and Myburg, A.A. (2015). Genome-wide mapping of histone H3 lysine 4 trimethylation in *Eucalyptus grandis* developing xylem. BMC Plant Biol. *15*, 117. http://dx.doi.org/10.1186/s12870-015-0499-0

Kamioka, M., Takao, S., Suzuki, T., Taki, K., Higashiyama, T., Kinoshita, T., and Nakamichi, N. (2016). Direct repression of evening genes by CIRCADIAN CLOCK-ASSOCIATED1 in the *Arabidopsis* circadian clock. Plant Cell *28*, 696–711. http://dx.doi.org/10.1105/tpc.15.00737

Ko, D.K., Rohozinski, D., Song, Q., Taylor, S.H., Juenger, T.E., Harmon, F.G., and Chen, Z.J. (2016). Temporal shift of circadian-mediated gene expression and carbon fixation contributes to biomass heterosis in maize hybrids. PLOS Genet. *12*, e1006197. http://dx.doi.org/10.1371/journal.pgen.1006197

Kowar, T., Zakrzewski, F., Macas, J., Kobližková, A., Viehoever, P., Weisshaar, B., and Schmidt, T. (2016). Repeat composition of CenH3-chromatin and H3K9me2-marked heterochromatin in Sugar Beet (*Beta vulgaris*). BMC Plant Biol. *16*, 120. http://dx.doi.org/10.1186/s12870-016-0805-5

Landt, S.G., Marinov, G.K., Kundaje, A., Kheradpour, P., Pauli, F., Batzoglou, S., Bernstein, B.E., Bickel, P., Brown, J.B., Cayting, P., *et al.* (2012). ChIP-seq guidelines and practices of the ENCODE and modENCODE consortia. Genome Res. *22*, 1813–1831. http://dx.doi.org/10.1101/gr.136184.111

Li, C., Qiao, Z., Qi, W., Wang, Q., Yuan, Y., Yang, X., Tang, Y., Mei, B., Lv, Y., Zhao, H., *et al.* (2015). Genome-wide characterization of *cis*-acting DNA targets reveals the transcriptional regulatory framework of opaque2 in maize. Plant Cell *27*, 532–545. http://dx.doi.org/10.1105/tpc.114.134858

Li, D., Fu, X., Guo, L., Huang, Z., Li, Y., Liu, Y., He, Z., Cao, X., Ma, X., Zhao, M., *et al.* (2016). *FAR-RED ELONGATED HYPOCOTYL3* activates *SEPALLATA2* but inhibits *CLAVATA3* to regulate meristem determinacy and maintenance in *Arabidopsis*. Proc. Natl. Acad. Sci. U.S.A. *113*, 9375–9380. http://dx.doi.org/10.1073/pnas.1602960113

Liu, A.C., and Cheng, C.P. (2016). Pathogen-induced ERF68 regulates hypersensitive cell death in tomato. Mol. plant Pathol. Online. http://dx.doi.org/10.1111/mpp.12460

Liu, L., Missirian, V., Zinkgraf, M., Groover, A., and Filkov, V. (2014). Evaluation of experimental design and computational parameter choices affecting analyses of ChIP-seq and RNA-seq data in undomesticated poplar trees. BMC Genomics *15* (Suppl. 5), S3. http://dx.doi.org/10.1186/1471-2164-15-S5-S3

Liu, L., Ramsay, T., Zinkgraf, M., Sundell, D., Street, N.R., Filkov, V., and Groover, A. (2015a). A resource for characterizing genome-wide binding and putative target genes of transcription factors expressed during secondary growth and wood formation in *Populus*. Plant J *82*, 887–898. http://dx.doi.org/10.1111/tpj.12850

Liu, L., Zinkgraf, M., Petzold, H.E., Beers, E.P., Filkov, V., and Groover, A. (2015b). The *Populus* ARBORKNOX1 homeodomain transcription factor regulates woody growth through binding to evolutionarily conserved target genes of diverse function. New Phytol. *205*, 682–694. http://dx.doi.org/10.1111/nph.13151

Liu, S., Kracher, B., Ziegler, J., Birkenbihl, R.P., and Somssich, I.E. (2015c). Negative regulation of ABA signaling by

WRKY33 is critical for *Arabidopsis* immunity towards Botrytis cinerea 2100. Elife *4*, e07295. http://dx.doi.org/10.7554/eLife.07295

Liu, T., Carlsson, J., Takeuchi, T., Newton, L., and Farré, E.M. (2013). Direct regulation of abiotic responses by the *Arabidopsis* circadian clock component PRR7. Plant J. *76*, 101–114. http://dx.doi.org/10.1111/tpj.12276

Liu, T.L., Newton, L., Liu, M.J., Shiu, S.H., and Farre, E.M. (2016). A G-box-like motif is necessary for transcriptional regulation by circadian pseudo-response regulators in *Arabidopsis*. Plant Physiol. *170*, 528–539. http://dx.doi.org/10.1104/pp.15.01562

Luo, C., and Lam, E. (2014). Quantitatively profiling genome-wide patterns of histone modifications in *Arabidopsis thaliana* using ChIP-seq. Methods Mol. Biol. *1112*, 177–193. http://dx.doi.org/10.1007/978-1-62703-773-0_12

Muiño, J.M., Hoogstraat, M., van Ham, R.C., and van Dijk, A.D. (2011). PRI-CAT: a web-tool for the analysis, storage and visualization of plant ChIP-seq experiments. Nucleic Acids Res. *39*, W524–7. http://dx.doi.org/10.1093/nar/gkr373

Nagaki, K., Tanaka, K., Yamaji, N., Kobayashi, H., and Murata, M. (2015). Sunflower centromeres consist of a centromere-specific LINE and a chromosome-specific tandem repeat. Front Plant Sci. *6*, 912. http://dx.doi.org/10.3389/fpls.2015.00912

Nakamichi, N., Kiba, T., Kamioka, M., Suzuki, T., Yamashino, T., Higashiyama, T., Sakakibara, H., and Mizuno, T. (2012). Transcriptional repressor PRR5 directly regulates clock-output pathways. Proc. Natl. Acad. Sci. U.S.A. *109*, 17123–17128. http://dx.doi.org/10.1073/pnas.1205156109

Ouyang, X., Li, J., Li, G., Li, B., Chen, B., Shen, H., Huang, X., Mo, X., Wan, X., Lin, R., *et al.* (2011). Genome-wide binding site analysis of FAR-RED ELONGATED HYPOCOTYL3 reveals its novel function in *Arabidopsis* development. Plant Cell *23*, 2514–2535. http://dx.doi.org/10.1105/tpc.111.085126

Pfeiffer, A., Shi, H., Tepperman, J.M., Zhang, Y., and Quail, P.H. (2014). Combinatorial complexity in a transcriptionally centered signaling hub in *Arabidopsis*. Mol. Plant *7*, 1598–1618. http://dx.doi.org/10.1093/mp/ssu087

Ricardi, M.M., Gonzalez, R.M., Zhong, S., Dominguez, P.G., Duffy, T., Turjanski, P.G., Salgado Salter, J.D., Alleva, K., Carrari, F., Giovannoni, J.J., *et al.* (2014). Genome-wide data (ChIP-Seq) enabled identification of cell wall-related and aquaporin genes as targets of tomato ASR1, a drought stress-responsive transcription factor. BMC Plant Biol. *14*, 29. http://dx.doi.org/10.1186/1471-2229-14-29

Rowley, M.J., Böhmdorfer, G., and Wierzbicki, A.T. (2013). Analysis of long non-coding RNAs produced by a specialized RNA polymerase in *Arabidopsis thaliana*. Methods *63*, 160–169. http://dx.doi.org/10.1016/j.ymeth.2013.05.006

Sacharowski, S.P., Gratkowska, D.M., Sarnowska, E.A., Kondrak, P., Jancewicz, I., Porri, A., Bucior, E., Rolicka, A.T., Franzen, R., Kowalczyk, J., *et al.* (2015). SWP73 subunits of *Arabidopsis* SWI/SNF chromatin remodeling complexes play distinct roles in leaf and flower development. Plant Cell *27*, 1889–1906. http://dx.doi.org/10.1105/tpc.15.00233

Shamimuzzaman, M., and Vodkin, L. (2013). Genome-wide identification of binding sites for NAC and YABBY transcription factors and co-regulated genes during soybean seedling development by ChIP-Seq and RNA-Seq. BMC Genomics *14*, 477. http://dx.doi.org/10.1186/1471-2164-14-477

Smith, K.M., Phatale, P.A., Sullivan, C.M., Pomraning, K.R., and Freitag, M. (2011). Heterochromatin is required for normal distribution of *Neurospora crassa* CenH3. Mol. Cell. Biol. *31*, 2528–2542. http://dx.doi.org/10.1128/MCB.01285-10

Solomon, M.J., Larsen, P.L., and Varshavsky, A. (1988). Mapping protein-DNA interactions *in vivo* with formaldehyde: evidence that histone H4 is retained on a highly transcribed gene. Cell 53, 937–947.

van Dijk, K., Ding, Y., Malkaram, S., Riethoven, J.J., Liu, R., Yang, J., Laczko, P., Chen, H., Xia, Y., Ladunga, I., *et al.* (2010). Dynamic changes in genome-wide histone H3 lysine 4 methylation patterns in response to dehydration stress in *Arabidopsis thaliana*. BMC Plant Biol. *10*, 238. http://dx.doi.org/10.1186/1471-2229-10-238

Wei, W., Zhang, Y.Q., Tao, J.J., Chen, H.W., Li, Q.T., Zhang, W.K., Ma, B., Lin, Q., Zhang, J.S., and Chen, S.Y. (2015). The Alfin-like homeodomain finger protein AL5 suppresses multiple negative factors to confer abiotic stress tolerance in *Arabidopsis*. Plant J *81*, 871–883. http://dx.doi.org/10.1111/tpj.12773

Xing, D., Wang, Y., Xu, R., Ye, X., Yang, D., and Li, Q.Q. (2013). The regulatory role of Pcf11-similar-4 (PCFS4) in *Arabidopsis* development by genome-wide physical interactions with target loci. BMC Genomics *14*, 598. http://dx.doi.org/10.1186/1471-2164-14-598

Zhang, L., Hu, W., Wang, Y., Feng, R., Zhang, Y., Liu, J., Jia, C., Miao, H., Zhang, J., Xu, B., *et al.* (2015). The MaASR gene as a crucial component in multiple drought stress response pathways in *Arabidopsis*. Funct. Integr. Genomics. *15*, 247–260. http://dx.doi.org/10.1007/s10142-014-0415-y

Zhong, X., Hale, C.J., Law, J.A., Johnson, L.M., Feng, S., Tu, A., and Jacobsen, S.E. (2012). DDR complex facilitates global association of RNA polymerase V to promoters and evolutionarily young transposons. Nat. Struct. Mol. Biol. *19*, 870–875. http://dx.doi.org/10.1038/nsmb.2354

Zhu, J., Sun, Y., and Wang, Z. (2012). Genome-wide identification of transcription factor-binding sites in plants using chromatin immunoprecipitation followed by microarray (ChIP-chip) or sequencing (ChIP-seq). Methods Mol. Biol. *876*, 173–188. http://dx.doi.org/10.1007/978-1-61779-809-2_14

Zong, W., Zhong, X., You, J., and Xiong, L. (2013). Genome-wide profiling of histone H3K4-tri-methylation and gene expression in rice under drought stress. Plant Mol. Biol. *81*, 175–188. http://dx.doi.org/10.1007/s11103-012-9990-2

# Cataloguing Plant Genome Structural Variations

11

Xingtan Zhang[1,2], Xuequn Chen[1,2], Pingping Liang[1,2] and Haibao Tang[1,2*]

[1]Center for Genomics and Biotechnology, Fujian Provincial Key Laboratory of Haixia Applied Plant Systems Biology, Fujian Agriculture and Forestry University, Fuzhou, China.
[2]Haixia Institute of Science and Technology, Fujian Agriculture and Forestry University, Fuzhou, China. Email: zhangxt@fafu.edu.cn; chenxq@fafu.edu.cn; lppleiyu@gmail.com; and tanghaibao@gmail.com

*Correspondence: tanghaibao@gmail.com

https://doi.org/10.21775/9781910190654.12

## Abstract

Structural variation (SV) is a type of genetic variation identified through the comparison of genome structures which often have direct and significant associations with phenotypic variations. Building on the next-generation sequencing (NGS) technologies, research on plant structural variations are gaining momentum and have revolutionized our view on the functional impact of the 'hidden' diversity that were largely understudied before. Herein, we first describe the current state of plant genomic SV research based on NGS and in particular focus on the biological insights gained from the large-scale identification of various types of plant SVs. Specific examples are chosen to demonstrate the genetic basis for phenotype diversity in model plant and major agricultural crops. Additionally, development of new genomic mapping technologies, including optical mapping and long read sequencing, as well as improved computational algorithms associated with these technologies have helped to pinpoint the exact nature and location of genomic SVs with much better resolution and precision. Future direction of plant research on SVs should focus on the population level to build a comprehensive catalogue of SVs, leading to full assessment of their impact on biological diversity.

## Introduction

Structural variations (SVs) are a collection of complex genomic DNA mutations that differentiate among individuals in a certain population. In contrast to single nucleotide polymorphisms (SNPs) and short insertions and deletions (indels), SVs typically consist of DNA changes that are relatively long in size. The initial detection of SVs before the advent of the sequencing era is often based on detection of large scale chromosomal changes from karyotypic observation under microscope, including abnormal number of chromosomes such as aneuploidies (Jacobs *et al.*, 1959; Edwards *et al.*, 1960), chromosomal rearrangement (Bobrow *et al.*, 1971) and copy number variations (CNVs) (Bailey and Eichler 2006). Karyotypic mutations larger than 3 Mb in size can often be observed with *in situ* hybridization. Building on the next-generation sequencing (NGS) technology, we are moving towards a new phase of variant discovery that focus on identification of SVs with their boundaries mapped at single-base resolution. Extensive studies on the human genome have revealed that SVs play important roles in human health (Spielmann and Klopocki 2013; Weischenfeldt *et al.*, 2013). For example, Alzheimer's disease (AD) is a chronic neurodegenerative disease that usually badly affects the elder with symptoms such as problems with language, disorientation, mood swings and loss of motivation (Burns and Iliffe, 2009). Recent genetic research links the duplications of dosage-sensitive *APP* genes and Alzheimer's disease (Rovelet-Lecrux *et al.*, 2006). Similarly, higher *C4* copy number leads to more *C4* expression in the brain, contributing to a 1.4-fold risk to develop schizophrenia (Sekar *et al.*, 2016). These disease studies in human have indicated that abnormal copy number variation of certain genes significantly impact health and traits.

In addition to their leading role in causing a wide spectrum of human disease, structural variations are pervasive when comparing plant genomes and these changes are involved in many important biological processes (Saxena *et al.*, 2014; Żmieńko *et al.*, 2014). Structural variations represent an essential part in plant adaptive evolution and functional diversity. For example, re-sequencing of 50 rice accessions identified 21 genes with variable copy numbers that were annotated as potential resistance genes, suggesting that CNVs were involved in response to environmental cues and essential in their adaptability (Xu *et al.*, 2012). Additionally, presence or absence variations (PAVs) underlie the diversification in molecular functions that are shown to be beneficial in respective ecological niches. Sequencing of many individual ecotypes and cultivars in the population have vastly enriched our knowledge about the nature and impact of SVs over the past few years. Herein, we first describe the current state of plant SV research based on NGS as well as the biological implications of these genetic variations, categorized under several major types of structural variations. We then review several specific examples to show how SVs influence phenotypic diversity in plants in a case-study fashion. We then provide an overview of newly developed technologies, especially single molecular sequencing and genomic mapping that have become popular in recent years as powerful tools for the discovery of SVs with increasing resolution and accuracy. We conclude this review with several important directions in future SV studies in plants.

## Large-scale detection of structural variations in plants

While different types of structural variations have been reviewed before, often with a human-centric view (Alkan *et al.*, 2011; Weischenfeldt *et al.*, 2013; Saxena *et al.*, 2014) – in this review we go through a list of SVs that represent the most abundant types in plant genomes, including copy number variations (CNVs), presence and absence variations (PAVs), mobile element insertions (MEIs) and homeologous exchanges (HEs).

### Copy number variation

Copy number variations are variations of number of repeated sequences in the genome between individuals. Growing evidence support that CNVs are prevalent in plant genomes, which makes it the most extensively studied types of SVs in plants. CNV is defined as unbalanced changes in the genome structure and cover deletions (which is the same as copy number of zero) as well as duplications of > 50 bp in size (Girirajan *et al.*, 2011). Based on the data from the genomic hybridization (CGH) array and NGS technologies, a couple of plant CNVs (Table 11.1) were identified at genome-wide level (DeBolt, 2010; Swanson-Wagner *et al.*, 2010; Zheng *et al.*, 2011; Muñoz-Amatriaín *et al.*, 2013; Ping Yu *et al.*, 2013; Boocock *et al.*, 2015; Zhang *et al.*, 2015; Zhou *et al.*, 2015; Bai *et al.*, 2016; Cardone *et al.*, 2016; Hardigan *et al.*, 2016).

CNVs have been shown to play important roles in the biodiversity in plants, including adaptation to abiotic and biotic stress and yield-related traits that could be important targets for crop improvement. DeBolt performed CGH to detect CNVs in *Arabidopsis* grown under different temperature and stress regimes for the fifth-generation offspring from a common ancestor (DeBolt, 2010). Comparison between the siblings identified numerous CNVs, which account for approximately 0.38% of *Arabidopsis* annotated genes. The study found that many CNVs have effect on genome instability under environment with different abiotic inputs. Structural variations analysis from four soybean genotypes identified a list of genes overlapping with CNV regions. Functional annotation revealed that these genes were associated with disease resistance genes, suggesting that CNVs have contributed to the host defence against pathogens (McHale *et al.*, 2012). A recent report displayed that 30.2% of the potato genome were contained within CNV regions with nearly 30% of protein-coding genes affected. These CNVs are implicated in the expansions of stress-related gene families that vary between species and within species (Hardigan *et al.*, 2016). CNVs have

**Table 11.1** Selected examples of genome-wide CNV studies in plants

| Species | CNVs identification | Methods | Implication | Reference |
|---|---|---|---|---|
| *Arabidopsis* | Identified intersibling CNVs in *Arabidopsis* grown under different temperature and SA treatment for the fifth-generation plants from one common ancestor | CGH | This study uncovered that CNVs contributed to genome instability and shaped genome diversity in *Arabidopsis* | DeBolt, 2010 |
| Rice | Identification of CNVs from 20 Asian cultivated rice comprising 6 *indica*, 3 *aus*, 2 *rayada*, 2 aromatic, 3 tropical *japonica* and 4 temperate *japonica* cultivars. Found 2886 CNV regions in total | CGH | Some rice CNVs might arise independently within different groups and contribute to group differences | Yu *et al.*, 2013 |
| Rice | Identified 9916 deletions compared to the reference genome Nipponbare and 2806 genes were affected by CNVs | NGS | Experimentally validated 28 functional CNV genes including *OsMADS56*, *BPH14*, *OsDCL2b* and *OsMADS30*, implying that CNVs might contributed to phenotypic variations in rice | Bai *et al.*, 2016 |
| Maize | Identified 479 genes exhibiting higher copy number in some genotypes and 3410 genes that have either fewer copies or are missing in specific genomes | CGH | Although 10% of Maize annotated genes exhibit CNV/PAV events relative to the B73 reference genome, the majority of SVs were observed in both maize and teosinte genomes, suggesting these variants predate maize domestication | Swanson-Wagner *et al.*, 2010 |
| Cucumber | Identified 26,788 SVs based on deep resequencing of 115 diverse accessions | NGS | These SVs affect the coding regions of 1676 genes, some of which are associated with cucumber domestication. Genome-wide association analysis revealed that a CNV defined the *Female* locus | Zhang *et al.*, 2015 |
| Potato | Identified 219.8 (30.2%) of potato genomes with nearly 30% of annotated genes were affected by CNVs | NGS | This study showed that CNV was the major component of the significant genomic diversity of clonally propagated potato and CNV-affected genes highly enriched in functions related to adaption | Hardigan *et al.*, 2016 |
| Barley | CNVs in Barley were account for 14.9% of genome and affected 9.5% of coding genes from 14 different accessions | CGH | CNV-affected genes are overlapping R genes and might contribute to the molecular mechanism for adaptation to biotic and abiotic stress in barley | Muñoz-Amatriaín *et al.*, 2013 |
| Apple | 876 CNV regions were detected from 30 domesticated apple accessions, which spanned 3.5% of the apple genome | NGS | Resistance genes were significantly enriched in CNV regions | Boocock *et al.*, 2015 |
| Sorghum | Identified 17,111 CNVs in two sweet and one grain sorghum inbred lines, affecting 2600 genes | NGS | CNVs were significantly enriched in genes encoding cellulose synthases, pectinesterases, GRAS transcription factors, BTB/POZ and auxin-responsive proteins, suggesting that these genes might be involved in differentiation | Zheng *et al.*, 2011 |
| Soybean | A total of 302 soybean accessions, including 62 wild soybeans (*G. soja*), 130 landraces and 110 improved cultivars | NGS | (1) 162 CNVs were potentially involved in sorghum domestication and improvement. (2) GWAS showed that CNVs contributed to important traits, such as hilum colour and plant height | Zhou *et al.*, 2015 |
| Soybean | Four soybean genotype were used to detect CNVs and PAVs | CGH | This study identified a full set of CNV-affected genes were associated with disease resistance genes, and provided insight into the mechanisms of soybean resistance gene evolution | McHale *et al.*, 2012 |
| Grapevine | Identified 746 CNVs from four grapevine varieties and more than 2000 genes were affected by CNVs | NGS and CGH | Genes with polymorphisms were related to auxin/hormone response, berry growth and development | Cardone *et al.*, 2016 |

also conduced to population differentiation. A total of 2886 CNV regions were detected from 20 Asian cultivated rice comprising six *indica*, three *aus*, two *rayada*, two *aromatic*, three tropical *japonica* and four temperate *japonica* cultivars. Comparison of various CNV events observed in *ssp. indica* and ssp. *japonica* expounded that 57.8% of the CNVs were only present in either ssp. *indica* or ssp. *japonica* (Yu *et al.*, 2013), suggesting that CNVs arose independently within different groups and might contribute to population differentiation.

## Presence or absence variation

Gene contents often vary among even closely related individuals or cultivars, the union of which is termed 'pan-genome' (Vernikos *et al.*, 2015). The gene set present in all individuals is often called the 'core genome'. In contrast to core genome, the 'dispensable genome' consists of the sequences present only in a subset of individuals. The dispensable genome often contains a large number of PAVs. It has been hypothesized that the core genome is critical in the fundamental pathways that control plant growth and development, while the dispensable genome played important roles in phenotypic variation and adaptive evolution.

A robust way to compare PAV variation in the population is to perform sequencing on a large collection of individuals. A couple of recent publications that utilized NGS methods to detect PAV genes with important functions (Table 11.2). The largest data collected thus far come from the '*Arabidopsis* 1001 Genomes' project (Alonso-Blanco *et al.*, 2016) and the 'Rice 3000 Genomes' project (Zhikang *et al.*, 2014). Multiple *Arabidopsis* genomes were sequenced in the *Arabidopsis* 1001 Genomes project to identify both the core and dispensable genomes (Weigel and Mott, 2009; Cao *et al.*, 2011; Gan *et al.*, 2011). Tan and colleagues surveyed 80 *Arabidopsis* accessions and reported 2407 genes that were not present in reference genome (Tan *et al.*, 2012). Functional investigation revealed that largest proportions of these genes were related to membranes and binding, suggesting that the associated PAVs were involved in pattern recognition in molecular function. In addition, 31% PAV genes showed uneven distribution in chromosomes, which might be resulted from unequal recombination as the likely genetic mechanism underlying the origin of PAVs. A pan-genome study of three rice accessions showed that PAV genes were common in the *Oryza* species and ~10% of protein-coding genes were genome specific (Schatz *et al.*, 2014). Another method based on metagenome assembly was used to perform large-scale identification of dispensable genomes based on low-coverage NGS data (Yao *et al.*, 2015). Re-sequencing data from 1483 cultivated rice accessions were divided into two subpopulations: *indica* and *japonica*. Unmapped reads from each set of individuals were *de novo* assembled separately, with *indica*-related accessions or *japonica*-related accessions merged to generate a metagenome assembly. By mapping the respective 'population' assemblies to reference genome, this approach was able to identify PAVs associated with agronomic traits and specific metabolite profiling (Yao *et al.*, 2015). A recent pan-genome study in soybean re-sequenced and *de novo* assembled seven wild soybeans (Li *et al.*, 2014). Comparisons among wild soybean (*G. soja*) genomes and the reference genome (*G. max*) revealed 2.3 to 3.9 Mb dispensable sequences in wild soybean. These PAV genes typically had significantly higher *dN/dS* ratios, indicating that dispensable genes had undergone weaker purifying selection and/or greater positive selection compared to the core genes that are more conserved (Li *et al.*, 2014).

## Mobile element insertion and deletion (MEI)

Transposable element (TE) insertions and deletions are also one of the most prevalent genetic variations in plants. Sequencing of 80 *Arabidopsis* accessions from eight populations explicated that ~80% of TEs were partially or completely absent from the genomes of at least one of the 80 individuals (Cao *et al.*, 2011). Crops with larger and more complex genomes typically contain a lot more TEs than the small genome of *Arabidopsis* or rice. Although first considered as selfish DNA or 'genome parasites', transposable elements have shown a dramatic impact on genome evolution and regulation of gene expression (Bennetzen and Wang, 2014). Accumulation of TEs, in particular LTR retrotransposons, were responsible for genome size variation (Vitte and Panaud, 2005; Hawkins *et al.*, 2006; Zedek *et al.*, 2010). For example, a study in genus *Eleocharis* revealed a positive correlation between *Ty1-copia* densities and genome size (Zedek *et al.*, 2010). El Baidouri and colleagues investigated 40 sequenced

**Table 11.2** Selected examples of genome-wide PAV studies in plants

| Species | PAVs identification | Methods | Implication | Reference |
|---|---|---|---|---|
| *Arabidopsis* | The authors investigated 80 re-sequenced *Arabidopsis* accessions and identified 2407 PAV genes | NGS | A large number of PAV genes were related to membranes, suggesting that these genes were involved in pattern recognition. In addition, 31% of PAV genes were unevenly distributed in chromosomes and showed a cluster pattern, suggesting that unequal recombination was a major mechanism for PAVs | Tan *et al.*, 2012 |
| Soybean | *De novo* genome assembly of seven wild soybean identified 2.3–3.9 Mb of *G. soja*-specific dispensable genome | NGS | The PAV genes had significantly higher *dN*/*dS* ratios than the core genes, indicating that dispensable genes have undergone weaker purifying selection and/or greater positive selection than core genes. In addition, these PAV genes were related to receptor activity, structural molecule activity and antioxidant activity | Li *et al.*, 2014 |
| Rice | A metagenome-like assembly strategy was used to identify rice dispensable genome from 1483 rice accessions | NGS | The new approach is powerful to identify dispensable genome using low-coverage NGS data. In addition, association mapping of rice dispensable genome found genes related to grain width and metabolic traits were located in dispensable genome, linking PAVs to important rice traits | Yao *et al.*, 2015 |
| Potato | Re-sequencing of 12 potato monoploid and doubled monoploid potato clones identified ~7000 genes with PAVs | NGS | These dispensable genes were associated with limited transcription and/or a recent evolutionary history, with lower deletion frequency observed in genes conserved across angiosperms | Hardigan *et al.*, 2016 |
| Maize | Transcriptome sequencing of seedlings from 503 maize uncovered 8681 representative transcript assemblies not present in the reference genome (B73) | NGS | This study investigated maize pan-transcriptome and showed that RNA-seq was also useful to uncover genetic variations and important phenotype related genes | Hirsch *et al.*, 2014 |

plant genomes and observed at least one case of horizontal TE transfer in 26 genomes and most of these TEs had remained functional after the transfer. Therefore, TE-driven genetic material exchange played an important role in genome evolution (El Baidouri *et al.*, 2014). It would be natural to expect that horizontal TE transfer might also be frequent between different individuals within species.

In addition to the impact on genome structure, TEs also have the capability to alter gene structure and expression. TE insertion could invade the space occupied by the protein-coding genes and disrupt the open reading frame (ORF), resulting in abnormal phenotypes (Chen *et al.*, 2005). TEs are also frequently observed to rewire existing regulatory networks by inserting into *cis*-regulatory elements. Study of DNA transposons *mPing* in rice genome showed that insertion of TEs have affected expression of 710 genes (Naito *et al.*, 2009). Once inserted into near-gene region, specific TE amplification could trigger the epigenetic silencing pathway through methylation or small RNAs (Hollister and Gaut, 2009; Joly-Lopez and Bureau, 2014), which regulated the gene expression in close proximity. Another study in soybean showed that a 5.7-kb transposon (*Tgm-Express I*) was inserted into intron 2 of the flavanone 3-hydroxylase gene (*F3H*), which converted purple flowers to pink (Zabala and Vodkin, 2005). Interestingly, the transposon consisted of five unrelated host gene fragments and the gene fragments were processed through a complex alternative splicing mechanism as added exons (Zabala and Vodkin, 2007).

## Homeologous exchange (HE)

One type of structural variation has been relatively understudied until recently and is relatively unique to the plant genomes. In polyploid plant genomes, there has been sequence exchanges between homeologous chromosomes (often in different subgenomes in the same polyploid cell) leading to simultaneous gene deletions and amplification, or Homeologous Exchanges (HEs). For example, gene deletions and HEs between subgenomes in *Brassica napus* have led to the reduction of seed glucosinolate (GSL) content (Chalhoub *et al.*, 2014). Genes responsible for the flowering time (FLC) were greatly expanded from a single copy in *Arabidopsis* to nine copies in the oilseed genome with several expansion events due to HEs. The identified HEs also co-localize with known QTLs for vernalization requirements and flowering time. Interestingly, such HEs vary across different re-sequenced *B. napus* lines and even re-synthesized lines (Chalhoub *et al.*, 2014). Indeed, we can link much of the SVs that have occurred in *B. napus* to its unique adaptive and agronomic traits. Similarly, HEs occur frequently in the tetraploid cotton (*G. hirsutum*) (Li *et al.*, 2015). Re-sequencing of polyploid varieties or individuals could reveal additional genome restructuring events that might have been consciously or unconsciously selected for during the history of human cultivation and crop improvements.

## Case studies: genetic basis for phenotype diversity contributed by plant SVs

Structural variations play important roles in phenotype diversity. In this part, we select a couple of examples that collectively demonstrated that how SVs influence a wide array of phenotypic variations, highlighting their critical role in the context of agricultural crops.

### Sex determination

Cucumber is a major vegetable crop consumed worldwide and also serves as a model system for sex determination studies (Tanurdzic and Banks, 2004). Most cultivated cucumbers are monoecious, which contain male and female flowers on a single plant. In contrast, another type, gynoecious cucumber, bears only female flowers (Fig. 11.1A). Previous studies showed that gynoecy was associated with Mendelian locus *Female* (*F*-locus), which contained four protein-coding genes including aminocyclopropane-1-carboxylic acid synthase gene (*ACS1*), a truncated MYB transcription factor, a branched-chain amino acid aminotransferase and a gene with unknown function (Tanurdzic and Banks, 2004). Comparison between the gynoecious and monoecious cucumber genomes led to the discovery of a 30.2-kb duplicated segment that was significantly associated with gynoecy. This particular CNV event is the underlying genetic cause for the variation at the *Female* locus that has led to the development of gynoecy in cucumber (Zhang *et al.*, 2015).

### Response to abiotic stress response

Algerian barley landrace Sahara showed superior boron-toxicity tolerance (Fig. 11.1B), when compared to intolerant genotypes (Sutton *et al.*, 2007). It has been reported that boron-tolerance is associated with the ability to maintain a low level of boron concentration in the shoot. Sutton and colleagues identified an efflux transporter (*Bot1*) as the candidate gene for the boron-toxicity tolerance. Comparison between Sahara and Clipper genomes revealed that Sahara contained 4x more copies of *Bot1* than Clipper, explaining their differential response to the toxic boron. Moreover, mRNA expression of this gene in Sahara were substantially higher than that in Clipper, consistent with the higher copy number of *Bot1*. Expression of the Sahara *Bot1* gene in yeast indicated that *Bot1* provided boron tolerance in yeast under high level treatment of $H_3BO_3$. These results, starting from the identification of the CNV at the *Bot1* locus to the functional validation, demonstrated that higher copy number of *Bot1* gene conferred boron tolerance in barley Sahara (Sutton *et al.*, 2007). Similarly, genic copy number variation that contributes to aluminium (Al) tolerance is found in maize (Maron *et al.*, 2013). Three tandem multidrug and toxic compound extrusion 1 (*MATE1*) copies were identified in a maize recombinant inbred line population. The expression of *MATE1* conferred a significant increase in Al tolerance and root citrate exudation in response to Al. Consequently, *MATE1* copy number was associated with its mRNA expression, which in turn resulted in superior Al tolerance in maize (Maron *et al.*, 2013).

### Response to biotic stress

Soybean cyst nematode (SCN) is one of the most economically damaging pathogens for soybean.

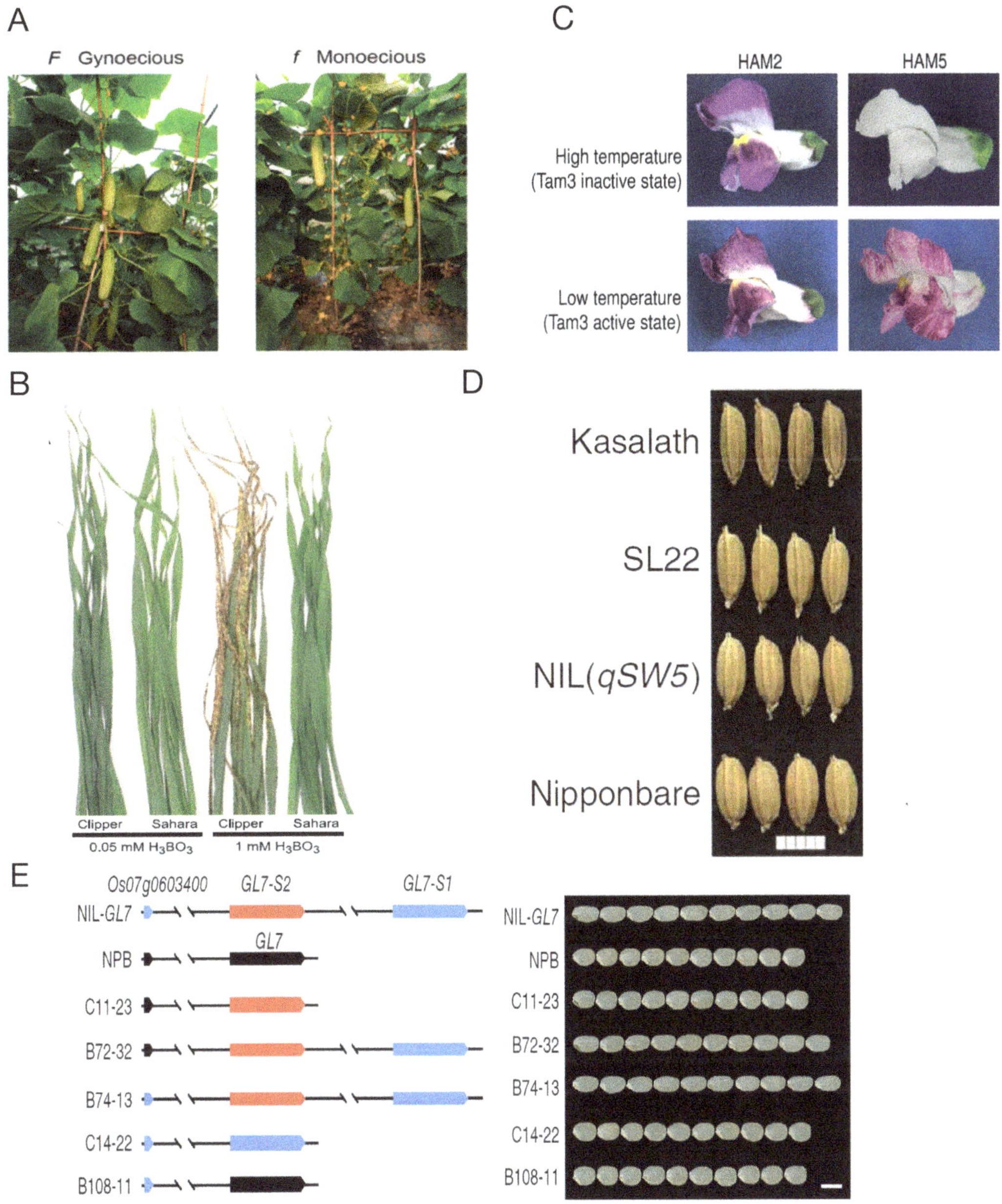


**Figure 11.1** Functional impact of structural variations in plant genomes. (A) Phenotypes of gynoecious (WI1983G) and monoecious (WI1983GM) cucumbers. Monoecious cucumbers contain both male and female flowers on a single plant, while gynoecious cucumbers contain 30.2-kb duplication in their genomes that lead to the development of female flowers only (Zhang *et al*., 2015). (B) Boron-toxicity symptoms in leaf blades of boron-intolerant (Clipper) and boron-tolerant (Sahara) barley plants. Compared with intolerant genotypes, Sahara contains four times more copies of *Bot1*, which encodes an efflux transporter and confers tolerance to boron-toxicity (Sutton *et al*., 2007). (C) Different flower pigmentation patterns in *Antirrhinum majus* lines HAM2 and HAM5. HAM2 only contains one copy of transposon *Tam3* in the upstream of *niv* gene, which is responsible for pale-red petal colour. In contrast, HAM5 contains two copies of *Tam3* in the upstream and downstream of *niv* gene, respectively. The paired *Tam3* inhibits the expression of *niv* and results in white petal colour (Uchiyama *et al*., 2013). (D) Phenotypic variation among rice Kasalath, SL22, NIL (*qSW5*) and Nipponbare accessions. Nipponbare contains a 1,212-bp deletion in the *qSW5* region and shows increased grain size (Shomura *et al*., 2008). (E) Rice grain length diversity is associated with copy number variations in *GL7* locus (Wang *et al*., 2015). Images were based on the respective studies (Sutton *et al*., 2007; Shomura *et al*., 2008; Uchiyama *et al*., 2013; Wang *et al*., 2015; Zhang *et al*., 2015), and reproduced with permission.

QTL analysis located that *Rhg1* regions consistently show a large contribution to SCN resistance in SCN-resistant cultivated soybean (Concibido *et al.*, 2004). *Rhg1* locus is a 31-kb segment which encodes multiple gene products. In susceptible varieties, only one copy of this 31-kb segment was observed. In contrast, ten tandem copies were identified in SCN-resistant soybeans. Overexpression of the cluster genes in *Rhg1* regions conferred enhanced SCN resistant, therefore confirming that copy number variation of *Rhg1* is associated with SCN resistance in soybean (Cook *et al.*, 2012).

### Flower colour

Flowers attract much attention due to their diversity in plants, and also for their critical role in agricultural production. A recent study shows that transposon is one of the major factors that contributes to the floral pigmentation and flower colour in *Antirrhinum* (Clegg and Durbin, 2003; Wei and Cao, 2016). The *nivea* (*niv*) gene encodes chalcone synthase (*CHS*) and is responsible for flower pigmentation (Sommer *et al.*, 1985). Expression of *niv* gene resulted in pale-red petal colour; while suppression of *niv* mRNA level lead to white colour. In HAM5 accession, *Tam3* transposons are inserted into both upstream and downstream of *niv* gene, and the corresponding inserted sequences paired and formed a loop structure under high temperature, thus preventing the binding of correlated transcription factor. Therefore, the transposon insertion inhibited the expression of *niv* gene, resulting in white petal colour. In contrast, only one *Tam3* transposon was observed in the promoter of *niv* gene in HAM2 line. The absence of the loop structure maintained an active *niv* expression and led to pale-red petal colour (Fig. 11.1C) (Uchiyama *et al.*, 2013).

### Grain size

Grain size is one of the most agronomically important trait for cereal domestication. Recent reports demonstrated that both CNV and PAV contributed to rice grain dimension as well as increased yields (Shomura *et al.*, 2008; Wang *et al.*, 2015). In an $F_2$ cross population between Nipponbare (*japonica*) and Kasalath (*indica*) cultivars, Izawa and his team identified several QTLs responsible for grain width. One of the QTLs, *qSW5* explained 38.5% of variation in the $F_2$ population (Shomura *et al.*, 2008). Compared to Kasalath, Nipponbare has increased grain size and its genome contained a specific 1212-bp deletion that appeared to be associated with grain width. Two rice lines were tested in confirm this association – one line with substitution of Kasalath chromosome 5 in a Nipponbare genetic background (SL22) and another nearly isogenic line (NIL) that both contained around 90 kb of Kasalath fragments of the *qSW5* region in a Nipponbare background. Both rice lines, in which *qSW5* region was present, showed relatively reduced rice grain size compared to Nipponbare (Fig. 11.1D). For grain length, another rice study revealed that tandem duplications of 17.1-kb segment containing *GL7* locus contributed to the increase in grain length and improvement of appearance quality (Wang *et al.*, 2015). Rice lines (NIL-GL7, B72–32 and B74–13) harbour multiple copies of *GL7* as well as increased grain width. In contrast, rice lines (NPB, C11–23, C14–22 and B108–11) that only harbour one copy of *GL7* displayed relatively narrow grain (Fig. 11.1E). *GL7* encodes a homologous protein to the LONGIFOLIA proteins, which is known to regulate longitudinal cell elongation in *Arabidopsis*. These two studies – with PAV controlling grain width and CNV controlling grain length – provide classical examples that specific SVs were probably selected for during the breeding history of rice.

## New experimental and computational approaches to detect SVs

Structural variations attracted much attention in the genomics field in the past decade and a number of methods have been developed on the basis of large amounts of NGS data. Most of these efforts focus on the careful analyses of alignment of short reads – such as read depth, split reads (reads that map to multiple distinct locations), and discordant read pairs (paired-end reads that show an abnormal distance between the two ends) (Zhao *et al.* 2013). Many of these alignment-based approaches have been extensively reviewed elsewhere (Escaramís *et al.*, 2015; Tattini *et al.*, 2015). We instead focus on the more recent genome mapping and sequencing technologies, including optical mapping and PacBio sequencing that show promise in the discovery of SVs that were intrinsically difficult to do with only short reads.

## Optical mapping

Optical mapping is an approach that combines DNA tagging with imaging to produce an ordered map of restriction sites or sequence motifs from a single linearized DNA molecule (Chaney *et al.*, 2016). There are two commercial vendors with slightly different variation of conceptually similar technologies – OpGen and BioNano (Tang *et al.*, 2015). We will briefly review the OpGen method here. Long DNA molecules are linearized and stretched first, followed by restriction enzyme digestion. Images of the resulting digested fragments are then captured with a camera, allowing a precise measurement of the distance between restriction sites. The resulting restriction map can be viewed as unique 'barcodes' for different regions in the genome (Lam *et al.*, 2012). Since optical mapping has the ability to barcode long DNA single molecules, this technology was initially used for the improvement of *de novo* genome assemblies (Tang *et al.*, 2015). Naturally, optical mapping is also a very powerful tool to detect large structural variations among individuals in a population. For example, Mak and colleagues generated genome maps for a trio family with long, fluorescently labelled DNA molecules using the BioNano Irys platform (Mak *et al.*, 2016). Comparison of these maps between closely related individuals and the reference human genome led to a comprehensive catalogue of relatively large structural variations (> 5 kb in size), including insertion and deletion, inversion and CNVs. Although so far very few studies plant structural variations were studied using optical mapping, there have been emerging proposals and pilot studies to map many individuals using optical mapping related technologies (Chaney *et al.*, 2016).

## PacBio single-molecule sequencing

Although the second-generation sequencing technologies are widely used in genomics, including *de novo* genome assembly and detection of small-scale genetic variations, their apparent disadvantages, namely the GC bias and short read length, make them unsuitable for calling variations that are larger scale or more complex in nature (Rhoads and Au, 2015). Repetitive sequences in complex genomes can often lead to multiple, ambiguous mapping of short reads that misidentifies structural variations. The alternative PacBio sequencing technology offers long, single-molecule sequencing and produces much more robust data for *de novo* assembly and the characterization of structural variations. Chaisson and colleagues sequenced and analysed a haploid human genome (CHM1) using PacBio platform (Chaisson *et al.*, 2015). Comparative analysis of the PacBio assembled genome and human GRCh37 reference genome resolved the complete sequences of 26,079 euchromatic structural variants at the base-pair level, most of which were not previously published (Chaisson *et al.*, 2015). Combining Illumina and PacBio technologies, Korbinian and colleagues generated *Arabidopsis thaliana* Landsberg erecta (Ler) assembly at chromosome-level (Dong *et al.*, 2016). Whole-genome comparison of the Ler assembly against the TAIR10 reference genome identified a number of SVs, including transpositions, inversions, rearranged regions as well as PAVs. In addition, the authors highlighted one case of 1.2 Mb inversion and postulated that this inversion might have reduced the level of meiotic recombination that could lead to the genetically isolated haplotypes in the worldwide population of *A. thaliana* (Dong *et al.*, 2016).

## Algorithmic advancement to detect structural variations

It is important to point out that the increased power and sensitivity achieved with the PacBio technology to detect SVs would be impossible without the algorithmic advances in the analyses of read alignments. Ritz *et al.* proposed a new algorithm, MultiBreak-SV, that tolerated high error rates of PacBio reads and employed a probabilistic approach to consider all possible solutions for break-reads alignments at the same time (Ritz *et al.*, 2014). Another mapping-based SV detection method (PB Honey) proposed two strategies, interrupted long-read mapping (PBHoney-Tails) and intra-read discordance (PBHoney-Spots) (English *et al.*, 2014). PBHoney-Tails first extracts and re-aligns unmapped tails from mapped long reads. By analysing piece-alignments with similarly mapped tails based on their respective location and orientation, one can then annotate each cluster as either a deletion, insertion, or translocation and predict breakpoints using the average interrupted position of each read. By leveraging the experimentally determined 15% per-base error rate, PBHoney-Spots was able to identify discordant 'spots' within the reference where the error rate was higher than expected (intra-read discordance).

Due to the overlapping goals of genome assembly and SV discovery, software used to assess the quality of genome assemblies could also be useful to discover SVs. For instance, Assemblytics (Nattestad and Schatz, 2016) was able to identify six classes of variants based on distinct alignment signatures between related genome assemblies.

## Future perspective

With recent technological and algorithmic advancements, much of the previously hidden genetic variations could be discovered across different plant taxa and crop varieties. Future studies of SVs should focus on the integration of various studies and databases, and associate the SVs with specific traits and phenotypes, in order to gain a comprehensive understanding of the genetic underpinnings of a rich set of biological functions.

### Population analysis on a grand scale

Population analysis of SNPs have been extensively studied in plants over past two decades since SNP is perhaps the easiest form of genomic variation to compare across individuals. However, most SV studies have been only limited to a small number of individuals and their population genetics is very much in its infancy. Recent CNV research in human sequenced 236 individuals, representing 125 distinct human populations and observed that duplications exhibit drastically different genetic and selective signatures between populations. For example, comparative analysis of CNVs at the population level found that large duplications that introgressed from the extinct Denisova lineage were exclusively present in Oceanic populations (Sudmant *et al.*, 2015). In plants, population analysis of 302 soybean re-sequencing lines showed that many CNVs could be targets of artificial selection in addition to the SNPs. Using a metric called relative frequency difference (RFD) to prioritize CNVs, the authors identified 162 potentially CNVs that are potential targets for human selection during domestication and improvement. Almost all of the discovered CNVs overlap with known domestication-related QTL regions (Zhou *et al.*, 2015).

### More and better GWAS studies

Building on the ever-increasing catalogue of SVs, it is natural to extend the statistical tests of Genome-Wide Association Studies (GWAS) to SVs. It has been suggested that many SVs could well account for the 'missing heritability' (Manolio *et al.*, 2009). In 2008, McCarroll already proposed extension of GWAS to CNVs since SNP-based GWAS only explain a modest fraction (2–15%) of heritable variation for certain disease risk (McCarroll, 2008). With more re-sequencing data, it has become more feasible to extend the gene–trait association across all genetic variants discovered – from simple SNPs to the more complex SVs. In a pilot study, Zhou and colleagues performed GWAS tests using CNV data identified by re-sequencing 302 soybeans and uncovered genes with varying copy numbers to be associated with cyst nematode resistance and hilum colour (Zhou *et al.*, 2015). GWAS with SVs was also performed in cucumber and seven SV loci were identified to be significantly related with the tuberculate fruit trait, with one matching locus from a previous study (Zhang *et al.*, 2015). While still rudimentary and relatively small-scale in terms of sample size, these studies did show promising applications of GWAS tests that are directly performed on SVs. Better SV-related GWAS analyses should become routinely performed at the population level for all re-sequencing projects, with the ultimate goal to identify much more gene–trait associations than previously suggested by SNPs alone in plants.

## Acknowledgements

We thank the Fujian provincial government for a Fujian '100 Talent Plan' award to HT. This work was supported by the National Key Research and Development Program of China (2016YFD0100305).

## References

1001 Genomes Consortium. Electronic address: magnus.nordborg@gmi.oeaw.ac.at. (2016). 1,135 Genomes Reveal the Global Pattern of Polymorphism in *Arabidopsis thaliana*. Cell *166*, 481–491. http://dx.doi.org/10.1016/j.cell.2016.05.063

Alkan, C., Coe, B.P., and Eichler, E.E. (2011). Genome structural variation discovery and genotyping. Nat. Rev. Genet. *12*, 363–376. http://dx.doi.org/10.1038/nrg2958

Bai, Z., Chen, J., Liao, Y., Wang, M., Liu, R., Ge, S., Wing, R.A., and Chen, M. (2016). The impact and origin of copy number variations in the *Oryza* species. BMC Genomics *17*, 261. http://dx.doi.org/10.1186/s12864-016-2589-2

Bailey, J.A., and Eichler, E.E. (2006). Primate segmental duplications: crucibles of evolution, diversity and disease. Nat. Rev. Genet. 7, 552–564.

Bennetzen, J.L., and Wang, H. (2014). The contributions of transposable elements to the structure, function, and evolution of plant genomes. Annu. Rev. Plant Biol. *65*, 505–530. http://dx.doi.org/10.1146/annurev-arplant-050213-035811

Bobrow, M., Joness, L.F., and Clarke, G. (1971). A complex chromosomal rearrangement with formation of a ring 4. J. Med. Genet. *8*, 235–239.

Boocock, J., Chagné, D., Merriman, T.R., and Black, M.A. (2015). The distribution and impact of common copy-number variation in the genome of the domesticated apple, Malus x domestica Borkh. BMC Genomics *16*, 848. http://dx.doi.org/10.1186/s12864-015-2096-x

Burns, A., and Iliffe, S. (2009). Alzheimer's disease. BMJ *338*, b158. http://dx.doi.org/10.1136/bmj.b158

Cao, J., Schneeberger, K., Ossowski, S., Günther, T., Bender, S., Fitz, J., Koenig, D., Lanz, C., Stegle, O., Lippert, C., *et al.* (2011). Whole-genome sequencing of multiple *Arabidopsis thaliana* populations. Nat. Genet. *43*, 956–963. http://dx.doi.org/10.1038/ng.911

Cardone, M.F., D'Addabbo, P., Alkan, C., Bergamini, C., Catacchio, C.R., Anaclerio, F., Chiatante, G., Marra, A., Giannuzzi, G., Perniola, R., *et al.* (2016). Inter-varietal structural variation in grapevine genomes. Plant J. *88*, 648–661. http://dx.doi.org/10.1111/tpj.13274

Chaisson, M.J., Huddleston, J., Dennis, M.Y., Sudmant, P.H., Malig, M., Hormozdiari, F., Antonacci, F., Surti, U., Sandstrom, R., Boitano, M., *et al.* (2015). Resolving the complexity of the human genome using single-molecule sequencing. Nature *517*, 608–611. http://dx.doi.org/10.1038/nature13907

Chalhoub, B., Denoeud, F., Liu, S., Parkin, I.A., Tang, H., Wang, X., Chiquet, J., Beleram, H., Tong, C., Samans, B., *et al.* (2014). Early allopolyploid evolution in the post-Neolithic *Brassica napus* oilseed genome. Science *345*, 950–953.

Chaney, L., Sharp, A.R., Evans, C.R., and Udall, J.A. (2016). Genome mapping in plant comparative genomics. Trends Plant Sci. *21*, 770–780. http://dx.doi.org/10.1016/j.tplants.2016.05.004

Chen, J.M., Stenson, P.D., Cooper, D.N., and Ferec, C. (2005). A systematic analysis of LINE-1 endonuclease-dependent retrotranspositional events causing human genetic disease. Hum. Genet. *117*, 411–427.

Clegg, M.T., and Durbin, M.L. (2003). Tracing floral adaptations from ecology to molecules. Nat. Rev. Genet. *4*, 206–215. http://dx.doi.org/10.1038/nrg1023

Concibido, V.C., Diers, B.W., and Arelli, P.R. (2004). A decade of QTL mapping for cyst nematode resistance in soybean. Crop Sci. *44*, 1121–1131.

Cook, D.E., Lee, T.G., Guo, X., Melito, S., Wang, K., Bayless, A.M., Wang, J., Hughes, T.J., Willis, D.K., Clemente, T.E., *et al.* (2012). Copy number variation of multiple genes at Rhg1 mediates nematode resistance in soybean. Science *338*, 1206–1209. http://dx.doi.org/10.1126/science.1228746

DeBolt, S. (2010). Copy number variation shapes genome diversity in *Arabidopsis* over immediate family generational scales. Genome Biology and Evolution *2*, 441–453.

Dong, J., Feng, Y., Kumar, D., Zhang, W., Zhu, T., Luo, M.C., and Messing, J. (2016). Analysis of tandem gene copies in maize chromosomal regions reconstructed from long sequence reads. Proc. Natl. Acad. Sci. U.S.A. *113*, 7949–7956. http://dx.doi.org/10.1073/pnas.1608775113

Edwards, J.H., Harnden, D.G., Cameron, A.H., Crosse, V.M., and Wolff, O.H. (1960). A new trisomic syndrome. Lancet *1*, 787–790.

El Baidouri, M., Carpentier, M.C., Cooke, R., Gao, D., Lasserre, E., Llauro, C., Mirouze, M., Picault, N., Jackson, S.A., and Panaud, O. (2014). Widespread and frequent horizontal transfers of transposable elements in plants. Genome Res. *24*, 831–838. http://dx.doi.org/10.1101/gr.164400.113

English, A.C., Salerno, W.J., and Reid, J.G. (2014). PBHoney: identifying genomic variants via long-read discordance and interrupted mapping. BMC Bioinf. *15*, 180. http://dx.doi.org/10.1186/1471-2105-15-180

Escaramís, G., Docampo, E., and Rabionet, R. (2015). A decade of structural variants: description, history and methods to detect structural variation. Brief. Funct. Genomics. *14*, 305–314. http://dx.doi.org/10.1093/bfgp/elv014

Gan, X., Stegle, O., Behr, J., Steffen, J.G., Drewe, P., Hildebrand, K.L., Lyngsoe, R., Schultheiss, S.J., Osborne, E.J., Sreedharan, V.T., *et al.* (2011). Multiple reference genomes and transcriptomes for *Arabidopsis thaliana*. Nature *477*, 419–423. http://dx.doi.org/10.1038/nature10414

Girirajan, S., Campbell, C.D., and Eichler, E.E. (2011). Human copy number variation and complex genetic disease. Annu. Rev. Genet. *45*, 203–226. http://dx.doi.org/10.1146/annurev-genet-102209-163544

Sommer, H., Carpenter, R., Harrison, B.J., and Saedler, H. (1985). The transposable element Tam3 of *Antirrhinum majus* generates a novel type of sequence alterations upon excision. Mol. Gen. Genet. *199*, 225–231.

Hardigan, M.A., *et al.* (2016). 'Genome reduction uncovers a large dispensable genome and adaptive role for copy number variation in asexually propagated Solanum tuberosum.' Plant Cell

Hawkins, J.S., Kim, H., Nason, J.D., Wing, R.A., and Wendel, J.F. (2006). Differential lineage-specific amplification of transposable elements is responsible for genome size variation in Gossypium. Genome Res. *16*, 1252–1261.

Hirsch, C.N., Foerster, J.M., Johnson, J.M., Sekhon, R.S., Muttoni, G., Vaillancourt, B., Peñagaricano, F., Lindquist, E., Pedraza, M.A., Barry, K., *et al.* (2014). Insights into the maize pan-genome and pan-transcriptome. Plant Cell *26*, 121–135. http://dx.doi.org/10.1105/tpc.113.119982

Hollister, J.D., and Gaut, B.S. (2009). Epigenetic silencing of transposable elements: a trade-off between reduced transposition and deleterious effects on neighboring gene expression. Genome Res. *19*, 1419–1428. http://dx.doi.org/10.1101/gr.091678.109

Jacobs, P.A., Baikie, A.G., Court Brown, W.M., and Strong, J.A. (1959). The somatic chromosomes in mongolism. Lancet *1*, 710.

Joly-Lopez, Z., and Bureau, T.E. (2014). Diversity and evolution of transposable elements in *Arabidopsis*. Chromosome. Res. *22*, 203–216. http://dx.doi.org/10.1007/s10577-014-9418-8

Lam, E.T., Hastie, A., Lin, C., Ehrlich, D., Das, S.K., Austin, M.D., Deshpande, P., Cao, H., Nagarajan, N., Xiao, M., *et al.* (2012). Genome mapping on nanochannel arrays for structural variation analysis and sequence assembly. Nat. Biotechnol. *30*, 771–776.

Li, F., Fan, G., Lu, C., Xiao, G., Zou, C., Kohel, R.J., Ma, Z., Shang, H., Ma, X., Wu, J., *et al.* (2015). Genome sequence of cultivated Upland cotton (*Gossypium hirsutum* TM-1) provides insights into genome evolution. Nat. Biotechnol. *33*, 524–530. http://dx.doi.org/10.1038/nbt.3208

Li, Y.H., Zhou, G., Ma, J., Jiang, W., Jin, L.G., Zhang, Z., Guo, Y., Zhang, J., Sui, Y., Zheng, L., *et al.* (2014). De novo assembly of soybean wild relatives for pan-genome analysis of diversity and agronomic traits. Nat. Biotechnol. *32*, 1045–1052. http://dx.doi.org/10.1038/nbt.2979

Mak, A.C., Lai, Y.Y., Lam, E.T., Kwok, T.P., Leung, A.K., Poon, A., Mostovoy, Y., Hastie, A.R., Stedman, W., Anantharaman, T., *et al.* (2016). Genome-wide structural variation detection by genome mapping on nanochannel arrays. Genetics *202*, 351–362. http://dx.doi.org/10.1534/genetics.115.183483

Manolio, T.A., Collins, F.S., Cox, N.J., Goldstein, D.B., Hindorff, L.A., Hunter, D.J., McCarthy, M.I., Ramos, E.M., Cardon, L.R., Chakravarti, A., *et al.* (2009). Finding the missing heritability of complex diseases. Nature *461*, 747–753. http://dx.doi.org/10.1038/nature08494

Maron, L.G., Guimarães, C.T., Kirst, M., Albert, P.S., Birchler, J.A., Bradbury, P.J., Buckler, E.S., Coluccio, A.E., Danilova, T.V., Kudrna, D., *et al.* (2013). Aluminum tolerance in maize is associated with higher MATE1 gene copy number. Proc. Natl. Acad. Sci. U.S.A. *110*, 5241–5246. http://dx.doi.org/10.1073/pnas.1220766110

McCarroll, S.A. (2008). Extending genome-wide association studies to copy-number variation. Hum. Mol. Genet. *17*, R135–42. http://dx.doi.org/10.1093/hmg/ddn282

McHale, L.K., Haun, W.J., Xu, W.W., Bhaskar, P.B., Anderson, J.E., Hyten, D.L., Gerhardt, D.J., Jeddeloh, J.A., and Stupar, R.M. (2012). Structural variants in the soybean genome localize to clusters of biotic stress-response genes. Plant Physiol. *159*, 1295–1308. http://dx.doi.org/10.1104/pp.112.194605

Muñoz-Amatriaín, M., Eichten, S.R., Wicker, T., Richmond, T.A., Mascher, M., Steuernagel, B., Scholz, U., Ariyadasa, R., Spannagl, M., Nussbaumer, T., *et al.* (2013). Distribution, functional impact, and origin mechanisms of copy number variation in the barley genome. Genome Biol. *14*, R58. http://dx.doi.org/10.1186/gb-2013-14-6-r58

Naito, K., Zhang, F., Tsukiyama, T., Saito, H., Hancock, C.N., Richardson, A.O., Okumoto, Y., Tanisaka, T., and Wessler, S.R. (2009). Unexpected consequences of a sudden and massive transposon amplification on rice gene expression. Nature *461*, 1130–1134. http://dx.doi.org/10.1038/nature08479

Nattestad, M., and Schatz, M.C. (2016). Assemblytics: a web analytics tool for the detection of variants from an assembly. Bioinformatics 32, 3021–3023. http://dx.doi.org/10.1093/bioinformatics/btw369

Yu, P., Wang, C.H., Xu, Q., Feng, Y., Yuan, X.P., Yu, H.Y., Wang, Y.P., Tang, S.X., and Wei, X.H. (2013). Genome-wide copy number variations in Oryza sativa L. BMC Genomics *14*, 649. http://dx.doi.org/10.1186/1471-2164-14-649

Rhoads, A., and Au, K.F. (2015). PacBio sequencing and its applications. genomics. proteomics. Bioinformatics. *13*, 278–289. http://dx.doi.org/10.1016/j.gpb.2015.08.002

Ritz, A., Bashir, A., Sindi, S., Hsu, D., Hajirasouliha, I., and Raphael, B.J. (2014). Characterization of structural variants with single molecule and hybrid sequencing approaches. Bioinformatics *30*, 3458–3466. http://dx.doi.org/10.1093/bioinformatics/btu714

Rovelet-Lecrux, A., Hannequin, D., Raux, G., Le Meur, N., Laquerriere, A., Vital, A., Dumachin, Cl, Feuillette, S., Brice, A., Vercelletto, M., *et al.* (2006). APP locus duplication causes autosomal dominant early-onset Alzheimer disease with cerebral amyloid angiopathy. Nat. Genet. *38*, 24–26.

Saxena, R.K., Edwards, D., and Varshney, R.K. (2014). Structural variations in plant genomes. Brief. Funct. Genomics. *13*, 296–307. http://dx.doi.org/10.1093/bfgp/elu016

Schatz, M.C., Maron, L.G., Stein, J.C., Hernandez Wences, A., Gurtowski, J., Biggers, E., Lee, H., Kramer, M., Antoniou, E., Ghiban, E., *et al.* (2014). Whole genome de novo assemblies of three divergent strains of rice, Oryza sativa, document novel gene space of aus and indica. Genome Biol. *15*, 506.

Sekar, A., Bialas, A.R., de Rivera, H., Davis, A., Hammond, T.R., Kamitaki, N., Tooley, K., Presumey, J., Baum, M., Van Doren, V., *et al.* (2016). Schizophrenia risk from complex variation of complement component 4. Nature *530*, 177–183. http://dx.doi.org/10.1038/nature16549

Tan, S., Zhong, Y., Hou, H., Yang, S., and Tian, D. (2012). Variation of presence/absence genes among *Arabidopsis* populations. BMC Evol. Biol. *12*, 86. http://dx.doi.org/10.1186/1471-2148-12-86

Shomura, A., Izawa, T., Ebana, K., Ebitani, T., Kanegae, H., Konishi, S., and Yano, M. (2008). Deletion in a gene associated with grain size increased yields during rice domestication. Nat. Genet. *40*, 1023–1028. http://dx.doi.org/10.1038/ng.169

Spielmann, M., and Klopocki, E. (2013). CNVs of noncoding cis-regulatory elements in human disease. Curr. Opin. Genet. Dev. *23*, 249–256. http://dx.doi.org/10.1016/j.gde.2013.02.013

Sudmant, P.H., Mallick, S., Nelson, B.J., Hormozdiari, F., Krumm, N., Huddleston, J., Coe, B.P., Baker, C., Nordenfelt, S., Bamshad, M., *et al.* (2015). Global diversity, population stratification, and selection of human copy-number variation. Science *349*, aab3761. http://dx.doi.org/10.1126/science.aab3761

Sutton, T., Baumann, U., Hayes, J., Collins, N.C., Shi, B.J., Schnurbusch, T., Hay, A., Mayo, G., Pallotta, M., Tester, M., *et al.* (2007). Boron-toxicity tolerance in barley arising from efflux transporter amplification. Science *318*, 1446–1449.

Swanson-Wagner, R.A., Eichten, S.R., Kumari, S., Tiffin, P., Stein, J.C., Ware, D., and Springer, N.M. (2010). Pervasive gene content variation and copy number variation in maize and its undomesticated progenitor. Genome Res. *20*, 1689–1699. http://dx.doi.org/10.1101/gr.109165.110

Tang, H., Lyons, E., and Town, C.D. (2015). Optical mapping in plant comparative genomics. GigaScience *4*, 3. http://dx.doi.org/10.1186/s13742-015-0044-y

Tanurdzic, M., and Banks, J.A. (2004). Sex-determining mechanisms in land plants. Plant Cell *16*, S61–71. http://dx.doi.org/10.1105/tpc.016667

Tattini, L., D'Aurizio, R., and Magi, A. (2015). Detection of genomic structural variants from next-generation sequencing data. Front. Bioeng. Biotechnol. *3*, 92. http://dx.doi.org/10.3389/fbioe.2015.00092

Uchiyama, T., Hiura, S., Ebinuma, I., Senda, M., Mikami, T., Martin, C., and Kishima, Y. (2013). A pair of transposons coordinately suppresses gene expression, independent of pathways mediated by siRNA in Antirrhinum. New Phytol. *197*, 431–440. http://dx.doi.org/10.1111/nph.12041

Vernikos, G., Medini, D., Riley, D.R., and Tettelin, H. (2015). Ten years of pan-genome analyses. Curr. Opin. Microbiol. *23*, 148–154. http://dx.doi.org/10.1016/j.mib.2014.11.016

Vitte, C., and Panaud, O. (2005). LTR retrotransposons and flowering plant genome size: emergence of the increase/decrease model. Cytogenet. Genome. Res. *110*, 91–107.

Wang, Y., Xiong, G., Hu, J., Jiang, L., Yu, H., Xu, J., Fang, Y., Zeng, L., Xu, E., Xu, J., *et al.* (2015). Copy number variation at the GL7 locus contributes to grain size diversity in rice. Nat. Genet. *47*, 944–948. http://dx.doi.org/10.1038/ng.3346

Wei, L., and Cao, X. (2016). The effect of transposable elements on phenotypic variation: insights from plants to humans. Sci. China Life Sci. *59*, 24–37. http://dx.doi.org/10.1007/s11427-015-4993-2

Weigel, D., and Mott, R. (2009). The 1001 genomes project for *Arabidopsis thaliana*. Genome Biol. *10*, 107. http://dx.doi.org/10.1186/gb-2009-10-5-107

Weischenfeldt, J., Symmons, O., Spitz, F., and Korbel, J.O. (2013). Phenotypic impact of genomic structural variation: insights from and for human disease. Nat. Rev. Genet. *14*, 125–138. http://dx.doi.org/10.1038/nrg3373

Xu, X., Liu, X., Ge, S., Jensen, J.D., Hu, F., Li, X., Dong, Y., Gutenkunst, R.N., Fang, L., Huang, L., *et al.* (2011). Resequencing 50 accessions of cultivated and wild rice yields markers for identifying agronomically important genes. Nat. Biotechnol. *30*, 105–111. http://dx.doi.org/10.1038/nbt.2050

Yao, W., Li, G., Zhao, H., Wang, G., Lian, X., and Xie, W. (2015). Exploring the rice dispensable genome using a metagenome-like assembly strategy. Genome Biol. *16*, 187. http://dx.doi.org/10.1186/s13059-015-0757-3

Zabala, G., and Vodkin, L. (2007). Novel exon combinations generated by alternative splicing of gene fragments mobilized by a CACTA transposon in Glycine max. BMC Plant Biol. *7*, 38.

Zabala, G., and Vodkin, L.O. (2005). The wp mutation of Glycine max carries a gene-fragment-rich transposon of the CACTA superfamily. Plant Cell *17*, 2619–2632.

Zedek, F., Smerda, J., Smarda, P., and Bureš, P. (2010). Correlated evolution of LTR retrotransposons and genome size in the genus Eleocharis. BMC Plant Biol. *10*, 265. http://dx.doi.org/10.1186/1471-2229-10-265

Zhang, Z., et al. (2015). 'Genome-Wide Mapping of Structural Variations Reveals a Copy Number Variant That Determines Reproductive Morphology in Cucumber.' Plant Cell **27**(6): 1595–1604.

Zhao, M., Wang, Q., Wang, Q., Jia, P., and Zhao, Z. (2013). Computational tools for copy number variation (CNV) detection using next-generation sequencing data: features and perspectives. BMC Bioinf. *14* (Suppl. 11), S1. http://dx.doi.org/10.1186/1471-2105-14-S11-S1

Zheng, L.Y., Guo, X.S., He, B., Sun, L.J., Peng, Y., Dong, S.S., Liu, T.F., Jiang, S., Ramachandran, S., Liu, C.M., *et al.* (2011). Genome-wide patterns of genetic variation in sweet and grain sorghum (Sorghum bicolor). Genome Biol. *12*, R114. http://dx.doi.org/10.1186/gb-2011-12-11-r114

Zhikang, L., Fu, B.Y., Gao, Y.M., Wang, W.S., Xu, J.L., Zhang, F., Zhao, X.Q., Zhao, T.Q., Zheng, T.Q., Zhou, Y.L., *et al.* (2014). The 3,000 rice genomes project. Gigascience 3, 7.

Zhou, Z., Jiang, Y., Wang, Z., Gou, Z. Lyu, J., Li, W., Yu, Y., Shu, L., Zhao, Y., Ma, Y., *et al.* (2015). Resequencing 302 wild and cultivated accessions identifies genes related to domestication and improvement in soybean. Nat. Biotechnol. 33, 408–414.

Żmieńko, A., Samelak, A., Kozłowski, P., and Figlerowicz, M. (2014). Copy number polymorphism in plant genomes. Theor. Appl. Genet. *127*, 1–18. http://dx.doi.org/10.1007/s00122-013-2177-7